设施园艺创新与进展

——2011第二届中国·寿光国际设施园艺高层学术论坛论文集

Protected Horticulture Advances and Innovations
—— Proceedings of 2011 the 2nd High-level International Forum on Protected Horticulture（Shouguang·China）

杨其长　Toyoki Kozai（日本）　Gerard P.A. Bot（荷兰）　主编
Edited by　Qichang Yang　Toyoki Kozai（Japan）　Gerard P.A. Bot（the Netherlands）

中国农业科学技术出版社

图书在版编目（CIP）数据

设施园艺创新与进展——2011第二届中国·寿光国际设施园艺高层学术论坛论文集/杨其长，（日）古在豊樹（Toyoki Kozai），（荷）伯特（Gerard P. A. Bot）主编
—北京：中国农业科学技术出版社，2011.4
ISBN 978-7-5116-0427-9

Ⅰ.①设…　Ⅱ.①杨…②古…③伯…　Ⅲ.①园艺—保护地栽培—文集　Ⅳ.①S62-53

中国版本图书馆 CIP 数据核字（2011）第 055368 号

责任编辑　　张孝安
责任校对　　贾晓红

出 版 者	中国农业科学技术出版社 北京市中关村南大街 12 号　邮编：100081
电　　话	（010）82109708（编辑室）　　（010）82109704（发行部） （010）82109703（读者服务部）
传　　真	（010）82109700
网　　址	http://www.castp.cn
经 销 者	新华书店北京发行所
印 刷 者	北京科信印刷厂
开　　本	787 mm×1 092 mm　1/16
印　　张	20.75
字　　数	500 千字
版　　次	2011 年 4 月第 1 版　2011 年 4 月第 1 次印刷
定　　价	86.00 元

版权所有·翻印必究

2011 中国·寿光国际设施园艺高层学术论坛组织委员会名单

组委会主席：
　　刘　旭　中国工程院院士　中国农业科学院副院长
　　朱兰玺　山东省寿光市人民政府市长

组委会副主席：
　　杨德峰　山东省寿光市人大常委会党组书记　菜博会组委会常务副主任
　　梅旭荣　中国农业科学院农业环境与可持续发展研究所所长　研究员

组委会成员：
　　桑文军　山东省寿光市人民政府副市长
　　刘继芳　中国农业科学院办公室主任
　　袁龙江　中国农业科学院科技局副局长
　　栗金池　中国农业科学院农业环境与可持续发展研究所党委书记
　　杨其长　中国农业科学院农业环境与可持续发展研究所研究员
　　隋子龙　山东省寿光市蔬菜高科技示范园管理处主任　菜博会组委会办公室主任
　　孙德华　山东省寿光市蔬菜高科技示范园管理处党支部书记
　　王启龙　中国（寿光）国际蔬菜科技博览会组委会办公室常务副主任
　　张志斌　中国农业科学院蔬菜花卉所研究员，中国园艺学会设施园艺分会会长
　　陈青云　中国农业大学教授，中国农业工程学会设施园艺工程专委会主任
　　何启伟　山东省园艺学会理事长　山东省农业专家顾问团蔬菜分团团长
　　王秀峰　山东农业大学园艺科学与工程学院院长

2011 第二届中国·寿光国际设施园艺高层学术论坛学术委员会名单

学术委员会主席：
　　Toyoki Kozai　千叶大学（日本）
　　杨其长　中国农业科学院
　　Gerard P. A. Bot　瓦赫宁根大学（荷兰）

学术委员会成员（按姓氏字母顺序）

A. P. Papadopoulos 教授	加拿大温室与农作物加工研究中心（加拿大）
艾希珍教授	山东农业大学
白义奎教授	沈阳农业大学
别之龙教授/主任	华中农业大学
Carl – Otto Ottosen 教授	奥尔胡斯大学（丹麦）
陈青云教授/副院长	中国农业大学
Constantinos Kittas 教授	塞萨罗尼斯大学（希腊）
Eddie Schrevens 教授	勒芬大学（比利时）
Eldert van Henten 教授	瓦赫宁根大学（荷兰）
方炜教授	台湾大学（中国台湾）
Gerrit van Straten 教授	瓦赫宁根大学（荷兰）
郭世荣教授	南京农业大学
郝秀明 教授	加拿大温室与农作物加工研究中心（加拿大）
何启伟研究员	山东农业科学院
黄丹枫教授/副院长	上海交通大学
李萍萍教授/副校长	南京林业大学
李建设教授/院长	宁夏大学
李天来教授/副校长	沈阳农业大学
李亚灵教授	山西农业大学
刘士哲教授	华南农业大学
罗卫红教授	南京农业大学
马承伟教授	中国农业大学
毛罕平教授/院长	江苏大学
Ryo Mastuda 教授	东京大学（日本）
Stephania de Pascale 教授	那不勒斯大学（意大利）
孙振元/研究员	中国林业科学院
孙治强教授/副院长	河南农业大学
孙忠富研究员	中国农业科学院
Takehiko Hoshi 教授	东海大学（日本）
仝雨欣博士	千叶大学（日本）
王铁良教授/处长	沈阳农业大学
王秀峰教授/院长	山东农业大学
魏珉教授	山东农业大学
温祥珍教授	山西农业大学
宋卫堂教授	中国农业大学
须晖教授/副院长	沈阳农业大学
徐志刚教授	南京农业大学

徐志豪研究员/主任	浙江省农业科学院
肖玉兰教授	首都师范大学
杨仁全研究员/处长	北京市科学技术委员会
喻景权教授/副院长	浙江大学
于贤昌教授	中国农业科学院
张明毅教授	宜兰大学（中国台湾）
张启翔教授/副校长	北京林业大学
张振贤教授	中国农业大学
张志斌研究员	中国农业科学院
周长吉研究员/副所长	农业部规划设计院
邹志荣教授/院长	西北农林科技大学

Organizing Committee of 2011 High–level International Forum on Protected Horticulture

Chairman of Organizing Committee

Xu Liu（Chinese Academy of Engineering, Chinese Academy of Agricultural Sciences）
Lanxi Zhu（The People's Government of Shouguang City）
Defeng Yang（The Standing Committee of Shouguang People's Congress）
Xurong Mei（The Institute of Environment and Sustainable in Agriculture）

Members of Organizing Committee

Wenjun Sang（The People's Government of Shouguang City）
Jifang Liu（Chinese Academy of Agricultural Sciences）
Longjiang Yuan（Science & Technology Bureau of Chinese Academy of Agricultural Sciences）
Jinchi Li（the Institute of Environment and Sustainable in Agriculture）
Qichang Yang（the Institute of Environment and Sustainable in Agriculture）
Zilong Sui（Organizing Committee of China International Vegetable Fair）
Dehua Sun（administrative office of Shouguang Vegetable Hi–Tech Demonstration Park）
Qilong Wang（Organizing Committee of China International Vegetable Fair）
Zhibin Zhang（Institute of Vegetable Crops and Flowers）
Qingyun Chen（China Agricultural University）
Qiwei He（Society of Horticultural Science of Shandong Province）
Xiufeng Wang（Shandong Agricultural University）

Scientific Committee of 2011 High-level International Forum on Protected Horticulture

Chairman of Scientific Committee

Toyoki Kozai (Japan)
Qichang Yang
Gerard P. A. Bot (The Netherlands)

Members of Scientific Committee

A. P. (Tom) Papadopoulos (Canada), Xizhen Ai, Yikui Bai, Zhilong Bie, Carl Otto-Ottosen (Danmark), Qingyun Chen, Constantinos Kittas (Greece), Eddie Schrevens (Belgium), Eldert van Henten (The Netherlands), Wei Fang (Taiwan), Gerrit van Straten (The Netherlands), Shirong Guo, Xiuming Hao (Canada), Qiwei He, Danfeng Huang, Pingping Li, Jianshe Li, Tianlai Li, Yaling Li, Shizhe Liu, Weihong Luo, Chengwei Ma, Hanping Mao, Roy Mastuda (Japan), Stephania de Pascale (Italy), Zhengyuan Sun, Zhiqiang Sun, Zhongfu Sun, Takehiko Hoshi (Japan), Yuxin Tong, Tieliang Wang, Xiufeng Wang, Min Wei, Xiangzhen Wen, Weitang Song, Hui Xu, Zhigang Xu, Zhihao Xu, Yulan Xiao, Renquan Yang, Jingquan Yu, Xianchang Yu, Mingyi Zhang (Taiwan), Qixiang Zhang Zhenxian Zhang, Zhibin Zhang, Changji Zhou, Zhirong Zou

前　言

2009 年 4 月 20～22 日，在第十届中国（寿光）国际蔬菜科技博览会期间，中国农业科学院和寿光市人民政府共同举办了"设施园艺与现代科技"为主题的首届"中国·寿光国际设施园艺高层学术论坛"（2009 High-level International Forum on Protected Horticulture, HIFPH2009），邀请了数十位国内外知名设施园艺专家和 150 余位参会代表，就设施园艺科技进展、节能与新能源利用、环境优化控制、高效栽培、新材料及新技术等内容进行了深入研讨，取得了圆满成功。会后多位代表认为，论坛的专家层次高、报告精彩有深度、对行业发展影响力大，希望能形成一定的机制，定期举办下去。经中国农业科学院和寿光市人民政府商定，"中国·寿光国际设施园艺高层学术论坛"每两年举办一届，时间与菜博会同期。2011 年第二届论坛在有关领导、专家的大力支持下，在寿光市如期举办。

近年来，设施园艺产业发展迅速，仅中国的设施栽培面积就已达 350 万公顷。设施园艺的快速发展为改善城乡居民的生活质量、增加农民收入，做出了巨大贡献。但随着人口的不断增长、耕地的不断减少、淡水资源的日趋匮乏以及化石能源的日益枯竭，设施园艺产业也面临着诸多亟待解决的难题。如何利用现代科技成果解决设施园艺生产中面临的资源、环境和可持续发展问题，是摆在世界设施园艺专家面前的重大课题。为此，本届论坛选择以"节能高效、绿色安全"为主题，安排 40 位国内外专家作大会主题报告和专题报告，并围绕节能与新能源利用、设施高效栽培、绿色安全生产、数字化调控技术等内容进行交流与研讨，探讨实现设施园艺节能、高效、安全生产的技术途径，并汇集了与会专家的 43 篇论文，正式编辑出版。

在论坛组织过程中，得到了中国农业科学院、山东省寿光市人民政府、荷兰瓦赫宁根大学（Wageningen UR）、日本千叶大学（Chiba University）、山东农业大学、山东省农业科学院、中国园艺学会设施园艺分会、中国农业工程学会设施园艺工程专业委员会以及北京中环易达设施园艺科技有限公司等单位的大力支持，在此表示衷心感谢！

由于时间仓促，论文集中难免会有错漏之处，恳请各位同仁和读者批评指正。

编　者
2011 年 4 月

目 录

综 述

Improving Utilization Efficiency of Electricity, Light energy, Water and CO_2 of a Plant Factory with Artificial Light ························· Toyoki Kozai(2)

发展设施蔬菜低碳生产技术对策 ························· 张志斌(9)

Optimal Greenhouse Cultivation Control: How to Get There? ············· Gerrit van Straten(14)

植物工厂与垂直农业及其资源替代战略构想 ························· 杨其长(15)

High Temperature Control in Mediterranean Greenhouse Production: the Constraints and the Options
 ························· Stefania De Pascale and Cecilia Stanghellini (20)

"Effects of Greenhouse Climate Control Equipment on Greenhouse Microclimate and Crop Response"
 ························· C. Kittas, N. Katsoulas1, T. Bartzanas (34)

设施无土栽培叶菜中硝酸盐和维生素 C 的累积调控 ············· 刘文科,杨其长,魏灵玲(62)

设施园艺工程技术

日光温室热环境分析及设计方法研究 ············· 马承伟,徐凡,赵淑梅,李睿,刘洋(70)

空气—空气热泵技术在温室环境综合控制中的应用 ························· 仝宇欣(80)

山地日光温室性能分析 ························· 邹志荣,张勇(86)

Designing a Greenhouse Plant: Novel Approaches to Improve Resource Use Efficiency in
 Controlled Environments ············· A. Maggio, S. De Pascale and G. Barbieri (93)

下沉式机打土墙结构的日光温室性能与适应性分析 ············· 丁小明,周长吉,魏晓明(100)

雪灾中钢骨架结构日光温室倒塌原因及对策 ············· 白义奎,李天来,王铁良,刘文合(108)

日光温室浅层土壤水媒蓄放热增温效果 ············· 方慧,杨其长,梁浩,王烁(113)

华北地区几种日光温室墙体热流量测定与分析
 ············· 胡彬,马承伟,王双瑜,阳萍,徐凡,曹晏飞(123)

日光温室墙体材料对墙体温度分布及室内温度的影响 ························· 佟国红(131)

山东寿光日光温室冬季热环境测试
 ············· 曹晏飞,张建宇,赵淑梅,马承伟,蒋程瑶,魏家鹏,桑毅振(138)

基于 ZigBee、3G 网络的温室无线远程植物生理生态监测系统
 ············· 杨玮,杨仁全,周增产,商守海,李东星,董明明(145)

蔬菜变量喷药研究与试验 ············· 马伟,王秀,郭建华(150)

相变蓄热技术应用于温室节能中的技术分析 ············· 梁浩,杨其长,方慧(159)

MATLAB 和 VB 在温室环境模型构建中的混合编程 …………… 孟力力,张 义,杨其长(165)
下挖式节能日光温室采光优化设计 ………………………… 李清明,艾希珍,于贤昌(175)
植物工厂炼苗系统及其配套装备
　　………………………… 刘文玺,张晓慧,周增产,卓杰强,商守海,李东星,李秀刚(183)
日光温室结构与环境因子相关性研究初探
　　……………………………………… 王克安,杨 宁,吕晓惠,王 伟,张卫华,柴秀乾(190)
LED 照明的发展潜力 ……………………………………………… Johann Buck,王 贺(译)(198)

设施栽培理论技术

植物照明中绿光比例对不同莴苣生长之影响 ………………… 张明毅,方 炜,邬家琪(202)
Production Pattern and Water Use Efficiency of Tomato Crops ……………… Li YaLing(209)
设施番茄砂培营养液配方筛选试验 …………………………… 高艳明,李建设,卜燕燕(213)
Reducing Nitrate Concentration in Lettuce by Continuous Light Emitted by Red and Blue LEDs
　　before Harvest ……………………… Zhou Wanlai,Liu Wenke,Yang Qichang(225)
不同红蓝 LED 对生菜形态与生理品质的影响
　　……………………… 闻 婧,杨其长,魏灵玲,程瑞锋,刘文科,孟力力,鲍顺淑,周晚来(232)
Effects of Silicon on Plant Growth and Antioxidan Enzyme Activities in Leaves of Cucumber
　　Seedlings under NO_3^- Stress
　　………………… Song Yunpeng,Wang Xiufeng,Wei Min,Shi Qinghua,Yang Fengjuan(240)
采前硝酸钙和氯化钾对水培生菜硝酸盐含量的影响 ………………… 刘文科,杨其长(243)
EM 菌对水培生菜生长和品质的影响 ………… 琚志君,刘厚诚,陈日远,孙光闻,宋世威(249)
Crop Management in Greenhouses:Adapting the Growth Conditions to the Plant Needs or
　　Adapting the Plant to the Growth Conditions? ……… L. F. M. Marcelis S. De Pascale(254)
温室番茄营养诊断研究初报
　　………………………… 郭建华,武新岩,王钟扬,王 秀,马 伟,张瑞瑞,徐 刚(267)
Influences of Air Inflated Film Covering and CO_2 Enrichment on Greenhouse Microclimate,
　　Growth and Yield of Tomato (*Lycopersicon esculentum* Mill)
　　……………… Min Wei,Yuki Tanoue,Toru Maruo,Masaaki Hohjo and Yutaka Shinohara(273)
影响日光温室 CO_2 浓度变化的因素与增施效果研究
　　……………………………………… 马 俊,贺超兴,闫 妍,张志斌,尹宏峰(276)
Exogenous Polyamines Enhance Cucumber (*Cucumis sativus* L.)Resistance to NO_3^- Stress,
　　and Affect the Nitrogen Metabolism and Polyamines Contents in Leaves of Cucumber
　　Seedlings ……… Wang xiuhong,Yang fengjuan,Wei Min,Shi QingHua,Wang xiufeng(284)
微生物有机肥在辣椒育苗中的应用效果研究 ………… 杨伟国,孙光闻,刘厚诚,宋世威(287)
新型超声波雾培装置及其系统设计简报 ……………… 程瑞锋,杨其长,魏灵玲,闻 婧(292)
设施内蔬菜水旱轮作治理连作障碍新模式 …………… 江解增,缪旻珉,曾晓萍,曹光亮(297)
LED 光源光质比对甘薯组培苗生长及电能消耗的影响 ……… 杨雅婷,杨其长,肖 平(304)
甜樱桃促成设施栽培调查报告 ………………… 孙玉刚,魏国芹,李芳东,秦志华,安 森(310)

Improving Utilization Efficiency of Electricity, Light energy, Water and CO_2 of a Plant Factory with Artificial Light

Toyoki Kozai

(Toyoki, Center for Environment, Health and Field Sciences, Chiba University, Kashiwa-no-ha, Kashiwa, Chiba 277-0822, Japan)

Key words: Closed system; COP; Electric energy utilization efficiency; Electricity consumption; Heat pump

There are two types of plant factory: one with solar light with or without supplementary light from lamps, and the other with artificial light only.

This paper deals with plant factory with artificial light only, and discusses methods of improving utilization efficiencies of the plant factory with respect to resources such as electric and light energy, water and CO_2.

It is thought by many people that electricity cost for lighting and cooling in the plant factory should be very high and that the plant factory will not be economically feasible for plant production. This paper shows that the electricity consumption and consumptions of other resources for plant production can be considerably reduced by improving the current environmental control method and cultivation systems, and by choosing plant species suitable for production in the plant factory.

This review paper is a reduced form of Kozai (2011).

1 Aims of plant factory

A simplest plant factory with artificial light is schematically shown in Figure 1, indicating that essential resources for the photosynthetic growth of a germinated seed with unfolded cotyledonary leaves or a transplant are water, CO_2, light, inorganic nutrients and a certain range of temperatures only.

Among the essential resources, light energy is the most costly resource for plant production in the plant factory with artificial light. Thus, achieving the highest light utilization efficiency, LUE, is of a primary importance in plant production in the plant factory.

Other environmental factors such as temperature, CO_2 concentration, water vapor pressure deficit, air current speed, nutrient solution composition, etc. are controlled primarily to maximize the LUE.

The second aim of environmental control in the plant factory is to utilize essential resources other than light (i.e., water, CO_2 and inorganic fertilizers) at their highest utilization efficiencies to obtain a maximum photosynthetic growth of plants with minimum consumptions of those resources and minimum emission of pollutants (Kozai et al., 2000; Kozai, 2005; Kozai et al., 2005). Consumptions of non-essential resources such as fossil fuel-derived plastics must be minimized.

The third aim is to obtain a highest quality and yield of produce and the fourth is to maximize the highest value of produce with minimum resource consumption and pollution.

2 Light utilization efficiency, LUE

2.1 Definitions of LUE and EUE

LUE for a certain period (hour, day or days required for one harvest), LUE_d, is expressed by Eq. (1) (Yokoi et al., 2005). Electric energy utilization efficiency, EUE_d, is expressed by Eq. (2).

"Light" in this paper means "photosynthetically active radiation" or, shortly, PAR (wave band: 400~700nm) (MJ m^{-2}). PAR energy accounts for about 98% of light energy emitted by standard fluorescent lamps, although it accounts for only about 50% of solar radiation (wave band: 300~3000nm) which spectrum varies with solar altitude, cloudiness and atmospheric transmissivity.

$$LUE_d = k \times D_n / PAR_c \qquad (1)$$
$$EUE_d = h \times LUE_d \qquad (2)$$

where k is the conversion factor from dry mass to chemical energy fixed in dry mass of plants (ca. 20 MJ kg^{-1}). The k value is slightly affected by weight percentages of carbohydrates, proteins and lipids in dry mass, but it is considered to be constant in this paper. D_n is dry mass increase in plants, its harvested part or its product, according to purpose (kg m^{-2}). h is the conversion coefficient of lamps from electric to PAR energy. Hourly LUE_p and EUE_p can also be expressed by Eqs. (3) and (4).

$$LUE_p = b \times C_f / PAR \qquad (3)$$
$$EUE_p = h \times LUE_p \qquad (4)$$

where b is the conversion factor from CO_2 fixed in plants to chemical energy fixed in dry mass of plants (MJ kg^{-1}). C_f is net photosynthetic rate of plants (kg CO_2 m^{-2}).

2.2 Representative Values of LUE

The average LUE_d over tomato seedling production in the closed system with artificial light was 0.027 (Yokoi et al., 2003), compared with the average LUE_d of 0.017 over tomato seedling production in the greenhouse (Shibuya and Kozai, 2001). Namely, the LUE in the closed system was 1.7 times the LUE_d in the greenhouse. Ohyama et al. (2000) show similar values for LUE_d and EUE_d for the plant factory.

2.3 Process of Energy Conversion

Figure 2 shows a scheme of electric energy conversion process to light, heat, and chemical energy in the plant factory.

As shown in Figure 2, only around 36% (100×8.9/25) of light energy emitted by lamps is generally absorbed by leaves, the rest is absorbed by culture beds, floor, walls, etc. and converted to heat energy in vain. On the other hand, this low percentage of 36% means that there is much room to improve the LUE by improving the lighting system of the plant factory. It is noted that approximately 99% of electric energy is converted to heat energy in the plant factory during photoperiod, which must be removed to the outside.

2.4 Factors Affecting LUE

Commercial production from plant factories is currently limited to value-added plants because electricity consumption for lighting to increase dry mass of plants is significant. Factors affecting LUE is elaborated in Kozai (2011), which basically correspond to the process shown in Figure 2 showing the issues requiring further improvement in lighting system of plant factory.

By improving the factors shown in Figare 2, LUE as for D_n can be expected to increase to 3.3% as shown in Figure 4 from 0.76 in Figure 2, being 4.3 times (=3.3/0.76).

3 Electricity consumption by components

Eqs. (5) and (6) are also relevant to consider the suitability of plant production in a plant factory. A_A is electricity consumption for air conditioning per floor area, not per cultivated area.

$$A_A = (e \times A_L + A_M + H_V) / COP \tag{5}$$

$$A_T = e \times A_L + A_A + A_M \tag{6}$$

where e is the cultivation area per floor area; $e \times A_L$) and A_M are, respectively, electricity consumption per floor area for lighting, and for air circulation fans, nutrient solution pump, etc. H_V is cooling load per floor area due to air infiltration and heat penetration through walls. COP is coefficient of performance of heat pump (or air conditioner) for cooling or ratio of heat energy absorbed by heat pump to its electricity consumption.

In Eq. (5), A_A will fall as COP of the heat pump rises. H_V increases with increasing the number of air exchanges of culture room per hour. The annual average COP for cooling in a closed system was 7.6 in Tokyo (Ohyama et al., 2002, Figure 5); approximately, 4 in summer and 10 in winter. COP was over 10 when the room air temperature is 25 C and outside air temperature is 5 C (Figure 5). On the average, ($e \times A_L$) accounts for 73% of A_T; A_A accounts for 12% of A_T; A_M accounts for 15% of A_T in Tokyo (Yokoi et al., 2003; Ohyama et al., 2003) (Table 1).

4 Electricity cost per plant and its percentage in total cost

Electricity consumption per plant can be estimated by dividing A_T in Eq. (6) by the number of plants per floor area. Electricity consumption for production of one tomato transplant is about 300~400 kJ and its cost in Japan is about 1 JPY (0.008 Euro or 1 US cent as of 2011) (Table 2).

In seedling or transplant production, the cost of electric power charges to the cost of producing seedlings in plant factories is only about 2%, but 25% for leafy vegetable production (Kozai 2007). This difference in cost is mainly due to much higher planting density in seedling production than in leafy vegetable production (around 500~1 000 and 50~100 plants per m^2, respectively).

Price of electricity per kWh differs from country to country. It is about 0.2 Euro in Japan but is 0.01 Euro in most Middle East countries.

By way of contrast, depreciation of the initial equipment/structure and labor costs account for approximately 30% each of production costs (Takatsuji and Mori, 2011). It follows that not only electric power charges and labor costs should be reduced, but also depreciation of the initial equipment/structure costs is essential because over-sized and inefficient equipment and structure are common at present.

5 Plants suited to production in plant factories

Electricity consumption for lighting per plant is proportional to E in Eq. (7).

$$E \propto PAR_C \times T \times P/J \quad (7)$$

where PAR_C is PAR flux received at culture bed; T is photoperiod per day; P is days required for cultivationt; J is the number of plants per cultivated area. Eq. (7) makes it clear that plants that can be produced with low light intensities, short (15 to 30 day) harvest cycles and high cropping densities will be suited to plant factory production. These include various types of grafted and rooted cuttings and seedlings, leafy vegetables, herbs and aromatic grasses and plants for herbal medicines (e.g., St. John's wort or *Hypericum perforatum*), small-rooted vegetables (*hatsuka daikon*, Japanese dwarf turnip and *wasabi* etc) and small high-end flowers (miniature roses and orchids).

On the other hand, plants suited to growing in greenhouses using sunlight rather than plant factories for improved quality and yields include fruit-type vegetables such as tomatoes, green peppers and cucumbers, leafy vegetables and herbs that contain large amounts of functional components, berries such as strawberries and blueberries, high-end flowers such as phalaenopsis, dwarf loquats, mangoes and grapes etc for growing in containers with trickle irrigation, and non-woody or annual medicinal plants such as angelica, medicinal dwarf dendrobium, Asian ginseng, saffron and *Swertia japonica*.

Plants that do not lend themselves to plant factory production are plants used primarily as sources of calories (carbohydrates, protein and fats) for people and livestock such as rice, wheat, corn and potatoes, plants such as sugarcane and rapeseed used primarily as fuel (energy) sources, larger fruit trees and trees used for timber such as cedar and pine and others including daikon, burdock and lotus. These plants require large areas for growth and have harvest cycles of several months to ten or more years, but they have relatively low value (prices) to mass.

6 Towards integrative environmental control

Set points of environmental factors need to be determined considering, in addition to LUE: (1) status of plant growth and development; (2) predicted yield, quality and value; (3) total costs of environmental control and resource inputs; (4) price of produce in the market; (5) weather forecast; (6) spread of pest insects and disease; (7) emission of pollutants including CO_2 gas;

(8) utilization efficiencies of water and CO_2. Thus, integrative environmental control of plant factory with use of a multi-purpose objective function is a challenging subject (Figare 6). For efficient integrative environmental control, use of heat pumps for cooling, air circulation and dehumidification is essential (Kozai et al., 2011).

7 Water and CO_2 utilization efficiencies

Water utilization efficiency, WUE, can be defined similarly to LUE, as shown in Eq. (8).

$$WUE = W_f/W_s = (W_s - W_c - W_r)/W_s \tag{8}$$

Where W_f is the water absorbed and kept in plants and substrate; W_s is water supplied or irrigated to the plant factory; W_c is water condensed at the cooling panel of heat pump for cooling and collected for its recycling use; W_r is the water vapor released to the outside through air gaps of the plant factory. In the plant factory with N of about $0.01~h^{-1}$, over 90% of W_s is condensed and collected as W_c. Thus, WUE is greater than 0.9. While WUE is lower than 0.02 in the greenhouse because W_c is zero and over 95% of W_s is released to the outside as W_r. This means that WUE is about 45 times greater in the plant factory than in the greenhouse (Water consumption in the plant factory is $1/45^{th}$ compared with that in the greenhouse).

CUE is defined by Eq. (9).

$$CUE = C_f/S_s = (C_s - C_r)/C_s \tag{9}$$

Where C_f is CO_2 fixed by plants or net photosynthetic rate; C_s is CO_2 supplied to the plant factory; C_r is CO_2 released to the outside through air gaps of the plant factory. W_r and C_r are, respectively, given in Eqs. (10) and (11).

$$W_r = d \times N \times V_a \times (AH_i - AH_o) \tag{10}$$

$$C_r = c \times N \times V_a \times (C_i - C_o) \tag{11}$$

Where d is specific weight of water vapor; N is the number of air exchanges per hour of the culture room; V_a is the air volume of the culture room; AH_i and AH_o are, respectively, absolute humidity inside and outside the culture room; c is specific weight of CO_2; C_i and C_o are, respectively, CO_2 concentration inside and outside the culture room.

Eqs. (10) and (11) show that reduction of N (increase in air tightness) is primarily important to reduce W_r and C_r and thus improve WUE and CUE (Figs. 7 and 8). N of the culture room in the plant factory should be preferably around $0.01 \sim 0.02~h^{-1}$. CUE is low at low LAI or low C_p (net photosynthetic rate) because C_r is constant and C_s increases with increasing LAI or C_p. Similarly, WAU is low at low LAI.

It should be noted that C_f in Eq. (3) can be estimated by Eq. (12). The C_p is an important variable for optimizing the set points of environmental factors. A method of continuous monitoring of N will be described elsewhere. In an integrative environmental control, monitoring and control of state variables such as C_p, C_r, W_s, A_T and maximizing LUE, WUE and CUE, in combination of monitoring and control of state variables such as temperature and humidity are essential.

$$C_p = C_s - C_r \tag{12}$$

8 Closed system

Understanding and introducing the concept of closed system with minimum N is essential in the design, construction and operation of plant factory. Its concept is schematically shown in Figure 8. In the "ideal" closed system, all resources inputted to the system are converted to produce with minimum generation of heat energy, resulting in no emission of pollutants to the outside and highest resource utilization efficiencies possible.

The closed system with minimum N is important not only for achieving maximum WAU and CUE, but also for preventing pest insects or pathogenic microorganisms from entering the closed system and for minimizing the environmental disturbance by weather outside. In order to minimize the thermal disturbance, the walls of the closed system need to be thermally well insulated (Kozai et al., 2000).

This closed system for plant production is schematically shown in Figure 8. This closed system was commercialized in 2005 in Japan (Kozai et al., 2006; Kozai 2007) and has been used for commercial plant production of transplants, leafy vegetables and herbs at about 150 locations in Japan.

Conclusion

Factors affecting utilization efficiencies of plant factory with artificial light are analyzed and methods to improve those efficiencies are discussed. It is suggested that current light energy utilization efficiency can be doubled or even tripled in the future. Electricity consumptions for lighting and cooling account for, respectively, about 73% and about 12% of total electricity consumption, in case that the plant factory is thermally well insulated. It is also indicated that the closed system is an important concept to design the plant factory having high utilization efficiencies of light, water, CO_2 and inorganic fertilizers. Plant factory technology will contribute to producing horticultural crops with minimum resource consumption and minimum emission of pollutants. Discussion is given in more detail in Kozai (2011).

References

[1] Kozai, T., Kubota, C., Chun, C., Afreen, F., and Ohyama, K. 2000. Necessity and concept of the closed transplant production system, p. 3~19, In: C. Kubota and C. Chun (eds.) Transplant Production in the 21st Century. Kluwer Academic Publishers, The Netherlands

[2] Kozai, T., 2005. Closed systems for high quality transplants using minimum resources (In: Plant Tissue Culture Engineering, SBN: 1-4020-3594-2, (eds. Gupta, S. and Y. Ibaraki, 480pp.), Springer, Berlin. 275~312

[3] Kozai, T., F. Afreen and S. M. A. Zobayed (eds.). 2005. Photoautotrophic (sugar-free medium) micropropagation as a new micropropagation and transplant production system, Springer, Dordrecht, The Netherlands, 315

[4] Kozai, T., K. Ohyama and C. Chun, 2006. Commercialized closed systems with artificial lighting for plant production, Acta Hort. 711, 61~70

[5] Kozai, T. 2007. Propagation, grafting, and transplant production in closed systems with artificial lighting for commercialization in Japan, J. Ornamental Plants. Vol. 7 (3): 145~149

[6] Kozai, T., K. Ohyama, Y. Tong, P. Tongbai, and N. Nishioka. 2011. Integrative environmental control using heat pumps for

reductions in energy consumption and CO₂ gas emission, humidity control and air circulation. Acta. Horti. (in press)

[7] Kozai, T. 2011. Improving light energy utilization efficiency for a sustainable plant factory with artificial light, Proceedings of Green Lighting Shanghai Forum 2011, Shanghai, China, May 11~13, 2011

[8] Ohyama, K., K. Yoshinaga, and T. Kozai. 2000. Energy and mass transfer of a closed-type transplant production system (Part 1) - Energy balance -. J. SHITA. 12 (3): 160~170 (JE)

[9] Ohyama, K., T. Kozai, C. Kubota, C. Chun. 2002. Coefficient of performance for cooling of a home-use air conditioner installed in a closed-type transplant production system. J. SHITA, 14 (3): 141~146 (JE)

[10] Ohyama, K., K. Manabe, Y. Ohmura, C. Kubota and T. Kozai. 2003. A comparison between closed-type and open type transplant production systems with respect of quality of tomato plug transplants and resource consumption during summer. Environment Control in Biology. 41 (1): 57~61 (JE)

[11] Salisbury, F. B. and C. W. Ross. 1991. Plant physiology, page 609, Wadsworth Publishing Company, USA, 682

[12] Sager, J., J. L. Edwards, and W. H. Klein. 1982. Light utilization efficiency for photosynthesis. Transactions of the ASAE. 1 737~1 746

[13] Shibuya, T. and T. Kozai, 2001. Light-use and water-use efficiencies of tomato plug sheets in the greenhouse. Environment Control in Biology. 39 (1): 35~42 (JE)

[14] Takatsuji, M. and Y. Mori, 2010. LED Plant factory, Nikkan Kogyo Co. (in Japanese) . p. 4. (141pp.)

[15] Yokoi, S., T. Kozai, K. Ohyama, T. Hasegwa, C. Chun and C. Kubota, 2003. Effects of leaf area index of tomato seedling population on energy utilization efficiencies in a closed transplant production system. J. SHITA, 15 (4): 231~238 (JE)

[16] Yokoi, S., T. Kozai, T. Hasegawa, C. Chun and C. Kubota. 2005. CO₂ and water utilization efficiencies of a closed transplant production system as affected by leaf area index of tomato seedling populations and the number of air exchanges. J. SHITA, 18 (3): 182181 (JE)

[17] Yoshinaga, K., K. Ohyama and T. Kozai, 2000. Energy and mass balance of a closed-type transplant production system (Part 3) - Carbon dioxide balance -, J. SHITA, (4): 225~231 (JE)

Note: JE in parenthesis denotes that the abstract and figure and table captions of the paper are written in English but text is written in Japanese.

发展设施蔬菜低碳生产技术对策

张志斌

(中国农业科学院蔬菜花卉研究所 北京中关村南大街12号,北京 100081)

摘要:中国是一个设施园艺大国,发展设施蔬菜低碳生产技术对适应全球变暖并减缓温室气体排放有重要意义。本文结合我国设施蔬菜低碳生产存在的问题和发展设施蔬菜低碳生产技术对策,从优化设施结构、研发节能减排技术、资源高效利用技术、建立设施蔬菜低碳生产技术体系研究等方面进行进行探讨,并提出有关建议,仅供参考。

关键词:设施蔬菜;低碳生产;节能减排

Development Countermeasure of Protected Vegetables Low Carbon Production Technology

Zhang zhibin

(Institute of vegetables and flowers Chinese Academy of Agricultural Sciences Zhongguancun Nandajie No. 12 Beijing 100081 China)

Abstract: China is a large country for protected Horticulture. It is important of vegetables low carbon production technology to adapt to slow global warming and greenhouse gas emissions. Some problems and developing technical countermeasures are discussed, including optimization facility structure, development of energy saving and emission reduction technology, resource efficient utilization technology, establishing protected vegetables low carbon production technology system etc.

Key words: Protected vegetable; Low coal production; Energy saving and emission reduction

气候变化对全球环境和生态系统已产生着重要影响,发展低碳农业的目标之一就是使农业生产系统适应全球变暖并减缓二氧化碳(CO_2)为主的温室气体排放。

低碳农业是以低能耗、低排放、低污染为基础的农业生产模式,其关键在于提高农业生态系统对气候变化的适应性并降低农业发展对生态系统碳循环的影响,维持生物圈的碳平衡,其根本目标是促进实现碳中性,即人为排放的CO_2与通过人为措施吸收的CO_2实现动态平衡。

中国是一个设施农业大国,2010年中国设施蔬菜(包括西甜瓜)种植面积达400多万hm^2,设施蔬菜占设施园艺面积的95%以上,占世界设施蔬菜总面积的80%以上。设施蔬菜年种植面积比2004年翻了近一番,占全国蔬菜种植总面积的20%左右。设施蔬菜总产值达7 000亿元,占蔬菜总产值的60%左右。实践证明,设施蔬菜产业在中国一些区域已成为农

---

* 张志斌,男,研究员,博士生导师,中国农业科学院蔬菜花卉研究所,北京中关村南大街12号,100081,电话:010 - 82109507, E-mail: zhangivf@ yahoo. com. cn。

业的支柱产业,也成为现代农业的重要标志。因此,在设施农业领域推行温室气体减排和适应气候变化措施,发展低碳生产技术,对提高设施农业应对气候变化能力,促进其可持续发展有重要意义。也为中国政府公布的《节能中长期专项规划》中,争取 2010~2020 年的 10 年中把年均节能率提高到 3% 做贡献[1]。

1 中国设施蔬菜低碳生产存在的问题

1.1 设施资源高效利用技术水平低

中国目前设施蔬菜资源利用率低,缺乏配套的资源高效利用和设施蔬菜低碳生产技术,主要表现在能源、水资源、土地资源及劳动力资源等方面。长期以来设施蔬菜生产多偏重于获得高产,不惜投入大量的资源,肥料、能源和水资源等浪费严重,中国设施农业单位面积水资源的利用率仅为以色列的 1/5 左右,而且肥料利用率更低,如氮素利用率只有 30%~35%,与先进国家相比低 20 个百分点左右,不仅造成资源浪费,还会引起面源污染,严重影响中国设施农业的持续高效发展。大型温室,特别是国产的大型温室环境的调控能力比较差,表现在耗能高、单产低、年利用率不高,与国际先进设施农业国家荷兰相比,荷兰地处北纬 52°,而中国北纬 40°左右温室燃煤消耗 40~100t/(667m^2·年),能源耗费占生产成本的 50% 以上,而荷兰能源耗费占生产成本的 20% 左右。中国日光温室虽节能性能较好,但日光温室的土地利用率仅 50% 左右,绝大多数日光温室设备比较简陋,生产设施和配套设备总体水平较低,环境控制的现代化水平和劳动生产率低。

1.2 产品安全质量急需进一步提高

中国的节能型日光温室和塑料大棚,棚室优化环境能力有限,与露地生态系统相比,棚室环境中具有温差大、高湿和弱光等特点,病虫害易于发生。由于设施结构不合理,加上环境调控能力差,造成病虫害大量发生,致使农药使用过量,给产品造成污染,从而降低了产品效益。有些温室蔬菜作物生产为了单纯追求产量,盲目过量施用化肥,重茬连作,产品和土壤污染严重。随着设施作物栽培年限的增加,引起土壤微生物种群的改变、土壤结构的破坏和次生盐渍化以及养分障碍的发生,造成土壤质量退化。设施连作障碍已经成为影响中国设施农业土壤资源持续高效利用的重要瓶颈。据估计,中国常年发生的重要设施园艺作物病虫害多达百种以上,而造成严重危害的约 50 余种,产量损失超过 25%,防治设施园艺作物病害药剂不合理施用,严重污染园艺产品和生态环境[2]。

1.3 设施蔬菜低碳生产配套栽培技术体系尚未建立

中国设施蔬菜低碳生产从品种到栽培技术和病虫害防治等栽培技术体系尚未建立,设施蔬菜专用品种的研究与利用相对较落后,特别是低能耗、抗逆强、高品质的设施专用品种的研究与开发力度不够,一些优质主栽品种如适宜设施环境种植的黄瓜、番茄和甜椒等还依赖进口,具有完全自主知识产权的专用品种尚不能占领主导市场。由于缺乏设施蔬菜低碳生产栽培技术体系,中国目前设施蔬菜的单位面积产量、质量、效益和劳动生产率与国外相比还有较大的差距。设施蔬菜平均年产量仅约为 15kg/m^2 左右,是国际先进国家设施蔬菜产量的 1/3~1/2。特别是单位能量蔬菜产出率较低,严重影响了设施蔬菜产品的高效益,影响中国设施蔬菜生产的持续性发展。要实现低碳发展和可持续发展,节能减排是一种重要的方式和手段。节能就是在尽可能地减少能源消耗量的前提下,获得与原来等效的经济产出,有效地利用能源，提高能源利用效率[3]。

2 设施蔬菜低碳生产技术对策

2.1 优化设施结构，提高节能效率

日光温室和塑料棚是中国设施蔬菜栽培的主要类型，日光温室本身节能减排优势突出，目前全国节能日光温室约70万hm^2，按每公顷年节煤750t计算，每年可节约5亿t煤炭，约占全国煤炭消耗量的1/5。按每吨标准煤燃烧放出CO_2 2.5t计算，每年减排二氧化碳12.5亿t。

应对节能减排，要进一步优化温室结构，研发合理采光、减少热损失和夜间保温技术，提高日光温室节能效果。开发节能性强的日光温室，要保持采光性、保温性优于目前温室的前提下，进行大跨度日光温室的自动化全季节安全利用研究，提高空间利用率、土地利用率和抗灾能力。研究开发不同纬度地区大跨度日光温室结构材料、优化设计技术原理与模型。优化的新型日光温室在日照60%以上地区冬季正常天气夜间最低温度应高于外界30℃以上。研究亚热带地区冬夏两用连栋温室结构性能优化设计技术原理。开发出具有节能、节水、节药、节肥功能，具有自主知识产权的工程技术装备。实现中国设施农业高技术的国产化。

2.2 研发设施综合节能减排技术

设施覆盖材料主要是塑料薄膜和夜间保温覆盖材料，研发高透光、高保温多功能覆盖材料，大幅度提高覆盖材料的透光率、增加太阳能的入射量。同时开发防止设施内部长波向外辐射，如对温室覆盖材料的内侧进行镀膜处理，以阻止长波向外辐射，减少热损耗。同时研发保温技术与材料设备，提高设施夜间保温能力。保温材料可研发墙体和后坡面的蓄热性能好、隔热性强的复合或相变材料等。进行便于机械化操作的多层覆盖技术、合理通风等量化节能管理技术提高节能效果。

研发热能的多用途利用和余热回收技术，开发浅层地能的利用，利用土壤作为蓄热源，夏季把低温冷源抽到地上，用于温室降温，把经过热交换的热量打到地下，冬季把高温热源抽上来，在热泵作用下，升温至45~50℃，只需要稍许加温就可以用于温室采暖，节能幅度达60%以上。

研究提高单位能量的作物产出率相关技术，重点研究设施逆境管理技术、作物防衰老技术等提高单位面积蔬菜产量，主要是低温、弱光、高湿、高温、强光等亚适宜环境作物的响应机理与代谢调控技术，研究散射光利用技术。通过配套机械工程和微电子技术，使设施内温度、湿度、光照、水分、营养、CO_2浓度等综合环境自动调控到作物生育所需的最佳状态，生产作业高度自动化和机械化，达到科学配置利用资源、能源，提高土地利用率、劳动生产率和优质农产品产出率，大幅度提高作物单产水平。荷兰温室蔬菜单位产量的耗能率大大降低原因之一，就是通过综合高新技术应用，温室番茄年平均产量可达40~50kg/m^2，黄瓜年产60kg/m^2以上，商品率高达90%以上，降低了温室蔬菜单位产量的耗能率。

2.3 研究开发设施蔬菜减药技术

减少化学农药使用，也是低碳农业的重要内容。随着全球经济的发展和社会的进步，人们对生活的质量和食品的品质产生了独特的要求，追求纯天然、无污染的健康食品已成为一种时尚，在设施农业发达的国家，利用生物防治技术、生态调控技术防治设施蔬菜病虫害越

来越普遍。荷兰设施蔬菜生物防治率已达90%以上,但中国这样一个设施农业大国,根据目前中国蔬菜生长的复杂自然环境以及目前中国的科技水平,防治蔬菜病虫害仍然要使用化学农药[4]。但要正确区分可以合理使用的农药和不可以使用的违禁农药。要有安全间隔期,以避免农产品农残超标。同时要加强农药使用技术的研究,就是要用最少量农药达到最佳防治效果目的,中国农药有效利用率只有10%～30%,远低于发达国家50%的平均水平,喷撒的大部分农药流失到环境中,造成了严重的环境污染和人畜中毒[5]。研究采用生物防治和物理防治手段相结合进行综合防控,尽量减少化学药剂的使用,努力实现蔬菜自身和对环境的零污染。

为保障设施蔬菜减药,需要研究设施条件下病虫害发生的规律和控制技术,重点研究主要设施病虫害的发生的规律和生化与分子早期诊断技术,设施毁灭性病虫害的高效专一性天敌和生物制剂的筛选与研制,天敌与生物制剂规模化生产技术及其在不同设施环境下的应用技术。

针对设施栽培中土壤线虫带来的毁灭性危害问题,研究土壤耕作强度、方式、作物茬口、土质、土壤pH值、土壤微生物和杀线虫剂对线虫发生、流行与控制的影响,研究综合治理技术。根据线虫流行、加重的生态规律,在化防大幅度抑制的基础上,寻求出初步的微生物防治手段,并最终使使土壤生物环境趋于平衡。

2.4 研究开发资源综合高效利用技术,建立设施蔬菜低碳生产技术体系

开展低能耗、资源高效利用、无污染等设施蔬菜低碳综合生产技术研发,研究利用农业废弃物等建立设施蔬菜低碳生产技术体系。

中国每年农产秸秆6亿t左右,若燃烧将造成CO_2排放和空气污染,研究利用技术意义重大。实验表明,大气中的CO_2浓度约为300μl/L左右,而当设施内CO_2浓度大于大气中的2～3倍时,大部分蔬菜产量可以显著提高。CO_2浓度充足可使蔬菜提早上市,减少农药用量,抑制硝态氮的反硝化,改善作物品质。一般瓜类、茄果类蔬菜CO_2饱和点浓度为1 500μl/L左右,而设施内白天在日出后至上午放风前CO_2浓度往往不足,甚至低于外界大气中的CO_2浓度,远远不能满足设施蔬菜光合作用对CO_2的需要。所以研究利用CO_2施肥技术和开发"碳基肥料"不仅可以提高农作物的产量,而且还可以大幅度减少CO_2来发展低碳农业。

研究沼气利用和设施环境土壤有机栽培关键技术,研究开发秸秆等农业废弃物无害化快速腐熟技术,栽培系统的施肥、灌溉等量化管理及病虫害全过程控制的绿色化技术。集成组装适合中国国情的土壤有机栽培生产体系。研究替代草炭和岩棉的无土栽培生态型基质和栽培基质理化性质的优化配比,研究作物营养与水分精准管理技术,基质与营养液的环境友好型消毒技术,建立园艺作物高产复合生态基质无土栽培的操作技术规范。

研究配方量化科学施用肥技术,通过测土壤养分,根据作物生长所需的养分提出营养配方。重点是提高化肥利用率和减少化肥用量技术。按提高氮肥利用率10%计算,中国每年可以比现在少损失氮肥250万t,折成尿素则为540万t。中国尿素约占氮肥的50%以上,按生产1t尿素需要1t标煤计算,可减少1 350万tCO_2排放。研发缓释肥料应用技术,使肥料中养分化合物在土壤中释放速度缓慢或者养分释放速度可以得到一定程度的控制、以减少肥料养分特别是氮素在土壤中的损失,减少施肥次数,节省劳力和费用。

研究开发节水模式和雨水收集利用技术，推广蔬菜节水灌溉技术和蔬菜渗灌、微喷灌、滴溉和雨水收集利用等技术。

总之，通过低能耗、抗逆和抗病性强的设施专用品种选育和低能耗、资源高效利用、无污染等设施蔬菜综合生产关键技术研发，研究建立适合中国国情的设施蔬菜低碳生产技术体系。为了尽快建立该栽培技术体系，提高设施蔬菜产品的科技含量和国际竞争力，建议国家应确立农业科技专项向设施蔬菜低碳技术研究倾斜，列入国家有关部委的国家"支撑计划"、"863"计划等重点项目，并将设施蔬菜低碳技术示范与产业化列入国家重大专项。

参考文献

[1] 邢继俊，赵刚. 中国要大力发展低碳经济. 中国科技论坛，2007（10）：87~92
[2] 张志斌. 我国设施蔬菜存在的问题及发展重点. 中国蔬菜，2008（5）：1~3
[3] 付允，马永欢，刘怡君，牛文元. 低碳经济的发展模式研究. 中国人口·资源与环境，2008（3）：14~19
[4] 李宝聚. 安全蔬菜是种出来的——对海南豇豆事件的思考. 中国蔬菜，2010（5）：1~2
[5] 我国农药使用技术存在的问题和认识"误区". 中国农药网，2010-4-15

Optimal Greenhouse Cultivation Control: How to Get There?

Gerrit van Straten*

(*Wageningen University, Systems & Control Group Wageningen, The Netherlands*)

Abstract: Greenhouses create a protected environment for the crop, and provide a means for the grower to steer his cultivation. Over the world, the technical options for adjustment of the inside conditions differ widely. The more advanced the greenhouse equipment, the more advanced are the control options, and the more complicated the control task becomes.

Classically, local on-off, P and PI controllers were added to each piece of equipment, such as ventilation control, heater control, control of CO_2 dosage, fogging control and screens, and more. As a result, the control algorithms have become overcrowded with user defined settings, and the coordination between the loops causes problems. Multivariable control methods, although available in theory, have found little application. In any case, the philosophy is to use control to maintain a pre-set climate. The settings defining the day-night and seasonal trajectories to create acceptable production originate from experience and blue-print advice.

The emergence of better crop models and more advanced computing methods allows, in principle, a paradigm shift. Not the indoor climate should be the target, but the cultivation as a whole. Optimal control methodology allows to maximize, on-line, an economic goal function that balances future income from harvesting the crop against current costs of operating the nursery. The 'Wageningen' method entails a decomposition of the problem in two parts: an open loop dynamic optimization over the season for nominal smooth weather, plus, on-line, in closed loop, a receding horizon optimization that exploits the short term weather forecast. The result of the dynamic optimization step is a shadow price for a unit of crop biomass, which can then be used as input for the feed-back step to allow the short term balancing of resource costs such as energy and CO_2 against the marginal value of the accrued crop.

Examples of applications for tomato and sweet pepper show that energy savings of 15% over already optimized classical control can be achieved, under equal production. The temperature and CO_2 in the greenhouse are steadily fluctuating; there is no more set-point tracking. Moreover, the optimal control results in much better compliance with hard constraints set by the grower for crop health and vulnerability to pests and diseases.

The advantages of this approach are, among others: far less settings by the grower; optimal exploitation of weather; easy adaptation to energy prices. Nevertheless, the method is not easily adopted in the market. Possible reasons might be: the limited scope of models; the feel with the grower to be out of control; perceived less transparent operation; problematic adjustment to equipment change and the level of knowledge required at the computer control suppliers. The talk aims at discussing possible remedies and pathways to promote further economic optimization in greenhouse horticulture operation.

References

Van Straten, G., Van Willigenburg, L. G., Van Henten, E. J., Van Ooteghem, R. J. C. *Optimal Control of GreenhouseCultivation*, CRC Press, 2011 (ISBN 978-1-4200-5961-8).

---

* E-mail: Gerrit. vanstraten@ wur. nl.

植物工厂与垂直农业及其资源替代战略构想

杨其长

(中国农业科学院农业环境与可持续发展研究所,北京 100081)

摘要:农业是资源消耗型产业,因传统农业资源(养分、水、土地等)的日益短缺和枯竭,严重限制了农业的可持续发展潜力。根据农业生产的因子综合作用律,辩证考虑环境要素(温度、光照和 CO_2 气体等)和传统养分、水、土地等资源在农业生产中的协同关系,提出了通过强化环境要素提高传统农业资源的利用率,增强农业可持续发展能力,实现"资源替代"的战略构想。文中还介绍了实现"资源替代"战略的技术途径,并着重介绍了"植物工厂"与"垂直农业"在实现"资源替代"战略方面的作用。

关键词:农业资源;环境;替代战略;植物工厂;垂直农业

Strategy of Resources Replacment with Plant Factory and Vertical Farming

Yang Qichang

(*Institute of Environment and Sustainable Development in Agriculture, Key Lab. for Agro-Environment & Climate Change, Ministry of Agriculture, Chinese Academy of Agricultural Sciences, Beijing 100081 P. R. China*)

Abstract: Shortage and exhaustion of traditional agriculture resources limits the sustainable development of agriculture for the heavy resources-consumption in agricultural production. Based on the synergistic relationship in agricultural production between environmental factors (temperature, lighting and CO_2 etc.) and traditional agricultural resources (nutrients, water and farmland etc.), a strategic idea, was put forward to strengthen the sustainability of agriculture. Technological strategy of resource replacement was introduced, emphasized on the effect of plant factory and vertical farming in resource replacement realizing.

Key words: Agricultural Resources; Environment; Replacement strategy; Plant factory; Vertical farming

农业是国民经济的基础,同时也是资源消耗型产业。传统的农业生产需要利用大量的土地、水和矿质养分等不可再生资源,同时由于受气候条件的限制,农业生产具有时空局限性以及众多的不确定性。当前,中国农业的资源环境问题极为突出,农业资源数量和质量的下降趋势明显,农用土地资源与水资源日益短缺,土壤质量不断退化,面源污染发生面积和程度不断扩大,极端气候及自然灾害频发。此外,城市化的快速发展,又大量占用土地等农业资源。这些问题严重制约了中国农业的可持续发展。因此,大力推进科技进步,挖掘农业资源潜力,突破农业的时空局限,实现资源替代,势在必行。

1 必要性和意义

中国是一个农业人口众多、人均资源极度匮乏的国家。耕地约占全国土地总面积的

14.2%,其中中低产田约占 2/3。截至 2009 年年底,中国的耕地实际保有量为 1.217 亿 hm^2,已逼近温家宝总理在《政府工作报告》中宣布的耕地 1.2 亿 hm^2 保护"红线"。此外,由于城市发展、基础设施建设等对耕地的占用,我国每年减少耕地量在 30 万 hm^2 左右,如果按这样的速度计算,我国耕地面积从 1.217 亿 hm^2 下降到 1.2 亿 hm^2 的红线,也就只有 5 年左右的时间。中国目前的人均耕地占有量不到 $0.09hm^2$,是世界平均水平的 1/3。中国的人均水资源占有量 2 300m^3,仅为世界平均水平的 1/4,排在世界第 88 位,是全球人均水资源最贫乏的国家之一。中国还是世界上最大的氮肥消费国,虽然氮肥来源于大气,但高生产能耗($2.4 \times 1\,010kJ/t$)和粗放施用所造成的严重环境污染(硝酸盐污染)问题也日趋严重。由于在矿质养分资源的管理和高效利用方面缺乏有效措施,粗放施用现象严重,磷、钾、钙和硫等矿物质养分资源面临枯竭危险。此外,由于养分资源的浪费以及农用化学品的过量施用,农产品品质和食品安全形势也日益严峻。

虽然中国实行了严格的计划生育政策,但人口的惯性增长仍将持续相当长的一段时期,预计到 2030 年中国人口将达到 15 亿。同时,随着人口的增长和社会经济的快速发展,城乡居民对农产品在数量和品质上的需求也将不断上升,农业资源紧张的态势日益显现,已经成为社会关注的热点问题。为此,加快科技进步,改进中国粗放的农业资源利用方式和生产模式,通过以农业环境工程为基础的高效农业的发展,大幅度提高资源利用效率,从而实现资源效益的倍增与替代,将是中国农业由资源消耗型向资源节约型转变的重要途径之一。

2 资源替代战略思路

根据因子综合作用律,作物的生长发育受到各种外部因子(水、肥、气、热、光及其他农业技术措施)的影响,只有在其他要素满足的前提下,才能充分发挥某一或多个因素的高效利用。因子综合作用律认为,作物产量形成是由影响作物生长发育诸因子综合作用的结果,但其中必然有一个起主导作用的限制因子,作物产量在一定程度上受该限制因子的制约。在传统的农业生产中,由于受自然气候的限制(温度和光照等),农业生产具有鲜明的地域和季节性特征。农民只能通过强化土壤肥力、大量用水、用肥和用药等措施来提高特定环境下农作物的产量。追求高产的欲望导致过量施肥、灌水现象相当普遍,造成不可再生资源的大量浪费,由此造成的环境污染等生态负效应更是长期的且难以修复。实际上,按照因子综合作用律理论,农业的资源要素包括两大类,即不可再生资源和可再生资源,前者如土地、水和矿质养分等,后者包括光照、温度和二氧化碳等。前者可称为传统的资源要素,而后者可称为环境要素。从时空角度而言,在传统农业生产中环境要素是利用程度很低的资源,同时也是限制传统农业种植制度、产量和当家品种的决定性因素。

综合上述思路,作者提出了"资源替代"的战略构想。该构想的理论核心是通过环境与工程技术手段,创建或优化农业生产中的环境要素(光照、温度、湿度、CO_2 等可再生资源),以达到提高不可再生资源(土地、水和矿质养分)的利用效率(相当于通过环境手段使不可再生资源大幅度增殖),实现资源高效利用的现代农业生产模式。资源替代的表现形式主要包括:①通过环境要素的优化,大幅提高单产;②环境控制技术的突破,实现作物的周年连续生产;③现代工程技术的突破,使作物栽培向立体空间拓展;④建筑工程与农业技术的结合,使农业向垂直空间发展。实施"资源替代战略"可以大幅度提高农业资源的利用效率，缓解中国农业资源短缺的压力，实现农业的可持续发展。

资源替代战略的实现途径将是以设施农业工程为基础、以环境与营养要素的综合调控为手段、以时空最大化利用为目标的一种超高效生产方式，其中最具有代表性的模式、也被认为是现代设施农业的最高级发展阶段的就是"植物工厂"与"垂直农业"。

3 资源替代的技术途径

"资源替代战略"是通过环境与工程技术手段，创建或优化农业生产的环境要素，以达到提高不可再生资源的利用效率，实现资源高效利用的一种现代农业手段。"资源替代战略"的技术核心是环境的相对可控性、时间上的持续性以及立体空间的可延伸性，主要技术途径包括：

（1）通过强化对光照、温度、湿度和 CO_2 等环境要素的优化，以及水肥资源的精准控制，大幅度地提高作物的单产水平和资源利用效率，变相地实现以环境替代资源；

（2）通过环境与工程技术的突破，实现作物的周年连续生产，使作物生产不受或很少受外界气候条件的制约，甚至可以在戈壁、沙漠、建筑屋顶、地下、废弃荒地、水面、太空、星球等场所进行作物生长，变相地实现耕地资源的增加；

（3）通过栽培工程、营养液循环控制与人工光源技术的突破，使作物栽培由单层逐渐向立体空间拓展，使单位土地的空间利用率成倍提高；

（4）通过垂直空间利用以及清洁能源与资源高效循环技术的突破，实现在摩天大楼进行农业生产，使耕地资源得到几何级数的增加。

当前，我国在实现"资源替代战略"的技术研究方面还相对薄弱，与日本、荷兰、美国等发达国家相比差距明显，植物工厂研究刚刚起步，垂直农业还未被认识。因此，要真正实现"资源替代"的战略构想，必须加大技术的研发力度，逐渐缩小与发达国家的差距。

4 "资源替代战略"的实现方式—"植物工厂"与"垂直农业"

4.1 植物工厂

植物工厂（Plant factory）是一种通过设施内高精度的环境控制实现作物周年连续生产的高效农业系统，是由计算机对作物生育过程的温度、湿度、光照、CO_2 浓度以及营养液等环境要素进行自动控制，不受或很少受自然条件制约的新型生产方式。植物工厂的突出优势：①作物生产计划性强，可实现周年均衡生产；②生长速度快，周期短，单位面积产量高，效益好；③机械化、自动化程度高，劳动强度低，用工少，工作环境舒适；④不施用农药，产品安全无污染；⑤多层立体栽培，节省土地和能源；⑥不受或很少受自然气候的影响。

植物工厂的技术核心主要包括：植物人工光源（LED）、制冷-加热调温控湿、光照-CO_2 耦联调控、空气均匀循环与流通、营养液（EC、pH、DO 和液温等）在线检测与控制、立体栽培、图像信息传输、环境数据采集与自动控制等，可实时对植物工厂的温度、湿度、光照、气流、CO_2 浓度以及营养液等环境要素进行自动监控，实现智能化管理。随着物联网技术的发展，植物工厂还可通过物联网技术的应用，实现在任何地点利用手机、笔记本电脑、PDA 等终端随时了解蔬菜长势，调整控制参数，进行在线管理与远程监控。在植物工厂系统栽培的叶用莴苣从定植到采收仅用16~18d时间，比常规栽培周期缩短40%，单位面积年产量为露地栽培的25~40倍。

植物工厂技术的突破彻底解决了人类发展面临的诸多困境，不仅能大幅度提高资源利用效率，为人类提供洁净安全的食品，而且还可以实现在戈壁、荒漠、海岛、极地等非可耕地，甚至在建筑屋顶、地下、居民家庭里进行作物生产。利用取之不尽的太阳能，加上一定的种子和矿质营养，就可源源不断地为人类生产所需要的食品。

植物工厂已经成为资源紧缺的国家如日本、韩国等解决资源问题的重要途径，也必将是我国未来农业"资源替代战略"的重要手段之一。

4.2 垂直农业

垂直农业（Vertical farming），也称摩天大楼农业或大厦农业（Skyscraper Agriculture），就像我们居住的房屋从平房向高楼发展一样，耕地也从农田向多层空间的农场发展。垂直农业是通过在人工构筑的多层建筑里模拟农业生物所需的环境，实现在高楼大厦内进行农业生产，大幅度提高土地利用效率和产量的农业发展模式、途径和方法。垂直农业不仅是未来农业发展的最高级生产方式，而且也是资源超高效利用的理想模式，其土地利用效率将达到露地的几百倍甚至数千倍，是未来解决土地紧缺问题的一种有效途径。

最早期的垂直农业设计方案是由美国设计师克里斯-雅克布斯和迪克森-戴斯波米尔共同完成的。在塔形建筑物顶部，设计有一个巨大的太阳能电池板，可以随着太阳位移而转动。窗户也很特别，具有防污染功能，不易凝结水珠，主要为了让其中生长的植物能够照射到更多的阳光。据有关专家估计，在纽约市建立150个左右这样的垂直农场就可满足全市1 800多万人口的粮食与蔬菜供给。

虽然垂直农业模式还没有真正在生产上应用，但世界各国的科学家与建筑学家都在积极研究最优的垂直农场模式，目前包括美国、英国、法国、加拿大、瑞典、比利时、荷兰、丹麦以及中国的设计师在内的相关专家，已经设计了几十种颇具创意的垂直农场或摩天大楼农业方案，相信在不久的将来一定会出现真正意义上的垂直农场。

5 结论

农业可持续发展的关键在于能否保证土地、水和矿质养分资源的高效利用，"资源替代战略"构想是基于作物发育的基本原理提出的。通过环境要素的优化与控制、立体种植以及垂直空间的拓展，完全可以实现资源的高效利用与变相替代，从而为解决资源短缺、人口膨胀、环境污染等人类面临的诸多问题作出贡献。

"资源替代战略"的技术核心是环境的相对可控性、时间上的持续性以及立体空间的可延伸性，植物工厂与垂直农业完全符合这种技术的要求，是实现"资源替代战略"的主要技术途径，将会在未来"资源替代"以及解决人类人口、资源、环境等诸多问题中发挥重要作用。

参考文献

[1] 张燕, 张洪. 农业转型与资源替代. 经济地理, 2001, 21 (6): 719~7
[2] 方小教. 农业资源多维替代路径的解析. 安徽教育学院学报, 2005, 23 (5): 36~40

[3] 张岳. 资源替代发展战略及其取向. 中国水利, 2004 (6): 22~23
[4] 张成波, 杨其长. 植物工厂现状与发展趋势. 2004, 华中农业大学学报（增刊）, 2004.12, 总第35期: 34~37
[5] 杨其长, 张成波. 植物工厂概论. 北京: 中国农业科学技术出版社, 2005
[6] 汪懋华. 设施农业的发展与工程科技创新. 北京: 北京出版社, 2000
[7] 王玉英. 激光植物工厂的现状与未来展望. 光机电信息, 2005 (1) 8~13
[8] 魏灵玲, 杨其长, 刘水丽. LED在植物工厂中的研究现状与应用景. 中国农学通报, 2007, 23 (11): 408-411
[9] 刘文科, 杨其长, 张成波. 中国农学通报专刊: 论设施农业在循环业中的运作模式//. 2006年中国农学会学术年会论文集, 2006: 282~284
[10] 王顺清, 刘文科. 农业可持续发展的有效途径一种养结合生态模式的探讨. 中国畜牧工程, 2005, 6 (7): 6~9
[11] 杨其长, 孙忠富, 魏玲灵等. 中国设施农业研究现状及发展战略. 2009中国寿光国际设施园艺高层学术论坛论文集, 北京: 中国农业科学技术出版社, 2009

High Temperature Control in Mediterranean Greenhouse Production: the Constraints and the Options

Stefania De Pascale [1]* and Cecilia Stanghellini [2]

(1. *Department of Agricultural Engineering and Agronomy, University of Naples Federico II, Via Università 10080055 Portici (Naples), Italy. E-mail: depascal@unina.it*; 2. *Wageningen UR Greenhouse Horticulture, Droevendaalsesteeg 1, 6708PB Wageningen, The Netherlands*)

Abstract: In open field, the environment is a critical determinant of crop yield and produce quality and it affects the geographical distribution of most crop species. In contrast, in protected cultivations environmental control allows to fulfill the actual needs more or less efficiently depending on the technological level. The economic optimum, however, depends on the trade-off between the costs of increased greenhouse control and increase in return, dictated by yield quantity, yield quality and production timing. Additional constraints are increasingly applied for achieving environmental targets. However, the diverse facets of greenhouse technology in the different areas of the world will necessarily require different approaches to achieve an improved utilization of the available resources. Although advanced technologies to improve resource use efficiency can be developed as a joint effort between different players involved in greenhouse technology, some specific requirements may clearly hinder the development of common "European" resource management models that, conversely should be calibrated for different environments. For instance, the quantification and control of resource fluxes can be better accomplished in a relatively closed and fully automated system, such as those utilized in the glasshouse of Northern-Central Europe, compared to Southern Europe, where different typologies of semi-open/semi-closed greenhouse systems generally co-exist. Based on these considerations, innovations aimed at improving resource use efficiency in greenhouse agriculture should implement these aspects and should reinforce and integrate information obtained from different research areas concerning the greenhouse production. Advancing knowledge on the physiology of high temperature adaptation, for instance, may support the development and validation of models for optimizing the greenhouse system and climate management in the Mediterranean. Overall, a successful approach will see horticulturists, plant physiologists, engineers and economists working together toward the definition of a sustainable greenhouse system.

Key words: Abiotic stress; Heat tolerance; Tomato; Mild-winter climate; Ventilation; Cooling; Shading

1 Introduction

Protected agriculture is a high-investment and a high-risk business that demands professional expertise, based on the synthesis of proven technologies and market requirements – along with economic objectives. Achieving high productivity in greenhouse is possible by exerting optimal control of both environmental parameters and cultural ractices. The economic optimum, however, depends on the trade-off between the costs of increased environment control and increase in return, dictated by

---

* E-mail: cecilia.stanghellini@wur.nl

yield quantity, yield quality and production timing. Additional constraints that are increasingly applied are aimed at improving resource use efficiency (water, nutrients, energy and soil) and decreasing environmental impact. Sustainability, however, must also be considered in the context of economic sustainability. In this respect, greenhouse crops are often the only form of economically sustainable agriculture that can be proposed in many marginal areas of the Mediterranean environment, where land abandonment is considerably expanding. Indeed, in contrast to what most may believe, the greenhouse system is in some respect quite a resource-efficient system compared to the open field agriculture (De Pascale and Maggio, 2005; 2008).

Nevertheless, the diverse facets of greenhouse technology in the various regions of Europe will necessarily require different approaches to achieve an improved utilization of the available resources. Among the most critical determinants that differentiate the North- Central vs. Southern Europe greenhouse agribusiness we must consider: 1) the climatic conditions; 2) the market dynamics; 3) the social/technical background of farmers and entrepreneurs. Besides technical innovations adapted to te relevant conditions, advancing knowledge on the physiology of plant stress adaptation may support decision in the management of crop, greenhouse and climate (De Pascale and Maggio, 2005; 2008).

The economic feasibility of these innovations highly depends on the yield increase that can be obtained. Therefore, in this paper we will present different research approaches that should integrated to *optimize* the greenhouse high temperature control and a few examples that may provide useful insights for defining the economic optimum of greenhouse control.

2 High temperature: From plant physiology to greenhouse management

In general, higher temperatures are associated with higher radiation and higher water use and it is relatively difficult to discriminate among the physiological effects (at the level of plants and plant organs) of the different factors. Except for transpiration cooling, plants are unable to adjust their tissue temperatures to any significant extent. On the other hand, plants have evolved several mechanisms that enable them to tolerate higher temperatures. These adaptive thermo-tolerant mechanisms reflect the environment in which a species has evolved and they largely dictate the environment where a crop may be grown (Wahid et al., 2007). Heat stress in plants may arise in leaves when transpiration is insufficient for maintaining lower leaf temperature or in organs with reduced capacity for transpiration such as fruits. Plants with high transpiration rates are more tolerant. Temperatures that inhibit cellular metabolism and growth for a cool season C3 species, such as spinach, may not inhibit warm-season C3 species, such as tomato, C4 and CAM species. In this section we will try to summarize the physiological mechanisms that are thought to be critical in plant heat stress tolerance. Most examples will refer to tomato, the most common greenhouse crop in warm environments.

2.1 Thermal dependence at the biochemical and metabolic levels

The relationship between the thermal environment for an organism and the thermal dependence of enzymes has been well established (Basra and Basra, 1997). Temperature effects on the rates of biochemical reactions may be modeled as the product of two functions, an exponentially increasing rate of the forward reaction and an exponential decay resulting from enzyme denaturation as temperatures increase. Failure of only one critical enzyme system can cause death of an organism. This fact may explain why most crop species survive sustained high temperatures up to a relatively narrow range, 40 to 45℃ (Burke, 1990). The shape of this function also describes temperature effects on most biological functions, including plant growth and development. Modelers frequently simplify the relationship into a stepwise linear function that has a plateau rather than an optimum temperature (Klueva et al., 2001).

The thermal dependence of the apparent reaction rate for selected enzymes may indicate the optimal thermal range for a plant (Burke et al., 1988). The identification of thermal kinetic windows (TKWs) for different species and cultivars can aid in the interpretation of the differential temperature stress responses for crop growth and development among genotypes in order to identify target traits for improving high temperature tolerance (Mahan et al., 1987).

For crop plants, the thermal kinetic window (TKW) is generally established as a result of thermally induced lipid phase changes, Rubisco activity and starch synthesis pathway in leaves and reproductive organs (Mahan et al., 1987). The light-dependent activation of Rubisco, which is mediated by Rubisco activase, is one of the most thermally labile reactions and is directly related to the inhibition of net photosynthetic rate with significant effect on plant growth and development (Law and Crafts-Brandner, 1999; Law et al., 2001). Activase sensitivity to high temperature differs both within species and among species and activase activity can acclimate during a relatively short period when the leaf temperature is increased in gradual increments (Law and Crafts-Brandner, 1999; Law et al., 2001). Acclimation of activase to high temperature may be associated with heat-stress-induced changes in the pools of ATP and ADP, which are substrates known to stabilize activase and/or to altered biosynthesis of the molecular forms of activase (Crafts-Brandner and Salvucci, 2000).

2.2 Thermal stability of cell membranes

In most crop species membrane stability limits growth at high temperature. The plasmalemma and membranes of cell organelles play a vital role in the functioning of cells. Any adverse effect of temperature stress on the membranes leads to disruption of cellular activity or death (Chaisompongopan et al., 1990). Injury to membranes from a sudden heat stress event may result from either denaturation of the membrane proteins or from melting of membrane lipids which leads to membrane rupture and loss of cellular contents measured by ion leakage (Raison, 1986). Heat stress may be an oxidative stress (Lee et al., 1983) and peroxidation of membrane lipids (Mishra and Singhal, 1992), observed at high temperatures is a symptom of cellular injury. Enhanced synthesis of an anti-oxidant by plant tissues may increase cell tolerance to heat stress (Upadhyaya et al., 1991).

A relationship between lipid composition and incubation temperature has been shown for algae, fungi and higher plants: increase in saturated fatty acids of membranes increases their melting temperature and thus increases heat tolerance (Hall, 1993). In *Arabiodopsis*, exposed to high temperatures, total lipid content decreases to about onehalf and the ratio of unsaturated to saturated fatty acids decreases to one-third of the levels at temperatures within the TKW (Somerville and Browse, 1991). *Arabiodopsis* mutants, deficient in activity of chloroplast fatty acid W-9 desaturase, accumulate large amounts of 16 : 0 fatty acids, resulting in greater saturation of chloroplast lipids and in higher optimal growth temperature (Kunst et al., 1989; Raison, 1986). In other species, however, heat tolerance or tolerance differences among cultivars have been unrelated to membrane lipid saturation (Rikin et al., 1993; Kee and Nobel, 1985). In such species, a factor other than membrane stability may be limiting growth at high temperature.

2.3 Heat Shock Proteins

Synthesis and accumulation of so-called 'Heat Shock Proteins' (HSPs) have been ascertained during a rapid heat stress. Increased production of these proteins also occurs when plants experience a gradual increase in temperature or other abiotic stresses including drought and low temperature. Three classes of proteins as distinguished by molecular weight account for most HSPs, namely HSP90, HSP70, and low molecular weight proteins of 15 to 30 kDa (LMW-HSP) (Vierling, 1991). In arid and semi-arid regions, plants may synthesize and accumulate substantial levels of HSP in response to elevated leaf temperatures (Burke et al., 1985). In general, HSP are induced by heat stress at any stage of development. The induction temperature for the synthesis, the accumulation of HSPs and the proportions of the three classes differ among species (Howarth, 1991). Levels were greater in the non-irrigated crops (Kimpel and Key, 1985). Correlation between synthesis and accumulation of heat shock proteins and heat tolerance suggests that the two are causally related. Further evidence for a causal relationship is that some cultivar differences in heat shock protein expression correlate with differences in thermo-tolerance (Pelham, 1986). In genetic experiments, HSP expression co-segregates with heat tolerance. Another evidence for the protective role of heat shock protein is that mutants unable to synthesize heat shock proteins, and cells in which HSP70 synthesis is blocked or inactivated, are more susceptible to heat injury (Abrol and Ingram, 1996).

The mechanism by which HSPs contribute to heat stress tolerance is still not certain. Many are chaperones (assist others proteins). In tomato it has been shown that (Heckathorn et al., 1998): Prevent proteins denaturation during stress; Play a structural role in maintaining cell membranes integrity during stress (some of them have been associated with particular organelles such as chloroplast, ribosome and mitochondria); Protect the Photosystem II from oxidative stress.

In addition to their protective effect under stress conditions, HSP play a role in plant development under normal growth conditions: promote carotenoids accumulation during fruit maturation (Neta-Sharir et al., 2005). HSPs provide a significant opportunity to increase heat tolerance of greenhouse crops (Siddique et al., 2003).

2.4 Photosynthesis

The temperature optimum for photosynthesis is broad, presumably because crop plants have adapted to a relatively wide range of thermal environments. A 1 to 2°C increase in average temperature is not likely to have a substantial impact on leaf photosynthetic rates (Abrol and Ingram,

1996). However, heat stress (from 5 to 10 °C above the optimum) significantly reduced net photosynthesis and increased night respiration and stomatal conductance on tomato (Sato et al., 2000). Differences between heat tolerant and heat sensitive genotypes have been observed: photosynthesis of genotypes adapted to higher temperature environments was less sensitive to high temperature than was photosynthesis of genotypes from cooler environments (Al-Khatib and Paulsen, 1990; Venema et al., 1999). Recent researches have shown significant variation among tomato cultivars with respect to reduction in photosynthesis at high temperature. Camejo et al. (2005) demonstrated that the reduction of net photosynthesis in heat-sensitive tomato Campbell 28 was associated with non-stomatal components such as the alteration of Photosystem II and injuries of the plasma membrane measured by ion leakage.

Relative humidity (Vapour Pressure Deficit), light, CO_2 concentration, irrigation (water availability) all affect the net photosythesis response to temperature and can be easily controlled in the greenhouse environment. For instance, the temperature optimum for net photosynthesis is likely to increase with elevated levels of atmospheric carbon dioxide (Aloni et al., 2001; Behboudian and Lai, 1994; Rawson, 1992).

2.5 Plant growth and development

Heat stress is one of the most important constraints on crop production in semiarid greenhouses and adversely affects the vegetative and reproductive stages of plants and ultimately reduces yield and fruit quality (Abdul-Baki, 1991). Several experiments have studied the effects of sub-optimal temperature on plant growth and development under controlled environment and field conditions (Chen et al., 1982). However, scattered examples of studies on the effects of high temperature on greenhouse crops are available in literature (El Ahamdi and Stevens, 1979; Khayat et al., 1985; Peet et al., 1996; 1997; 1998; Picken, 1984; Pressman et al., 2002; Rivero et al., 2004; Rylski, 1979; Sato et al., 2000; 2001; 2002; 2005; Sugiyama et al., 1966). The optimum temperature for vegetative growth of tomatoes is reported to be 18~25℃ (Heuvelink and Dorais, 2005). Temperature increases above 25°C (corresponding to day/night regime of 30/21℃) cause non-linear yield reductions depending on stress severity and duration (Table 1).

The major conclusions from these studies are:
— Reproductive stages (from anthesis to harvest) are more sensitive than vegetative stages;
— Fruit weight is less sensitive to heat stress than fruit number;
— Fruit yield reduction results from:
— reduced numbers of fruits formed;
— shorter fruit growth duration;
— inhibition of carbohydrate assimilation in fruits.

A number of explanations have been offered for the poor reproductive performances of tomato at high temperatures. These include reduced or abnormal pollen production, abnormal development of the female reproductive tissues, hormonal imbalances, low level of carbohydrates and lack of pollination. Unfortunately, only a few experiments have been conducted with a sufficient number of cultivars to assess the genetic variability in these traits. Sato *et al.* (2000) concluded that pollen release and pollen viability may be the major limiting factors for fruit set under high temperature stress. In this study, cultivar differences in high temperature sensitivity were related to differences in pollen release and germination under heat stress determining their ability to set fruit. Peet *et al.* (1998), comparing heat stress effects on male fertile and sterile tomatoes observed that fruit number declined sharply with increasing temperature, whereas fruit weight per plant declined only slightly in both groups of plants. In male fertile tomato cultivars there was a more drastic decline in seeds/fruit as temperatures increased, which was not paralleled by a decrease in average fruit weight, suggesting an increased tendency for parthenocarpic fruit-set at high temperatures. In this respect, parthenocarpy can be considered a verypromising plant genetic trait for fruit greenhouse crops. On the other hand, plant growth reduction is mainly associated to the above discussed reduction of the net photosynthetic rate, accompanied by impaired translocation and assimilate partitioning caused by high air temperatures (Dinar and Rudich, 1985).

2.6 Fruit Quality

It is well known that tomato fruits are sensitive to heat stress. Temperatures above 30°C suppress many of the parameters of normal fruit ripening including color, softening, respiration rate and Ethylene production. Moreover exposure of fruits to temperature approaching 40°C induces metabolic disorders and facilitates fungal and bacterial invasion (Mohammed *et al.*, 1996). Recently, Riga *et al.* (2008) studied the effects of cumulative air temperature and photosynthetically active radiation on tomato quality and concluded that cumulative air temperature has a stronger influence on tomato quality than cumulative photosynthetically active radiation. Growers could thus obtain tomatoes of similar quality under lower photosynthetically active radiation (shading) than that provided by natural sunlight.

3 Best management: "The lesser of two evils"

A preliminary assessment of the specific environmental parameters of a given area may support the choice of an optimal combination of greenhouse/environment to facilitate the greenhouse management in terms of natural resources use efficiency and energy costs. An analysis of the specific control requirements, necessary to maximize yield, in different areas of the world confirmed that light is the limiting factor in The Netherlands for fall/winter crops and heating is required for most of the year. High temperatures (also because of high radiation) may strongly reduce crop yield in Naples (South Italy) and cooling is necessary most of the time, whereas heating may provide some advantages for a short period but is not critical (Figure 1). On an annual basis, in the Netherlands inside a greenhouse about 2 800MJ/m^2 energy is received from the sun, which is almost 3 times more than the annual heating requirement, whereas in Naples this amount reaches 4 300MJ/m^2 per year.

In Europe, climatic differences have fostered the development of greenhouse systems based on simple structures and inexpensive climatic control devices in the warmer Southern regions compared to the Northern greenhouse systems. These simplified systems present some limitations for an efficient use of the natural resources: poor ventilation, inefficient humidity control and reduced light transmission of plastic coverage all pose serious constraints for timing productions and guaranteeing high yields and quality standards. As the result, the majority of the Mediterranean greenhouses currently under-uses the potential energy in the fall/winter period and is strongly limited during summer/spring (with two production peaks, in spring-early summer and in autumn and approximately 3 ~ 5 months of non-productive time) due to a costly control of high temperatures. On the other hand, approximately 20% of the production costs are attributable to heating in the high tech Dutch greenhouses (Bakker et al., 2009).

If we compare the typical trend of the mean monthly temperature inside and outside a poorly ventilated greenhouse in South Italy with the optimal temperature for some horticultural crops we can conclude that the high indoor temperatures that typically occur from May to September strongly limit the potential yield (Figure. 2). Shading of the greenhouse (by shading nets, screening or white wash) during this period is a common means for passive cooling. This reduces considerably the canopy photosynthesis (though not necessarily the one of the upper leaves, which may be saturated), Figure 3. The shading experiments of Cockshull et al., 1992 allow inferring that 30g of tomatoes are harvested for each MJ PAR radiation at the top of the canopy, a value corroborated by the experiments of Li et al. 2001. The results of Scholberg et al. (2000) imply quite a similar value for field crops. Each 10% shading of the 13 MJ $m^{-2} d^{-1}$, typical of May through July in Naples (Figure. 1) is then a potential yield loss of about 0.6 kg m^{-2} $month^{-1}$, under the conservative assumption that light use efficiency at high PAR levels is only half the value given above.

Nevertheless, if shading is not applied, the high temperature will cause plant growth and yield (in terms of both quantity and quality) to be reduced (Figure. 4). The worth of a cooling method that would allow a higher radiation without increasing the temperature can then be estimated. If we stick to tomato as an example, and as we are dealing with increasing production in summer, we must consider that this is traditionally a low-price period. Fig. 5 gives a worst-case of 0.30 €/kg. Each 10% more PAR into the greenhouse, without increase in temperature, is then worth at least 1 800€Ha^{-2} $month^{-1}$. Another very important advantage of cooling is reduced ventilation, coupled to carbon fertilization. Stanghellini et al. (2009) using a relatively simple model, have demonstrated that a fertilization rate of 70 kg $Ha^{-1} h^{-1}$ could increase net income by more than 100€$Ha^{-1} h^{-1}$, even at low tomato prices, under relatively high irradiation, up to ventilation rates around $10h^{-1}$. Unfortunately, without other means for temperature control, a higher ventilation rate than that would be required in a traditional Mediterranean greenhouses in sunshine.

Photo-selective plastics (with a higher transmissivity in the PAR than in the NearInfraRed range of sun radiation) are now available. Kempkes et al. (2009) concluded that a year-round photo-selective cover is bound to lower the winter-time mean temperature in passive greenhouses, therefore such materials have a potential only for application as movable screens, in most places. Stanghellini et al. (2009) have shown that implemented as movable screen, a commercially available photo-selective film could reduce ventilation by 10% and increase water use efficiency by 12%, even during a relatively cold and dark Dutch summer. Alternatively, photo-selective whitewash may combine functionality and flexibility. The materials commercially available presently, however, have a very poor selectivity.

Where water of good quality is available and not expensive, probably misting is a more effective mean to control the temperature during summer. The capital and maintenance costs of an installation is estimated to be 8 000€Ha^{-1} year^{-1} (Vermeulen, 2008), with maximal capacity of 5m^3 Ha^{-1} h^{-1} and electric power of 5 k · Wh Ha^{-1}. The break even point in a very worst-case scenario (only 500h of operation per year; value of tomato 0.30€/kg; price of electricity 0.4€/k · Wh and of water 1€m^{-3}) would be with misting delivering an increase in production of 3 kg m^{-2} over 500 hours. This is anything between 5 and 20% of production, depending on the greenhouse. Feasibility of misting, therefore, should be evaluated accounting for prices, greenhouse management and access to good water.

It is important to realize, however, that full advantage of all means for temperature control and reduction of the ventilation requirement can be had only with carbon fertilization. In its absence a lower ventilation rate will result in lower CO_2 concentration and lower crop yield.

4 Linking biotechnology to greenhouse environment

Nowadays, tomato heat stress tolerance has also been obtained by over-expressing the HsfA1, a transcription factor with critical function in thermo-tolerance (Mishra et al., 2002) by genetic engineering of Heat Shock Proteins such as HSP70, HSP110, HSP17.1 (Siddique et al., 2003). However, in most of these studies the improved thermo-tolerance was assessed only on a qualitative basis, whereas a clear quantification in terms of yield was not reported. One aspect that has not been sufficiently addressed in discussing sustainability is how biotechnology may contribute to identifying genetic target traits that are particularly important for greenhouse environment. In this respect, using genetic engineering to generate genotypes that may be functional to understand how plant would respond to specific environmental variables is an aspect that has a great potential (Maggio et al., 2008; Marcelis and De Pascale, 2009).

Reinforcing collaboration among different research areas will speed up the progress in developing new cultivars (both transgenic and not transgenic) by:

1. Isolating critical components and producing the relative genotypes.
2. Studying and assessing their relevance in greenhouse production.
3. Generating new cultivars or identifying the relative gene in wild type stress resistant plants.
4. Introducing them in commercial cultivars using assisted breeding.

5 Conclusions

Although advanced technologies to improve resource use efficiency can be developed as a joint effort between different players involved in greenhouse technology, some specific requirements may clearly hinder the development of common "European" resource management models that, conversely should be calibrated for different environments. Strategies to improve the resources use efficiency in greenhouse management should also look at improving the plant's ability to use the available resources and at identifying genetic traits that may be important in the specific environment. Today we do have a good idea of the mechanisms that need to be potentiated to improve heat stress tolerance, however, much more needs to be known and understood on the functional biology of heat stress adaptation of greenhouse vegetable crops. Advancing knowledge on the physiology of low/high temperature adaptation, for instance, may support the development and validation of models for optimizing the greenhouse system and climate management in the Mediterranean greenhouses. Based on these considerations, innovations aimed at improving resource use efficiency in greenhouse horticulture should implement and integrate all these aspects and should reinforce scientific collaboration.

It is our belief that much knowledge in this field could be gained by combining information from different research areas in an interdisciplinary approach to the greenhouse management. Overall, a successful approach will see horticulturists, plant physiologists, engineers and economists working together toward the definition of a sustainable greenhouse system.

References

[1] Abdul-Baki A. A. 1991. Tolerance of tomato cultivars and selected germplasm to heat stress. J. Amer. Soc. Hort. Sci. 116 (6): 1 113 ~ 1 116

[2] Abrol Y. P., Ingram K. T. 1996. Effects of higher day and night temperatures on growth and yields of some crop plants. In: Global climate change and agricultural production. Direct and indirect effects of changing hydrological, pedological and plant physiological processes. Bazzaz F., Sombroek W. (Eds.). FAO, Rome, Italy

[3] Al-Khatib K., Paulsen G. M. 1990. Photosynthesis and productivity during high temperature stress of wheat genotypes from major world regions. Crop Sci. 30: 1 127 ~ 1 132

[4] Aloni, B., Peet, M.M., Pharr M., Karni, L. 2001. The effect of high temperature and high atmospheric CO_2 on carbohydrate changes in bell pepper (*Capsicum* annum L.) pollen in relation to its germination. Physiologia Plantarum 112: 505 ~ 212

[5] Anonimous, 2008. Analísis de la Campana Hortofrutícola de Almería 2007/2008, Fundación Cajamar, Almeria, Monografía 17: 45

[6] Bakker J. C., Adams S. R., Boulard T., Montero J. I. 2009. Innovative technologies for an efficient use of energy. Acta Hort., 801: 49 ~ 62

[7] Basra A. S., Basra R. K. 1997. Mechanisms of Environmental Stress Resistance in Plants. Harwood Academic Publishers, Amsterdam, The Netherlands, 407

[8] Behboudian M. H., Lai R. 1994. Carbon dioxide in "Virosa" tomato plants: responses to enrichment duration and to temperature. Hort. Sci. 29: 1 456 ~ 1 459

[9] Burke J. J. 1 990. High temperature stress and adaptation in crops. In: Stress Response in Plants: Adaptation and Acclimation Mechanisms. Alscher R. G., Cummings J. R. (Eds.). Wiley-Liss, New York, 295 ~ 309

[10] Burke J. J., Hatfield J. L., Klein R. R., Mullet J. E. 1 985. Accumulation of heat shock proteins in field grown soybean. Plant Physiol. 78: 394 ~ 398

[11] Burke J. J., Mahan J. R., Hatfield J. L. 1988. Crop specific thermal kinetic windows in relation to wheat and cotton biomass production. Agron. J. 80: 553 ~ 556

[12] Camejo D., Rodrìguez P., Morales M. A., Dell'Amico J. M., Arturo Torrecillas A., Juan Josè Alarcòn J. J. 2005. High temperature effects on photosynthetic activity of two tomato cultivars with different heat susceptibility. J. Plant Physiol. 162 (3): 281 ~ 9

[13] Chaisompongapan, N., Li, P. H., Davis, D. W. and Mackhart, A. H. 1990. Photosynthetic responses to heat stress in

common bean genotypes differing in heat acclimation potential. Crop Sci. 30: 100~104

[14] Chen H. H., Shen Z-Y., Li P. H. 1982. Adaptability of crop plants to high temperature stress. Crop Sci. 22: 719~725

[15] Cockshull, K. E., Graves, C. J. and Cave, C. R. J. 1992. The influence of shading on glasshouse tomatoes. Journal of Horticultural Science. 67: 11~24

[16] Crafts-Brandner S. J., Salvucci M. E. 2000. Rubisco activase constrains the photosynthetic potential of leaves at high temperature and CO_2. Proc. Natl. Acad. Sci. USA 297: 13 430~13 435.

[17] De Pascale S., Maggio A. 2005. Sustainable protected cultivation at a Mediterranean climate. Perspectives and challenges. Acta Hort. 691: 29~42

[18] De Pascale S., Maggio A. 2008. Plant Stress Management in Semiarid Greenhouse. Acta Hort. 797: 205~215

[19] Dinar M., Rudich J. 1985. Effect of heat stress on assimilate partition in tomato. Ann. Bot. 56: 239~249

[20] El Ahamdi A. B., Stevens M. A. 1979. Reproductive responses of heat-tolerant tomatoes to high temperatures. J. Amer. Soc. Hort. Sci. 104 (5): 686~691

[21] Hall A. E. 1993. Breeding for heat tolerance. Plant Breed Res. 10: 129~168

[22] Heckathorn S. A., Downs C. A., Sharkey T. D., Coleman J. S. 1998. The Small, Methionine-Rich Chloroplast Heat-Shock Protein Protects Photosystem II Electron Transport during Heat Stress. Plant Physiol. 116: 439~444

[23] Heuvelink E., Dorais M., 2005. Crop growth and yield. In: Heuvelink, E. (Ed.) Tomatoes. CABI publishing, Wallingford, 85~144

[24] Howarth C. J. 1991. Molecular response of plants to an increased incidence of heat shock. Plant Cell Environ. 14: 831~841

[25] Kee S. C., Nobel P. S. 1985. Fatty acid composition of chlorenchyma membrane fractions from three desert succulents grown at moderate and high temperature. Biochim. Biophys. Acta 820: 100~106

[26] Kempkes K., Stanghellini C., Hemming S. 2009. Cover materials excluding Near Infrared radiation: what is the best strategy in mild climates? Acta Hort. 807: 67~72

[27] Khayat E., Ravad D., Zieslin N. 1985. The effect of various night-temperature regimes on the vegetative growth and fruit production of tomato plants. Scientia Horticulturae, 27: 9~13

[28] Kimpel J. A., Key J. L. 1985. Presence of heat shock mRNAs in field-grown soybean. Plant Physiol. 79: 622~678

[29] Klueva N. Y., Maestri E., Marmiroli N., Nguyen H. T. 2001. Mechanisms of Thermotolerance. in Crops. In: Basra A. S. (Ed.), Crop Responses and Adaptations to Temperature Stress. Haworth Press, Inc., New York, 177~181

[30] Kunst L., Browse J., Somerville. 1989. Enhanced thermal tolerance in a mutant of Arabidopsis deficient in palmitic acid unsaturation. Plant Physiol. 91: 401~408

[31] Law R. D., Crafts-Brandner S. J. 1999. Inhibition and acclimation of photosynthesis to heat stress is closely correlated with activation of ribulose-1, 5-bisphosphate carboxylase/ oxygenase. Plant Physiol. 120: 173~81

[32] Law R. D., Crafts-Brandner S. J., Salvucci M. E. 2001. Heat stress induces the synthesis of a new form of ribulose-1, 5-bisphosphate carboxylase/oxygenase activase in cotton leaves. Planta 214: 117~125

[33] Lee P. C., Bochner, B. R., Ames B. N. 1983. A heat shock stress and cell oxidation. Proc. Natl. Acad. Sci. USA 80: 7 496~7 500.

[34] Li, Y. L., Stanghellini, C., Challa, H. 2001. Effect of electrical conductivity and transpiration on production of greenhouse tomato. Scientia Horticulturae, 88: 11~29

[35] Maggio A., De Pascale S., Barbieri G. 2008. Designing a Greenhouse Plant: Novel Approaches to Improve Resource Use Efficiency in Controlled Environments. Acta Hort. 801: 1 235~1 242

[36] Mahan, J. R., Burke, J. J., Orzech, K. A. 1987. The "thermal kinetic window" as an indicator of optimum plant temperature. Plant Physiol. 82: 518~522

[37] Marcelis L. F. M., De Pascale S. 2009. Crop management in greenhouses: Adapting the Growth Conditions to the Plant Needs or Adapting the Plant to the Growth Conditions? Acta Horticulturae, 807: 163~173

[38] Mishra R. K., Singhal G. S. 1992. Function of photosynthetic apparatus of intact wheat leaves under high light and heat stress and its relationship with thylakoid lipids. Plant Physiol. 98: 1~6

[39] Mishra S. K., Tripp J., Winkelhaus S., Tschiersch B., Theres K., Nover L., Scharf D. K. 2002. In the complex family of heat stress transcription factors, HsfA1 has a unique role as master regulator of thermotolerance in tomato. Genes&Development 16: 1 555 ~ 1 567

[40] Mohammed M., Wilson L. A., Gomes P. I. 1996. Influence of High Temperature Stress on Postharvest Quality of Processing and Non-Processing Tomato Cultivars. Journal of Food Quality 19: 41 ~ 55

[41] Neta-Sharir I., Isaacson T., Lurie L., Weiss D. 2005. Dual Role for Tomato Heat Shock Protein 21: Protecting Photosystem II from Oxidative Stress and Promoting Color Changes during Fruit Maturation. The Plant Cell 17: 1 829 ~ 1 838

[42] Peet M. M., Batholemew M. 1996. Effect of night temperature on pollen characteristics, growth, and fruit set in tomato. J. Amer. Soc. Hort. Sci. 121 (3): 414 ~ 519

[43] Peet M. M., Sato S., Gardner R. G. 1998. Comparing heat stress effects on male fertile and sterile tomatoes. Plant, Cell and Environment 21: 225 ~ 231

[44] Peet M. M., Willits D. H., Gardner R. 1997. Response of ovule development and post- pollen production processes in male-sterile tomatoes to chronic, sub-acute high temperature stress. Journal of Experimental Botany 48: 101 ~ 111

[45] Pelham H. 1986. Speculations on the major heat shock and glucose regulated proteins. Cell 46: 959 ~ 961

[46] Picken A. J. F. 1984. A review of pollination and fruit set in the tomato (*Lycopersicon esculentum* Mill.). Journal of Horticultural Science 59: 1 ~ 13

[47] Pressman E., Peet M. M., Pharr D. M. 2002. The Effect of Heat Stress on Tomato Pollen Characteristics is Associated with Changes in Carbohydrate Concentration in the Developing Anthers. Annals of Botany 90: 631 ~ 636

[48] Raison J. K. 1986. Alterations in the physical properties and thermal response of membrane lipids: correlations with acclimation to chilly and high temperature. In: Frontiers of Membrane Research in Agriculture. St. John J. B., Berlin E., Jackson P. C. (Eds.). Rowman&Allanheld, Totoma, NJ, 383 ~ 401

[49] Rawson H. M. 1992. Plant responses to temperature under conditions of elevated CO_2. Aust. J. Plant Physiol. 40: 473 ~ 490

[50] Riga P., Anza M., Garbisu C. 2008. Tomato quality is more dependent on temperature than on photosynthetically active radiation. Journal of the Science of Food and Agriculture 88: 158 ~ 166

[51] Rikin A., Dillworth J. W., Bergman D. K. 1993. Correlation between circadian rythm of resistance to extreme temperature and changes in fatty acid composition in cotton seedlings. Plant Physiol. 101: 31 ~ 36

[52] Rivero R. M., Ruiz J. M., Romero, L. M. 2004. Importance of N source on heat stress tolerance due to the accumulation of proline and quaternary ammonium compounds in tomato plants. Plant Biology 6: 702 ~ 707

[53] Rylski, I. 1979. Fruit set and development of seeded and seedless tomato fruits under diverse regimes of temperature and pollination. J. Am. Soc. Hort. 104: 835 ~ 838

[54] Sato S., Peet M. M., Thomas J. F. 2000. Physiological factors limit fruit set of tomato (*Licopersicon esculentum* Mill.) under chronic mild heat stress. Plant, Cell and Environment 23: 719 ~ 726

[55] Sato S., Peet M. M., Gardner R. G. 2001. Formation of parthenocarpic fruit, undeveloped flowers and aborted flowers in tomato under moderately elevated temperatures. Scientia Horticulturae 90: 243 ~ 254

[56] Sato S., Peet M. M., Thomas J. F. 2002. Determining critical pre- and post- anthesis periods and physiological processes in *Lycopersicum esculentum* Mill. exposed to moderately elevated temperatures. Journal of Experimental Botany 53 (371): 1 187 ~ 1 195

[57] Sato S., Peet M. M. 2005. Effects of moderately elevated temperatures stress on the timing of pollen release and its germination in tomato (*Lycopersicon esculentum* Mill.) Journal of Horticultural Science & Biotechnology 80 (1): 23 ~ 28

[58] Scholberg, J., McNeal, B. L., Jones, J. W., Boote, K. J., Stanley, C. D., Obreza, T. A., 2000. Growth and Canopy Characteristics of Field-Grown Tomato Agron J, 92: 152 ~ 159

[59] Siddique M., Port M., Tripp J., Weber C., Zielinski D., Calligaris R., Winkelhaus S., Scharf K-D. 2003. Tomato heat stress protein Hsp16. 1-CIII represents a member of a new class of nucleocytoplasmic small heat stress proteins in plants. Cell Stress & Chaperones 8 (4): 381 ~ 394

[60] Somerville, C. and Browse, J. 1991. Plant lipids, metabolism and membranes. Science 252: 80 ~ 87

[61] Stanghellini, C., Heuvelink, E., 2007. Coltura e clima: effetto microclimatico dell'ambiente serra. *Italus Hortus*, 14 (1): 37~49

[62] Stanghellini, C., Kempkes, F. L. K., Hemming, S., Jianfeng D., 2009. Reflective materials for Near InfraRed radiation: effect on climate and effect on crop. FlowerTech, in press. Sugiyama T., Iwahori S., Takahashi K. 1966. Effect of high temperature on fruit setting of tomato under cover. Acta Hort. 4: 63~69

[63] Upadhyaya A., Davis T. D., Sankhla M. 1991. Heat shock tolerance and anti-oxidant activity in moth bean seedlings treated with tetayclasis. Plant Growth Regulation 10: 215~222

[64] Venema J. H., Posthumus F., Van Hasselt P. R. 1999. Impact of suboptimal temperature on growth, photosynthesis, leaf pigments and carbohydrates of domestic and high altitude wild Lycopersicon species. Journal of Plant Physiology 155: 711~718

[65] Vermeulen, P. C. M., 2008, Kwantitatieve Informatie voor de Glastuinbouw, rapport 185: Wageningen UR Greenhouse Horticulture Wageningen: 146

[66] Vierling E. 1991. The roles of heat shock proteins in plants. Annu. Rev. Plant Physiology Plant Mol. Biol. 42: 579~620

[67] Wahid A., Gelani S., Ashraf M., Foolad M. R. 2007. Heat tolerance in plants: An overview. Environmental and Experimental Botany 61: 199~223

Tables

Table 1 Compilation of literature over effect of high temperature on tomato. Production loss is given in % of potential yield at 25℃ (Stanghellini & Heuvelink, 2007)

Temperature(℃)	Duration	Production loss	Reference
42	6 hours	100	Heckathorn et al., 1998
40	2 days	100	Klueva et al., 2000
	4 h/9d before anthesis	no fruit set	Peet et al., 1998
	n. a.	endosperm damage	Sato et al., 2000
	30 hours	pollen damage	Sugiyama et al., 1966
35	30 days average	40	Rivero et al., 2004
32/36	33 days	25	Sato et al., 2001; 2002
27	daily mean	15	Peet et al., 1997
26	daily mean	5	Peet et al., 1998. 12

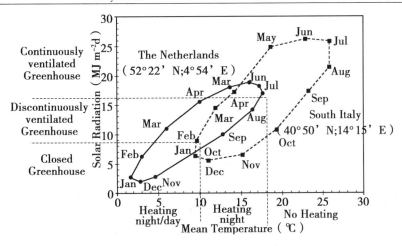

Figure 1 Climatic control requirements for greenhouse vegetable crops production in The Netherlands (Amsterdam) and in South Italy (Naples)

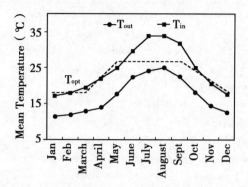

Figure 2 Typical trend of the mean monthly temperature inside (T_{in}) and outside (T_{out}) a poorly ventilated greenhouse in Italy (Naples, 40°49'N; 14°15'E; 30m. s. l.) and optimal temperature (T_{opt}) for some horticultural crops

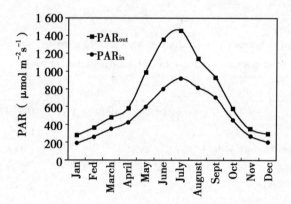

Figure 3 Typical trend of the mean monthly Photosynthetically Active Radiation inside (PAR_{in}) and outside (PAR_{out}) a poorly ventilated greenhouse in South Italy (Naples, 40°49'N; 14°15'E; 30m. s. l.) with shading screen from May to September

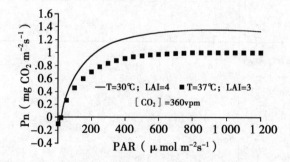

Figure 4 Light response curves for net photosynthesis (Pn) of a greenhouse tomato crop at 30℃ and at 37℃ and 360 vpm CO_2. Crop photosynthesis at 37℃ was calculated considering a reduction of 25% of the leaf area indices (LAI) (Sato et al., 2000)

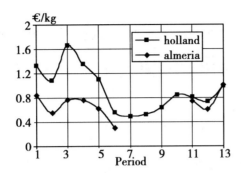

Figure 5 Evolution of the year of producer price of tomato (€/kg), in Holland and in Almeria. Mean of the years 2005~2008, for each 4-week period. Sources: Anomimous, 2008 and Vermeulen, 2008

"Effects of Greenhouse Climate Control Equipment on Greenhouse Microclimate and Crop Response"

C. Kittas[1], N. Katsoulas[1], T. Bartzanas[2]

(1 *School of Agricultural Sciences, Department of Agriculture Crop Production and Rural Environment, University of Thessaly, Fytokou St. N. Ionia Magnisias GR* – 38446, *Greece*;
2. *Centre for Research and Technology* – *Thessaly, Institute of Technology and Management of Agricultural Ecosystems, Technology Park of Thessaly,*
1st *Industrial Area,* 38500 *Volos, Greece*)

Abstract: An overview of the good agricultural practices for greenhouse horticulture in the Mediterranean and South East Europe regions is presented and discussed. Special emphasis is given on the effects of climate control equipment on: (a) greenhouse microclimate and (b) crop response. The main greenhouse climate control systems presented and discussed are the heating, dehumidification, ventilation, shading and cooling systems. These systems are the most widely used for greenhouse climate control especially in the Mediterranean area. In these areas, natural ventilation and whitewashing are the most common methods/systems used for greenhouse climate control during summer, since it require less energy, less equipment operation and maintenance and are much cheaper to install that other cooling systems. However, generally these systems are not sufficient for extracting the excess energy during sunny summer days and therefore, other cooling methods such as forced ventilation combined with evaporative cooling (mist or fog system, sprinklers, wet pads), are used. On the other hand, during winter period, heating and dehumidification are necessary for a standard quality production. Optimal greenhouse climate control become more important in the latter part of the twentieth century, when concerns about food safety, environmental pollution stimulated extensive research in the development of Integrated Production and Protection systems aiming at a significant reduction of pesticides use. The prospective and needs for future research and development in greenhouse climate control are presented in the conclusion. Special emphasis is given on the necessity to combine both physical and ecophysiological studies, because of the crucial role played by the crop in determining the microclimate in greenhouses.

1 Driving forces for greenhouse climate control and sustainable energy use in Mediterranean Greenhouse

All greenhouse cultivation systems, regardless of geographic location, consist of fundamental climate control components, and depending on their design and complexity, they can provide a greater or lesser amount of climate control, and subsequent plant growth and productivity.

Temperature is the most important variable of the greenhouse climate that can and needs to be controlled. The majority of plants grown in greenhouses are warm – season species and are adapted to average temperatures in the range 17 – 27℃, with approximate lower and upper temperature limits of 10℃ and 35℃ (von Zabeltitz, 1999). If the average minimum outside temperature is below 10℃ the greenhouse is likely to require heating, particularly at night. When the average maximum outside temperature is less than 27℃ ventilation will prevent excessive internal temperatures during the day;

however, if the average maximum temperature exceeds 27 – 28℃ then artificial cooling may be necessary. The maximum greenhouse temperature should not exceed 30 – 35℃ for prolonged periods. The climograph of some Mediterranean and North Europe regions against Beijing – China is shown in Figure 1. This shows that in temperate climates e. g. in the Netherlands, heating and ventilation enables the temperature to be controlled over the whole year, however, at lower latitudes, e. g. Almeria – Spain and Volos – Greece, the daytime temperatures are too high for ventilation to provide sufficient cooling during the summer. The attainment of suitable temperatures then requires positive cooling. In Beijing area heating during winter and positive cooling during summer are required.

The second important variable of the greenhouse climate is humidity, which has traditionally been expressed in terms of relative humidity. Relative humidity within the range 60% ~90% has little effect on plants (Grange and Hand, 1987). Values below 60% may occur during ventilation in arid climates, or when plants are young with small leaves, and this can cause water stress. Serious problems can occur if the relative humidity exceeds 95% for long periods, particularly at night as this favours the rapid development of fungus diseases e. g. *Botrytis cinerea* (O'Neill et al., 2002). The increased interest in maintaining adequate transpiration to avoid problems associated with calcium deficiency has resulted in humidity being expressed in terms of the vapour pressure deficit (VPD) or the moisture deficit both of which are directly related to transpiration. Maintaining the VPD above some minimum value helps to ensure adequate transpiration and also reduces disease problems. During the day humidity can usually be reduced using ventilation. However at night, unless the greenhouse is heated the internal and external temperatures may be similar and if the external humidity is high reducing the greenhouse humidity is not easy.

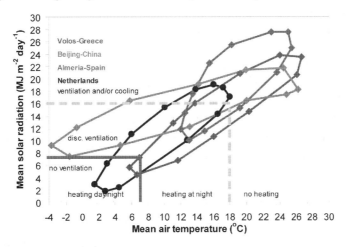

Figure 1　The mean solar radiation vs. mean air temperature for several locations around Europe: The climograph. Dotted lines indicate border lines for different control action in the greenhouse.

After the first "energy crisis" in the early 1980ies, where the limited supply of energy caused the first significant increase in energy prices, the energy use of greenhouses has become a major research issue. With the recently more pronounced interest in global warming and climate change, the use of fossil fuel is on the political agenda again and many governments have set maximum CO_2 emission levels for different industries including the greenhouse sector. There are mainly two ways to increase the energy efficiency: a) reduction of the energy input into the greenhouse system and b) increase the production per unit energy. The major challenge is to find ways which meet both needs: improved energy efficiency combined with an absolute reduction of the overall energy consumption and related CO_2 emission of the greenhouse industry. Technological innovations must focus on the energy consumption for the return on productivity, quality and societal satisfaction.

Clearly there are numerous technologies for greenhouse systems which can be adopted by the growers enabling a better and more efficient climate control and energy use. However, many obstacles and constraints remain to be solved. The existing technology and know – how developed in Northern Europe countries are generally not directly transferable to the Mediterranean growers: high – level technology is out of reach for most of the Mediterranean growers because their cost is too high compared to the modest investment capacity of these growers. Know – how from Northern Europe growers is often inappropriate to the problems encountered in the Mediterranean shelters.

In the case they could be used, an important effort in training and educating Mediterranean growers will be necessary. Taking into account this context, specific research and development tasks were initiated by the research institutes and extension services of the Mediterranean countries. The issues that are addressed in this paper concern the means and best practices by which Mediterranean growers can alleviate the climate – generated stress conditions that inhibit the growth and the development of the crops during the long extending warm season is a sustainable and energy friendly way.

2　Climate Control

2.1　Ventilation Cooling & Shading

Getting rid of the heat load is the major concern for greenhouse climate management in arid and semi – arid climate conditions. This can be realised by: (1) reducing the income of radiation; (2) removing the extra heat through air exchange; and (3) increasing the fraction of energy partitioned into latent heat (Figure 2).

Shade screens and whitewash are the major existing measures used to reduce the income of solar radiation; greenhouse ventilation is an effective way to remove the extra heat through air exchange between inside and outside, when outside air temperature is lower; and evaporative cooling is the common technique to reduce sensible heat load by increasing the latent heat fraction of dissipated energy. Other cooling technological solutions are available (heat pump, heat exchangers, ...), but are not yet widely used, especially in the Mediterranean countries, because investment cost is yet very high.

In order to explore means of relieving greenhouse heat load in hot summers, many studies on

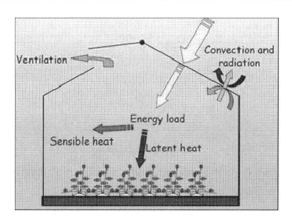

Figure 2 Greenhouse energy balance

shading, natural ventilation and evaporative cooling have been reported. Most of the existing reports focus on explaining how shading, ventilation and evaporative cooling affect greenhouse microclimate (Baille et al., 1994; 1999; Kittas et al., 2001a, Baille et al., 2001), demonstrating that shading (Bailey, 2000a; Baille et al., 2001; Castilla, 2001) and evaporative cooling (Arbel et al., 1999; Montero, 2001; Katsoulas et al., 2001; 2007; 2009) are effective means for relieving greenhouse heat load, under dry and sunny summer climate conditions in the Mediterranean area.

2.1.1 Ventilation

High summer temperatures result in the need for constant heat removal from the greenhouse. One of the simple and effective ways to reduce the difference between inside and outside air temperature is to improve ventilation. Natural or passive ventilation uses very little external energy. It is based on the difference in pressure, which the outside wind or the greenhouse temperature creates, between the greenhouse and the outside environment. If the greenhouse is equipped with ventilation openings (Figure 2), both near the ground and at the roof, then this type of ventilation replaces the internal hot air by external cooler one during the hot sunny days with weak wind. The external cool air enters the greenhouse through the lower side openings while the hot internal air exits through the roof openings due to density difference between air masses of different temperature causing the lowering of temperature in the greenhouse.

Sufficient ventilation is very important for optimal plant growth, especially in the case of high outside temperatures and solar radiation, which are common conditions during summer in Mediterranean countries. In order to study the variables, determining the greenhouse air temperature and to decide about the necessary measurements for greenhouse air temperature control, a simplified version of the greenhouse energy balance is formulated. According to Kittas et al. (2005), the greenhouse energy balance can be simplified to:

$$V_a = \frac{0.0003 \tau R_{s,o-\max}}{\Delta T} \qquad (2.1)$$

where:

V_a is the ratio Q/Ag, Q is the ventilation flow rate, in m³ [air] s⁻¹ and Ag is the greenhouse ground surface area, in m²

τ is the greenhouse transmission coefficient to solar radiation

$R_{s,o-max}$ is the maximum outside solar radiation W m⁻²,

- ΔT is the temperature difference between greenhouse and outside air, in ℃.

Using eqn. (2.1), it is easy to calculate the ventilation needs for several values of $R_{s,o-max}$ and ΔT. Figure 3 presents the variation of V_a for several values of $R_{s,o-max}$, τ and ΔT.

From Figure 3 it can be seen that for the area of Magnesia, Greece, where during the critical summer period, values of outside solar radiation exceed the value of 900 W m⁻² (Kittas et al., 2005), a ventilation rate of about 0.06 m³ s⁻¹ m⁻² (which corresponds, for a greenhouse with a mean height of 3 m, to an air-exchange of 60 h⁻¹) is needed in order to maintain a ΔT of about 4℃.

Figure 3　Greenhouses with different type of ventilation openings

The necessary ventilation rate can be obtained by natural ventilation by means of air exchange through the greenhouse vent openings or by forced ventilation by means of fans.

For effective ventilation, ventilators should, if possible, be located at the ridge, on the side walls and the gable. In some studies, the static ventilation systems for greenhouse cooling were described as less efficient and were not satisfactory during sunny days without wind as described by Verlodt et al. (1984) and Silva and Rosa (1987). Total ventilator area equivalent to 15% ~ 30% of floor area was recommended by White and Aldrich (1975). Above 30%, the effect of additional ventilation area on the temperature difference was very small.

Systems like exhaust fan; blower, etc. can supply high air exchange rates whenever needed. These are simple and robust systems and significantly increase the air transfer rate from the greenhouse and allow maintaining inside temperature to a level slightly higher than the outside temperature by increasing the number of air changes (Figure 4).

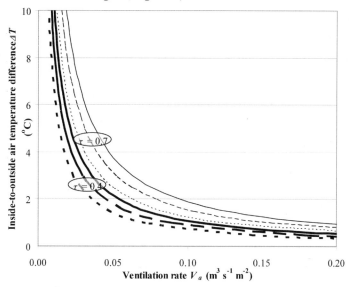

Figure 4 Inside-to-outside greenhouse air temperature difference ΔT as a function of ventilation rate V_a as calculated using eqn. (3), for a greenhouse with a regularly transpiring rose crop. Thin lines correspond to a greenhouse cover transmission to solar radiation τ of 0.7. Heavy lines correspond to a τ of 0.4. (——): $R_{s,o} = 900$ W m^{-2}, (——): $R_{s,o} = 750$ W m^{-2}, (– – –): $R_{s,o} = 600$ W m^{-2}. (Kittas et al., 2005)

The principle of forced ventilation is to create an air flow through the house. Fans suck air out on the one side, and openings on the other side let air in. Forced ventilation by fans is the most effective way to ventilate a greenhouse, but consumes electricity. It is estimated that the needs for electrical energy for greenhouse ventilation, for a greenhouse located in the Mediterranean are about 70000 kWh per greenhouse ha.

Some key elements are: Ventilation fans should develop a capacity of about 30 Pa static pres-

Figure 5 Fans for greenhouse forced ventilation

sure (3 mm on a water gauge), should be located on the leeside or the lee end of the greenhouse and the distance between two fans should not exceed 8 ~ 10 m. Furthermore, an inlet opening on the opposite side of a fan should be at least 1.25 times the fan area. The velocity of the incoming air must not be too high, in the plant area; the air speed should not exceed 0.5 m s^{-1}. The openings must close automatically when the fans are not in operation.

Willits (2003) presented a modelling study of a fan – ventilated greenhouse also equipped with fan – pad system. The results showed that with fan cooling alone (no evaporative cooling) little advantage could be derived from increasing airflow rates beyond $0.05 \text{ m}^3 \text{ m}^2 \text{ s}^{-1}$. The evapotranspiration coefficient was also compared with only fan and fan – pad cooling. Kittas *et al.* (2005) described the forced ventilation (using fan) in a multi – span greenhouse located near Karditsa Greece, for cooling and humidity regulation.

Kittas *et al.* (2001a) studied the influence of the greenhouse ventilation regime (natural or forced ventilation) on the energy partitioning of a well – watered rose canopy during several summer days in warm Mediterranean conditions (Eastern Greece). Two types of ventilation systems were evaluated: (i) fans and roof opening (forced ventilation) and (ii) roof openings only (natural ventilation). They found that significant differences existed in the microclimate created by the type of ventilation regime, as well as in the bulk aerodynamic conductance. These changes did not appear to modify the energy partitioning at the canopy level. Transpiration was not significantly enhanced under fan assisted ventilation. They mention that a high rate of ventilation is not a priori the best solution for alleviating the crop stress in greenhouse during summer conditions. When not limited by too low external wind speed, natural ventilation may be more appropriate as it creates a more humid and cooler environment, although less homogeneous, around the canopy. It appears that it exists a physiologically – based optimal value for the air exchange rate, and that this value depends on several factors: the outside conditions of wind, temperature and humidity, the internal air flow characteristics (which depends on the type of ventilation), the bulk aerodynamic conductance and the specific stomatal response of the species to humidity (Kittas *et al.*, 2001a). All these factors have to be taken into account when searching for the optimal (and time – variable) value of ventilation rate.

Considering the complexity of the mechanisms and the high number of interactions, the challenge is difficult, but might be met because the understanding and modelling of the coupling mechanisms between outside – inside atmospheres and canopy – inside atmosphere markedly progressed in the last decade.

Many researchers also studied the effects of insect – proof screens in roof openings (Figure 5) on greenhouse microclimate. Teitel (2001) experimented in a leeward roof openings greenhouse and revealed the obstruction offered by fine mesh screens to flow through the openings resulting in the higher temperature and humidity within the greenhouse. Demrati *et al.* (2001) studied the natural ventilation performance in a large scale, canarian type of greenhouse with a mature banana crop in Morocco.

Figure 6 Insect – proof screens in roof openings

The greenhouse ventilation performance was compared with those of other greenhouse types and discussed with respect to the improvement of greenhouse design, the use of insect – proof nets and irrigation control. Bartzanas *et al.* (2002) also analyzed the ventilation process in a tunnel greenhouse equipped with an insect – proof screen in the side openings using CFD. Insect screens significantly reduced airflow and increased thermal gradients inside the greenhouse.

Guidelines
- For a coastal area like Magnesia, Greece, where during the critical summer period, values of outside solar radiation exceed the value of 900 W m^{-2}, a ventilation rate of about 0.06 m^3 s^{-1} m^{-2} (which corresponds, for a greenhouse with a mean height of 3 m, to an air – exchange of 60 h^{-1}) is needed in order to maintain a ΔT of about 4℃. By natural ventilation an air exchange rate of about 40 h^{-1} could be obtained. For higher ventilation rate forced ventilation would be necessary.
- For effective ventilation, ventilators should, if possible, be located at the ridge, on the side walls and the gable.
- Total ventilator area equivalent to 15% ~ 30% of floor area is recommended. Above 30%, the effect of additional ventilation area on the temperature difference is very small.
- When not limited by too low external wind speed, natural ventilation may be more appropriate as it creates a more humid and cooler environment, although less homogeneous, around

the canopy.
- With roof ventilators, the highest ventilation rates per unit ventilator area are obtained when flap ventilators faced the wind (100%), followed by flap ventilators facing away from the wind (67%). The lowest rates of roof ventilation are obtained with the rolling ventilators (28%).
- Systems like exhaust fan; blower, etc. can supply high air exchange rates whenever needed. These are simple and robust systems and significantly increase the air transfer rate from the greenhouse and allow maintaining inside temperature to a level slightly higher than the outside temperature by increasing the number of air changes.
- Forced ventilation by fans is the most effective way to ventilate a greenhouse, but consumes electricity. It is estimated that the needs for electrical energy for greenhouse ventilation, for a greenhouse located in the Mediterranean are about 70000 kWh per greenhouse ha.
- Ventilation fans should develop a capacity of about 30 Pa static pressure (3 mm on a water gauge), should be located on the leeside or the lee end of the greenhouse and the distance between two fans should not exceed 8~10 m. Furthermore, an inlet opening on the opposite side of a fan should be at least 1.25 times the fan area. The velocity of the incoming air must not be too high, in the plant area; the air speed should not exceed 0.5 m s^{-1}. The openings must close automatically when the fans are not in operation.
- With fan cooling alone (no evaporative cooling) little advantage could be derived from increasing airflow rates beyond $0.05 \text{ m}^3 \text{ m}^2 \text{ s}^{-1}$.

2.1.2 Shading

Natural or forced ventilation is generally not sufficient for extracting the excess energy during sunny summer days (Baille, 1999). Therefore, other cooling methods have to be used in combination with ventilation. The entry of direct solar radiation through the covers into the greenhouse enclosure is the primary source of maximum heat gain. The entry of unwanted radiation (or light) can be controlled by the use of shading or reflection. Shading can be done by various methods such as by the use of paints, external shade cloths, use of nets (of various colors), partially reflective shade screens (Figure 6) and water film over the roof and liquid foams between the greenhouse walls. Shading is the ultimate solution to be used for cooling greenhouses, because it affects the productivity. However, in some cases, a better quality can be obtained from shading. More knowledge about the effect of shading on fruit or flower yield and quality must be obtained in order to determine the optimal intensity of shading (Cockshull et al., 1992). One of the most used methods adopted by growers due to the low cost is white painting, or whitening, the cover material. The use of screens has progressively been accepted by growers and has gained, through the last decade, a renewed interest as shown by the increasing area of field crops cultivated under screenhouses (Raveh et al. 2003; Cohen et al. 2005), while roof whitening, due to its low cost, is a current practice in the Mediterranean Basin.

In spite of the widespread use of whitening for alleviating the radiation load, the literature related to the influence of whitening on microclimate and crop behaviour is very sparse. Most studies were related to other shading techniques, like outside or inside permanent or movable shading screens (Nisen and Coutisse 1981, Miguel et al. 1994, 1997), and dealt mainly with the radiative modifications induced by the shading device. Baille et al. (1980) reported that glasshouse whitening reducing solar transmission coefficients by about 40% allowed inside air temperature to be maintained close that outside during summer periods in the South of France. In recent studies, Kittas et al. (1999) and Baille et al. (2001) reported that whitening applied onto a glass material enhanced slightly the PAR proportion of the incoming solar irradiance, thus reducing the solar infrared fraction entering the greenhouse. This characteristic of whitening could represent an advantage with respect to other shading devices, especially in warm countries with high radiation load during summer. Another advantage of whitening is that it does not affect the greenhouse ventilation, while internal shading nets affect negatively the performance of the roof ventilation. Whitening also significantly increases the fraction of diffuse irradiance, which is known to enhance the radiation use efficiency. With transmittance reduced to the half, the mean value of greenhouse to outside air temperature difference (average 8:00 ~ 19:00) was 4.4℃ lower than without application of whitewash at Volos, Greece (39.22 N) according to Baille et al. (2001).

Figure 7 **Thermal screen used for energy saving and greenhouse shading**

Whitening transforms a large part of the inside direct radiation into diffuse radiation (Baille et al., 2001). An enhancement of this ratio has been reported to increase the absorbed radiation by the crop (Hand et al., 1993) and has been suggested to increase crop productivity under shading conditions (Healey 1998). This characteristic of whitening could represent an advantage with respect to other shading devices, especially in warm countries with high radiation load during summer.

The screens mounted inside the greenhouse also contribute to decrease the inside wind speed, thus lessening the leaf boundary layer and restraining the availability of air CO_2 concentration near the leaf surface. It is not clear whether the shading nets should be used over the growth cycle or only during the most sensitive stages when the crops had a low leaf area and the canopy transpiration rate cannot significantly contribute to the greenhouse cooling (Seginer 1994).

Guidelines
- Natural or forced ventilation is generally not sufficient for extracting the excess energy during sunny summer days. The entry of unwanted radiation (or light) can be controlled by the use

of shading or reflection. Shading can be done by various methods such as by the use of paints, external shade cloths, use of nets (of various colors), partially reflective shade screens and water film over the roof and liquid foams between the greenhouse walls.

- Greenhouse whitening reducing solar transmission coefficients by about 40% allowed inside air temperature to be maintained close that outside during summer periods in the South of France.
- The advantage of whitening is that it does not affect the greenhouse ventilation, while internal shading nets affect negatively the performance of the roof ventilation. Whitening also significantly increases the fraction of diffuse irradiance, which is known to enhance the radiation use efficiency. This characteristic of whitening could represent one more advantage with respect to other shading devices.
- The screens mounted inside the greenhouse also contribute to decrease the inside wind speed, thus lessening the leaf boundary layer and restraining the availability of air CO_2 concentration near the leaf surface.

2.1.3　Evaporative cooling

One of the most efficient solutions for alleviating the climatic conditions is to use evaporative cooling systems, based on the conversion of sensible heat into latent heat by means of evaporation of water supplied directly into the greenhouse atmosphere (mist or fog system, sprinklers) or through evaporative pads (wet pads). Evaporative cooling allows simultaneous lowering of temperature and vapour pressure deficit (Cohen et al., 1983, Arbel et al., 1999, Willits, 1999) and can lead to greenhouse air temperatures lower than the outside air temperature. Its efficiency is higher in dry environments (Montero et al., 1990, Montero and Segal, 1993). The advantage of mist and fog systems over wet pad systems is the uniformity of conditions throughout the greenhouse, therefore eliminating the need for forced ventilation and airtight enclosure. Before installing a system, the air and water flow rates required must be calculated.

2.1.3.1　Fog system

This system is based on spraying the water as small droplets (in the fog range, 2～60 nm in diameter) with high pressure into the air above the plants in order to increase the water surface in contact with the air (Figure 7). Free fall velocity of these droplets is slow and the air streams inside the greenhouse easily carry the drops. This can result in high efficiency of water evaporation combined with keeping the foliage dry. Fogging is also used to create high relative humidity, along with cooling inside the greenhouse. Katsoulas et al. (2001) studied the effect of misting on transpiration and conductance of a greenhouse rose canopy in a greenhouse located in the coastal area of eastern Greece. Measurements were carried out during several days in the summer without air humidity control and with a mist system operating. The contribution of the mist system to total evaporative cooling was estimated to be about 20%, with only 40%～50% of the mist water being effectively used in cooling. Calculation of the crop water stress index (Jackson et al., 1981) confirmed that the crop was less stressed under misting conditions.

Figure 8 Fog system used for greenhouse cooling

A wide range for fog system cooling efficiency ($n_{f,cool}$) has been reported in the literature. According to the investigations of Gates et al. (1991) and Arbel et al. (2003), increased efficiency in the cooling process in relation to water consumption can be expected if fogging is combined with a reduced ventilation rate. Furthermore, a close relationship has been observed between $n_{f,cool}$ and system operation cycling (Abdel – Ghany and Kozai, 2006). Similar values for $n_{f,cool}$ have been reported by Li et al. (2006), who concluded that the fog cooling efficiency increases with spray rate and decreases with ventilation rate.

To improve the greenhouse cooling performance, Toida et al. (2006) used an upward air stream to enhance the fog evaporation rate. A vertically installed conventional nozzle with two small fans (100mm × 100 mm) to provide an upward air stream for enhancing fog evaporation was compared with a vertical nozzle and an approximately horizontal nozzle, both without any fan. The nozzle with fans provided a 1.5 times better evaporation ratio and three times wider cooling area than did the nozzle without fans under similar conditions. The nozzle with fans produced a lower and more uniform air temperature.

Guidelines
- Evaporative cooling allows simultaneous lowering of temperature and vapour pressure deficit and can lead to greenhouse air temperatures lower than the outside air temperature. Its efficiency is higher in dry environments.
- The advantage of mist and fog systems over wet pad systems is the uniformity of conditions throughout the greenhouse, therefore eliminating the need for forced ventilation and airtight enclosure. Before installing a system, the air and water flow rates required must be calculated.
- Fog systems can be high (40 bars) or low (5 bars) pressure systems. High pressure systems are more effective than low pressure.
- The nozzles of the fog system should be located at the highest possible position inside the greenhouse to allow water evaporation before the water drops to crop or the ground.
- During the operation of the fog system a vent opening of 20% of the maximum aperture should be maintained.

- Nozzle with fans provided 1.5 times better evaporation ratio and three times wider cooling area than nozzle without fans. Nozzles with fans produce a lower and more uniform air temperature.

2.1.3.2 Fan and pad cooling

The fan – and – pad cooling system (Figure 9) is most commonly used in horticulture. Air from outside is blown through pads with as large a surface as possible. The pads are kept permanently wet by sprinkling. The water from the pads evaporates and cools the air. For this reason, the outside air humidity must be low. There are basically two systems of fan – and – pad cooling: the negative – pressure system and the positive – pressure system: The negative – pressure system consists of a pad on one side of the greenhouse and a fan on the other side. The fans suck the air through the pad and through the greenhouse. The pressure inside the greenhouse is lower than the pressure outside. Thus hot air and dust can get into the greenhouse. There is a temperature gradient from pad to fan. The positive – pressure system consists of fans and pads on one side of the greenhouse and vents on the other side. The fans blow the air through the pads into the greenhouse. The pressure inside the greenhouse is higher than outside. Dust cannot get into the greenhouse. In order to achieve an optimal cooling the greenhouse should be shaded. The water flow rate, water distribution system, pump capacity, recirculation rate and output rate of the fan – and – pad cooling system must be carefully calculated and designed to provide a sufficient wetting of the pad and to avoid deposition of material in it.

Figure 9　Pad (a) and fan (b) greenhouse cooling system

The manufacturers' guidelines for pad selection and installation have to be observed. The following rules should be considered when designing a fan – and – pad cooling system:

– The cooling efficiency should provide inside air humidity of about 85% at the outlet. Higher air humidity slows down the transpiration rate of the plants. Plant temperature can then increase above air temperature.

– The pad material should have a high surface, good wetting properties and high cooling efficiency. It should cause little pressure loss, and should be durable. The average thickness of the pad is about 100 to 200 mm. It is very important that there are no leaks in the pad where the air can pass through without making contact with the pad. Different pad materials are available, such as wood wool, swelling clay minerals, and specially impregnated cellulose paper.

- The pad area depends on the air flow rate necessary for the cooling system and the permissible surface velocity over the pad. Average face velocities are $0.75 \sim 1.5$ m s^{-1}. Excessive velocities may cause problems with entrainment of drops into the greenhouse. The pad area should be about 1 m2 per $20 \sim 30$ m^2 greenhouse area. The maximum fan – to – pad distance should be $30 \sim 40$ m.

- Pads may be oriented horizontally of vertically, the latter being more common. Vertical pads are supplied with water from a perforated pipe along the top edge. In the case of horizontal pads, the water is sprayed over the upper surface. The water distribution must be such as to ensure an even wetting of the pad. Pads have to be protected from direct sunlight to prevent local drying – out. Salt and sand might clog them, if they become locally dry. In areas with frequent sandstorms it is recommended to protect the wet pad with a thin dry pad serving as a sand filter. The pads have to be located and mounted in a way which permits easy maintenance and cleaning. They should be located on the side facing the prevailing wind.

- Belt – driven or direct – driven propeller fans are used. Direct – driven fans are easier to maintain. Fans should be placed on the lee side of the greenhouse. If they are on the windward side, an increase of 10% in the ventilation rate will be needed. The distance between the fans should not exceed $7.5 - 10$ m, and the fans should not discharge towards the pads of an adjacent greenhouse less than 15 m away. All exhaust fans should be equipped with automatic shutters to prevent air exchange when the fans are not operating, and also to prevent back – draught when some of the fans are not being used.

- When starting the cooling system, the water flow through the pad should be turned on first to prevent the pads from clogging. The fans should be started only after the whole pad has been completely wetted. When stopping the cooling system in the evening, the fan should be turned off before the water flow through the pad. It is recommended to operate the cooling system by a simple control system depending on the inside temperature.

- The air flow rate depends on the solar radiation inside the greenhouse – that is, on the cladding material and shading – and on the evapotranspiration rate from the plants and soil. The air flow rate can be calculated by an energy balance. Generally, a basic air flow rate of $120 \sim 150$ m^3 per m^2 greenhouse area per hour will permit satisfactory operation of an evaporative cooling system.

Five factors mainly affect the temperature distribution along the greenhouse: ventilation rate; crop transpiration and soil evaporation, the latter being neglected in what follows; percentage of shading; water evaporation from the pads; and heat loss coefficient of the cover. The energy balance equation combines these factors and gives access to the temperature distribution along the greenhouse length.

Guidelines
- The pad material should have high surface, good wetting properties and high cooling efficiency. A suggested pad thickness is 200 mm. It is very important that there are no leaks in the pad where the air can pass through without making contact with the pad.

- The pad area depends on the air flow rate necessary for the cooling system and the permissible surface velocity over the pad. Average face velocities are $0.75 \sim 1.5$ m s^{-1}. The pad area should be about 1 m^2 per $20 \sim 30$ m^2 greenhouse area. The maximum fan – to – pad distance should be 40 m.
- Fans should be placed on the lee side of the greenhouse. If they are on the windward side, an increase of 10% in the ventilation rate will be needed. The distance between the fans should not exceed $7.5 - 10$ m, and the fans should not discharge towards the pads of an adjacent greenhouse less than 15 m away.
- When starting the cooling system, the water flow through the pad should be turned on first to prevent the pads from clogging. When stopping the cooling system in the evening, the fan should be turned off before the water flow through the pad.
- A basic air flow rate of $120 \sim 150$ m^3 per m^2 greenhouse area per hour will permit satisfactory operation of an evaporative cooling system.

2.1.3.3 Effect of evaporative cooling on Greenhouse Microclimate and Crop Response

Evaporation cooling is without doubt the most efficient way to cool the environment (Giacomelli et al., 1985; Montero 2006), especially if the outside atmosphere is dry. The technology is now available, but these techniques require a water of good quality. This is the main restriction to the use of fog – systems. As reported by Montero (2006), evaporative cooling techniques have recently become more popular in areas like the Mediterranean basin. This recent interest is associated with the incorporation of insect – proof screens that impose a strong reduction in the rate of air exchange. Hence, since greenhouses with evaporative cooling require less ventilation for given climatic conditions (Boulard and Baille, 1993), evaporative cooling can compensate for the reduced rate of air exchange during hot periods. Evaporative cooling allows simultaneous lowering of temperature and vapour pressure deficit (Arbel et al., 1999; Willits 1999; Katsoulas et al., 2001) (Figure 4a) and can lead to greenhouse air temperatures lower than the outside air temperature. The advantage of mist and fog systems over wet pad systems is the uniformity of conditions throughout the greenhouse, thereby eliminating the need for forced ventilation and airtight enclosure.

The fog system in a greenhouse is expected to lower air temperature and vapour pressure deficit, thereby lowering crop transpiration and thus crop irrigation needs. However, the transpiration water saving could be offset by the amount of water needed for evaporative cooling (Montero, 2006). Changes in crop transpiration and crop water needs, especially in soilless grown crops, may also alter plant nutrient requirements and prevent nutrient – related physiological disorders. In addition, fruit temperature, a parameter influencing several fruit quality characteristics (Jun et al., 1988; Agati et al., 1995), may be affected by fog cooling (Baille and Leonardi, 2001) through the fruit energy balance.

Values of air vapour pressure deficit, VPD, in excess of $1.5 \sim 2$ kPa are known to decrease the stomatal conductance of horticultural crops (Bakker 1991; Baille et al., 1994, Katsoulas et al., 2002), this decline being probably mediated by lower values of internal CO_2 concentration.

Katsoulas et al. (2009) studying an eggplant crop observed that the ratio of transpiration rate λE measured with fog to λE measured without fog cooling, was close to 0.70 indicating that fog cooling reduced crop water consumption about 30%. Montero (2006) pointed out that the above ratio was close to 0.85 during a whole growing season of a tomato crop in Almeria, Spain. Katsoulas et al. (2009) mention that the measured crop transpiration rate averaged at 3.09 kg m^{-2} day^{-1} with fog, and 4.41 kg m^{-2} day^{-1} without fog. Taking into account a 35% drainage rate, the amount of water consumed for irrigation averaged 4.75 kg m^{-2} day^{-1} with fog, and 6.79 kg m^{-2} day^{-1} without fog. Accordingly, Katsoulas et al. (2009) mention that fogging saved about 30% of irrigation water. However, the saved amount of water from crop irrigation was lower than the amount of water needed for fog system operation. It was found that mean value of water consumption for fog cooling was 3.3 kg m^{-2} day^{-1}. Adding this value to the water consumed for crop irrigation results in total water consumption for the fog cooled greenhouse compartment at 8.05 kg m^{-2} day^{-1}, 19% higher than the total water consumed in the non cooled compartment. Lorenzo et al. (2004) reported that the amount of water used by the fog system was equivalent to almost 30% of the total water uptake of a tomato crop.

A wide range for fog cooling efficiency has been reported in the literature. According to Gates et al. (1991) and Arbel et al. (2003), increased cooling efficiency in relation to water consumption can be expected if fogging is combined with low ventilation rate. Li et al. (2006) concluded that fog cooling efficiency increases with spraying rate and decreases with ventilation rate. Sase et al. (2006) concluded that control of the fog system by temperature set point (on/off control) and of vent openings by relative humidity set point (proportional control) resulted in 21% fog water saving compared to no control of vent openings (maximum opening). Perdigones et al. (2008) found that pulse width modulation strategy applied to fog system control reduced fog water consumption by 8% ~ 15% compared to a simple on/off control with fixed fog cycles. Finally, Toida et al. (2006) suggested incorporating small fans to fogging nozzles to increase (about 1.5 times) evaporation ratio of sprayed water and cooling system efficiency. The above reports indicate that fog system efficiency could easily increase, resulting in reduced greenhouse water consumption.

2.2 Heating

Greenhouse heating is essential even in countries with temperate climate, like the Mediterranean region, in order to maximize crop production in terms of quantity and quality and thus to increase the overall efficiency of greenhouse. Heating costs not only have a critical influence on the profitability, but in the long term may also determine the survival of the greenhouse industry. Apart from the costs problems associated with high energy consumption heating is associated with environmental problems through the emission of noxious gases.

Heating needs

There are number of formulations for calculating greenhouse heating needs (Hg) (W) among which the simplest is that proposed from ASHRAE (1978):

$$H_g = U A (T_i - T_o) \tag{2.2}$$

where,
U = heat loss coefficient ($W\ m^{-2}\ K^{-1}$), see table 1
A = exposed greenhouse surface area (m^2)
Ti = inside air temperature, (K)
To = outside air temperature (K)

Table 1 Total heat loss coefficient U at wind speed of n ms^{-1} (von Zabeltitz, 1986)

Covering materials	U value, $W\ m^{-2}\ K^{-1}$
Single glass	6.0 ~ 8.8
Double glass, 9mm air space	4.2 ~ 5.2
Double acrylic 16mm	4.2 ~ 5.0
Single plastic	6.0 ~ 8.0
Double plastic	4.2 ~ 6.0
Double plastic	
Single glass plus energy screen of	
Single film, non – woven	4.1 ~ 4.8
Aloumized single film	3.4 ~ 3.9

Note the estimation of greenhouse needs using equation 2.2 did not take into account the heat loss due to infiltration.

Heating systems

The heating system must provide heat to the greenhouse at the same rate at which it is lost. There are three popular types of heating systems for greenhouses. The most common and least expensive is the unit heater system.

Unit Heaters, In this systems, warm air is blown from unit heaters that have self – contained fireboxes. Heaters are located throughout the greenhouse, each heating a floor area of 180 to 500 m^2). The typical cost, including installation labour is 4 € to 8 €/m^2 of greenhouse floor. The low initial investment for the unit heater systems is suitable for greenhouse firms that start small and expand steadily, purchasing heaters as needed. These heaters are better suited for smaller greenhouses. The use of poly tubes to distribute heat more evenly minimizes temperature fluctuation. Unit heaters must be properly vented to the outdoors.

Central Heating, that produces steam or hot water, plus a radiating mechanism in the greenhouse to dissipate the heat (Figure 10). The typical cost of a central boiler system for 1 ha installation including the heat distribution components in the greenhouse and installation, can range from 30 € to 80 €/m^2 of greenhouse floor space, depending on the number of heat zones and the heat requirement.

Figure 10 Unit heaters and poly tubes to distribute heat

The materials cost proportions into 35% fro the boiler with controls and pumps and 65% for the pipe coils and/or unit heaters. Unlike unit heater systems a portion from the heat from central boiler systems is delivered to the root and crown zone of the crop. This can lead to improved growth of the crop and to a higher level of disease control. Hot water or steam heating systems can also be used to pre-heat irrigation water and prevent cold – water shock to plants. The heat is centrally provided by the boilers and transferred to the greenhouse environment either by pipes that are smooth or finned or by unit heat exchangers equipped with fans.

The amount of pipe needed in a greenhouse coil can be determined by referring to the heat supply values listed in Table 2 for various types of pipe. A greenhouse requiring 100.000 W would need 45 m of 51 – mm hot water pipe to provide this heat. This was determined by dividing the total heat requirement for the greenhouse by the amount of heat that 1 linear meter of pipe can provide.

Placement of heating pipes is very important. Considerable heat can be lost according to the placement of the pipes. The placement of pipes in the walls resulted in a significant amount of losses through greenhouse sides

Wall Pipes coils: Perimeter – wall heating can partially provide the additional heat required and contribute to a uniform thermal environment in the greenhouse. Both bare and finned pipe applications are common. Side pipes should have a few centimetres of clearance on all sides to permit the establishment of air currents and should be located low enough to prevent blockage of light entering through the side wall.

Figure 11 Pipe/Rail Heating Systems

Overhead pipes coils: Heat loss through the roofs and gables is supplied through an overhead coil of pipes that is situated across the entire greenhouse. The overhead coil is not the most desirable source of heat because it is located above the plants. Heat rises from the coil to the top of the greenhouse, where it serves no function and is quickly lost to the outside. Energy needs to be expended to drive the heat down to the plant zone. Overhead Heating Systems can provide the additional heat required for winter months. Such systems are compatible with hanging basket areas, snowmelt systems, and mono – or duo – rail systems. Both bare and finned pipe applications are common.

In bed – pipe coils: A better arrangement, when the greenhouse layout allows it, is the in – bed coil. By placing the heating pipes near the base of the plants, the roots and crown of the plants are heated better than in the overhead system. Air movement caused by the warmer underbench pipe reduces the humidity around the plant. Also heat is kept lower in the greenhouse resulting in better energy efficiency. Such systems are suitable for plants grown on benches, fixed tables, rolling, or transportable tables.

Floor pipe coil: Floor heating is more effective than in – bed pipe coil heating. Apart from the other advantages of in bed coils floor heating has the ability to dry the floor quickly. This is essential when flood floors are used for irrigation/fertilizations. In this system, plants are set on the floor, which makes drying the floor difficult. Air movement caused by the warmer floor reduces the humidity around the plant. Such systems are suitable for plants directly grown on the floor, flooded – floor areas or work areas.

Pipe/Rail Heating Systems: maintain uniform temperatures and positively affect the microclimate of plants. Air movement caused by the warmer pipe/rail reduces the humidity around the plant. Such systems are suitable for vegetable production systems.

Radiant heater systems: These heaters emit infrared radiation, which travels in a straight path at the speed of light. Objects in the path absorb this electromagnetic energy, which is immediately converted to heat. The air through which the infrared radiation travels is not heated. After objects such as plants, walks and benches have been heated, they will warm the air surrounding them. Air temperatures in infrared – radiant – heated greenhouses can be 3 ℃ to 6 ℃ cooler than in conventionally heated greenhouses with equivalent plant growth. Grower reports on fuel savings suggest a 30 to 50 percent fuel reduction with the use of low energy infrared – radiant heaters, as compared to the unit heater system.

Thermostats and Controls: There are several types of thermostats and environmental controllers that are available for commercial greenhouse production. Regardless of how sophisticated this equipment is, there are some very basic factors that must be considered if the system is to operate properly. Sensing devices should be placed at plant level in the greenhouse. Thermostats hung at eye level are easy to read but do not provide the necessary input for optimum environmental control. It is also important to have an appropriate number of sensors throughout the production area. Often time's environmental conditions can vary significantly within a small distance. Do not place thermostats in the direct rays of the sun. This will obviously result in poor readings. Mount thermostat so that they face

North or in a protected location. It is also sometimes necessary to use a small fan to pull air over the thermostat to get appropriate values.

Energy heaters and generators: The risk of electrical power is always present. If a power failure occurs during a cold period, such as a heavy snow or ice storm, crop loss due to freezing is likely. Heaters and boilers depend on electricity. Solenoid valves controlling fuel entry, safety control switches, thermostats and fans providing air to the fire box all depend on electricity energy.

A standby electrical generator is essential to any greenhouse operation. It may never used, but if required for even one critical cold night, it becomes a highly profitable investment. A minimum of 1 kilowatt (kW) of generator capacity is required per 200 m^2 of greenhouse floor area.

Heating for antifrost protection: In these greenhouses heating was used to protect crops from freezing. It is also used to keep air temperature in greenhouse in levels above critical thresholds for condensation control. These greenhouses are not equipped with heavy and complicated heating systems. Usually a unit heater is enough. Other useful recommendations for heating a greenhouse in order to avoid freezing of fruits are:

- To back the north wall to an existing structure such as a house or outbuilding. This rear wall offers extra wind protection and insulation.
- Use water to store heat (a simple passive solar heating system). We can use barrels or plastic tubes filled with water, inside the greenhouse to capture the sun's heat. The heat accumulated during the day will be released at night when temperatures drop.
- Insulate your greenhouse. If your greenhouse is constructed of plastic, insulate with a foam sheet. These sheets can easily be placed over the structure at night and removed during the day. Also install an additional layer of plastic to the interior of the green house for added insulation.

2.3 CO_2 enrichment

The lack of climate control in many greenhouses of Mediterranean countries results in an inadequate microclimate that negatively affects yield components and input – use efficiency. A better control of the greenhouse aerial environment can improve marketable yield and quality, and extend the growing season (Baille, 1999). Air CO_2 concentration is a relevant climate variable to be controlled in greenhouses as it has a marked effect on plant CO_2 assimilation. The atmospheric level limits the potential photosynthesis of most plant species and their productivity (Bowes, 1993). Inside an unenriched greenhouse, the CO_2 concentration drops below the atmospheric level whenever the CO_2 consumption rate by photosynthesis is greater than the supply rate through the greenhouse vents. CO_2 depletion depresses the daily photosynthetic rate, which is estimated to be about 15%, integrated over 29 days of simulation, when the concentration drops below 340 μmol mol^{-1} (Schapendonk and Gaastra, 1984). The poor efficiency of ventilation systems of the low – cost greenhouses in Mediterranean countries, coupled with the use of insectproof nets (Munoz et al., 1999) explains the relatively high CO_2 depletion (about 20% or more) reported in southern Spain (Lorenzo et al., 1990). The solution is to increase the ventilation rate through forced air, to improve design and management of the ventilation system, or to provide CO_2 enrichment. The latter is common in the greenhouse industry of Northern Europe as a means to enhance crop photosynthesis under the low radiation conditions that prevail during winter in those regions. Enrichment reportedly increases crop yield and quality under a CO_2 concentration of 700 ~ 900 μmol mol^{-1} (Kimball, 1986; Mortensen, 1987; Nederhoff, 1994). This situation explains why most of the present information on the effects of CO_2 enrichment on horticultural crops was gathered under climatic conditions and production systems (computerized climate – controlled greenhouses) typical of Northern Europe. Such knowledge and technology are not directly transferable to the environmental and socio – economic conditions of the Mediterranean countries, where CO_2 enrichment is not a common practice for several reasons.

One of the main restrictions is the short time duration available for an efficient use of CO_2 enrichment, due to the need to ventilate for temperature control (Enoch, 1984). The fact that greenhouses have to be ventilated during a large proportion of the daytime makes it uneconomical to maintain a high CO_2 concentration during the daytime. However, some authors advise supplying CO_2 even when ventilation is operating (Nederhoff, 1994) to maintain the same CO_2 concentration in the greenhouse as outside and enriching to levels of about 700 ~ 800 μmol mol^{-1} during the periods when the greenhouse is kept closed, usually in the early morning and the late afternoon.

In the absence of artificial supply, of carbon dioxide in the greenhouse environment, the CO_2 absorbed in the process of photosynthesis must ultimately come from the external ambient through the ventilation openings. This requires the CO_2 within the greenhouse to be lower than the external concentration, as there would be no flow inwards otherwise. Since potential assimilation is heavily dependent on carbon dioxide concentration, this implies that assimilation is reduced, whatever the light level or crop status. The ventilation of the greenhouse implies a trade – off between ensuring inflow of CO_2 and maintaining an adequate temperature within the greenhouse, particularly during sunny days. Stanghellini et al. (2007) applied a simple model for estimating the potential production loss, through data obtained in commercial greenhouses in Almeria Spain and Sicily Italy. They analysed the cost, potential benefits and consequences of bringing more CO_2 in the greenhouse either though increase in ventilation, at the cost of lowering temperature, or through artificial supply. They found that whereas the reduction in production caused by depletion is comparable to the reduction resulting from the lower temperature caused by ventilation to avoid depletion, compensating the effect of depletion is much cheaper than making up the loss by heating.

Sanchez – Guerrero et al. (2005) evaluated variable CO_2 enrichment under the autumn – winter climatic conditions prevailing in a coastal zone of southern Spain. Pure CO_2 was supplied to a greenhouse cucumber crop, maintaining in the greenhouse the air CO_2 concentration at close to 700 μmol mol^{-1} when the greenhouse was closed and 350 μmol mol^{-1} when the vents were open. CO_2 dynamics, efficiency of radiation and CO_2 use, and crop responses were determined over the growing season and compared to those of a similar cucumber crop grown in a non – enriched greenhouse. While the average diurnal concentration remained above 400 μmol mol^{-1} in the enriched greenhouse, significant CO_2 depletion was observed in the unenriched compartment, where the CO_2 concentration fell below 300 $\mu mol\ mol^{-1}$ during 60% of the daytime when the crop was fully developed. The average increase in fruit production for both dry and fresh matter due to CO_2 enrichment was 19%, which agrees well with previous results on the agricultural response of cucumber to moderate CO_2 enrichment.

Optimal CO_2 enrichment depends on the margin between the increase in crop value and the cost of providing the CO_2 gas. Attempting to establish the optimal concentration by experiment is not feasible because the economic value of enrichment is not constant but varies with solar radiation through photosynthesis rate, and with greenhouse ventilation rate through loss of CO_2 (Bailey & Chalabi, 1994). The optimal CO_2 set point depends on several influences: the effect of CO_2 on the photosynthetic assimilation rate, the partitioning to fruit and to vegetative structure, the distribution of photosynthate in subsequent harvests, and the price of fruit at those harvests, as well as the amount of CO_2 used, greenhouse ventilation rate and the price of CO_2. The effects of these individual factors can be described using mathematical models. These models can then be combined to produce a calculation procedure to determine the optimum CO_2 concentration (e.g. Aikman, 1996).

The two principal sources of CO_2 for enrichment are pure gas and the combustion gases from a hydrocarbon fuel such as low sulphur paraffin, propane, butane or natural gas.

Chalabi *et al.* (2001) showed that when tomato greenhouses are enriched with CO_2 obtained from the combustion of natural gas and the enrichment is controlled in an optimal way a considerable increase in economic performance is obtained. Compared with enriching only when the greenhouse requires heating, optimum control without heat storage increases the margin of crop value over expenditure on natural gas by 11%, and optimal control with heat storage increases the margin by 24%. A 30% increase in gas price reduces the financial margin by 11%, whereas a 30% increase in tomato price increases the margin by 40%.

2.4 Dehumidification

Condensation refers to the formation of liquid drops of water from water vapor. Condensation occurs when warm, moist air in a greenhouse comes in contact with a cold surface such as glass, fiberglass, plastic or structural members. The air in contact with the cold surface is cooled to the temperature of the surface. If the surface temperature is below the dew point temperature of the air, the water vapor in the air will condense onto the surface. Condensation will form heaviest in greenhouses during the period from sundown to several hours after sunrise. During the daylight hours, there is sufficient heating in the greenhouse from solar radiation to minimize or eliminate condensation from occurring except on very cold, cloudy days. The time when greenhouses are most likely to experience heavy condensation is sunrise or shortly before. Condensation is a symptom of high humidity and can lead to significant problems, including germination of fungal pathogen spores, including Botrytis and powdery mildew. Condensation can be a major problem and it is unfortunately one which, at least at certain times of the year, is almost impossible to avoid entirely.

The role of VPD

Relative humidity is still the most commonly used measurement for greenhouse control, even though it is not a perfect indication of what the plants 'feel'. Therefore, another kind of measurement, called the Vapour Pressure Deficit is often used to measure plant/air moisture relationships. VPD is the partial pressure difference (deficit) between the amount of moisture in the air and how much the air can hold when saturated. Key variables affect by VPD are temperature and relative humidity. VPD can be an excellent indicator for condensation control. High VPD increases plant transpiration and usually leads low chances for condensation whereas in low VPD environments transpiration is prohibited and condensation risk is high.

How to dehumidify your greenhouse

Combined used of heating and ventilation systems

By opening the windows, moist greenhouse air is replaced by relatively dry outside air. This is common practice to dehumidify a greenhouse. This method does not consume any energy when excess heat is available in the greenhouse and ventilation is needed to reduce the greenhouse temperature. Though, when the need for ventilation to reduce the temperature is less than the ventilation needed to remove moisture from the air, dehumidification consumes energy. The warm greenhouse air is replaced by cold dry outside air, lowering the temperature in the greenhouse.

Absorption using a hygroscopic material

The research on the application of hygroscopic dehumidification in greenhouses is minimal because the installation is too complex and the use of chemicals is not favourable in greenhouses. During the process, moist greenhouse air is in contact with the hygroscopic material releasing the latent heat of vaporisation as water vapour is absorbed. The hygroscopic material has to be regenerated at a higher temperature level. A maximum of 90% of the energy supplied to the material for regeneration can be returned to the greenhouse air with a sophisticated system involving several heat exchange processes including condensation of the vapour produced in the regeneration process.

In the above mentioned calculations the heat loss needed for regeneration of the hygroscopic material and the energy needed to circulate the air are not included in these calculations. The air treatment has to be done centrally because hygroscopic materials are toxic and dangerous for the greenhouse environment. An air distribution system is needed increasing the complexity of the system making it less attractive for application. In all, the installation is very costly. With a pay back time of 10 yr the system may cost between 4 and 21 m^{-2} which is not enough to install this system. But especially the extra environmental risk of the hygroscopic system is a disadvantage.

Condensation on cold surfaces

With this system wet humid air is forced to a clod surface which is located inside the greenhouse, different than the covering material. In the cold surface condensation occurs, the condensate water is collected and can be re – used and absolute humidity of the wet greenhouse air is reduced.

One metre of finned pipe used in their study at a temperature of 5 ℃ can remove 54 g of vapour per hour from air at a temperature of 20℃ and 80% relative humidity. The 3% light interception of the construction used in their experiment is not acceptable in a commercial greenhouse. By implementing the system directly in the construction of the greenhouse, part of this problem can be solved.

It can be concluded from the results that the fraction of latent heat removal over the total heat removal is less than 40%. For air with a relative humidity of 80% only one-third of the heat removed is latent heat. By using a heat pump the latent heat and the sensible heat gathered at the cold surface can be returned to the greenhouse together with the power needed to operate the heat pump.

Forced ventilation usually with the combined use of a heat exchanger

Mechanical ventilation is applied to exchange dry outside air with moist greenhouse air, exchanging heat between the two airflows. Albright and Behler (1984) tested an air – liquid – air heat exchanger for greenhouse humidity control. They concluded that around one-third of the enthalpy could be recovered from the ventilation air. De Hallaux and Gauthier (1998) studied this system and concluded that the use of heat exchangers would lower the energy consumption in direct proportion to the efficiency of the exchanger. The energy savings did not justify the costs of the equipment though. In a more recent study, Rousse et al. (2000) studied a heat recovery unit in Canada. The heat recovery unit had an efficiency of around 80%. The ventilation flux of 0.9 change/h was not enough to dehumidify the greenhouse. The coefficient of performance defined by dividing the recovered heat by the power consumed by the fan, ranged between 1.4 and 4.8 in their study. In a study with smaller sized heat exchangers that were placed in the gutter of the greenhouse, it was possible to retrieve between 60 and 70% of the sensible heat (Speetjens, 2001).

Based on the results of Campen et al. (2003) a ventilator capacity of 0.01 m^3 s^{-1} is sufficient for all crops. The energy needed to operate the ventilators is not considered because the experimental study by Speetjens (2001) showed that the energy consumption by the ventilators is less than 1% of the energy saved. In the experiments by Rousse et al. (2000) the energy consumption was around 20% ~ 40% of the energy saved. They used a central dehumidifying unit and a less optimal heat exchange system explaining this large discrepancy. In Table 5, the energy use, the amount of heat exchanged and the savings by the system compared to dehumidification by ventilation are given. The installation may cost between 3 and 13 h m^{-2} if the return time on investment is 10 yr.

Anti – drop covering materials

The use of anti-drop covering materials is an alternative technology for greenhouse dehumidification. The "anti-dripping" films that contain special additives which eliminate droplets and form instead a continuous thin layer of water running down the sides. The search for anti-drip cover materials has been mainly focused on the optical properties of the cover materials. Katsoulas et al. (2007) evaluated experimentally the effect of two anti-drip cover materials on the greenhouses microclimate and on the development and production of a cucumber crop. A standard greenhouse polyethylene (PE) cover film covered one of the three experimental greenhouses (C – PE). The other two greenhouses were covered: the first by a PE film with anti-drip (AD) and anti-fog (AF) properties (AD + AF – PE) and the second one by a PE film with AD properties (AD – PE). Air relative humidity levels were much higher in the AD – PE greenhouse than in the C – PE and the AD + AF – PE, with the AD + AF – PE covered greenhouse to present the lower values of air relative humidity and the higher values of air vapour pressure deficit. The temperature difference between the cover material and the dew point air temperature was more negative under the AD – PE covered greenhouse, leading in higher condensation rates over the PE film (Figure 12).

Nevertheless, in order to control fungus development, the greenhouses covered by the C – PE and the AD – PE film needed about double fungicide applications, than the greenhouse covered by the AD + AF – PE film. Accordingly, the AD + AF – PE film could be considered as an alternative technique for high humidity levels management in rudimentary equipment greenhouses for safer food production with less chemical residues.

3 Prospective and Needs For Future Research

There are several systems, methods, techniques and models for greenhouse climate control. In the future and towards greenhouse sustainability, climate control should focus on crop activity. The development of low cost and fast response biosensors and the integration of modelling and information technology could help to this direction. Distributed climate modelling should move from design and analysis to climate control, enabling more homogeneous climate conditions. Engineers should work more closely with plant scientists aiming on optimal plant production. Growers' education and training remains an important point to be solved in order to take advantage of the new technology

References

[1] Adams, S. R., Woodward, G. C., and Valdes, V. M. 2002. The effects of leaf removal and of modifying temperature set – points with solar radiation on tomato. Journal of Horticultural Science and Biotechnology, 77:733 ~ 738

[2] Abdel – Ghany, A. M., and T. Kozai. 2006. Cooling efficiency of fogging systems for greenhouses. Biosyst. Eng. 94(1): 97 ~ 109

[3] Agati, G., Mazzinghi P., Fusi, F., Ambrosini, I., 1995. The F685/F730 chlorophyll fluorescence ratio as a tool in plant physiology: response to physiological and environmental factors. J. Plant Physiol. 145, 228 ~ 238

[4] Antón, A. 2004. Utilización del análisis del ciclo de vida en la evaluación del impacto ambiental del cultivo bajo invernadero mediterráneo. Thesis Doctoral, 235 pp Universitat Politécnica de Catalunya

[5] Anton, A., Montero, J. I., Munoz, P., Pérez – Parra, J., Baeza, E. 2006. Environmental and economic evaluation of greenhouse cooling systems in Southern Spain. Acta Hort., 719: 211 ~ 214

[6] Arbel, A., M. Barak, and A. Shklyar. 2003. Combination of forced ventilation and fogging systems for cooling greenhouses. J. Agric. Eng. Res. 84(1): 45 ~ 55

[7] Arbel A, Yekutieli O, Barak M (1999) Performance of a fog system for cooling greenhouse. Journal Agricultural Engineering Research 72, 129 ~ 136

[8] Baeza, E. J. 2007. Optimización del diseño de los sistemas de ventilación en invernadero tipo parral. Tesis Doctoral 204 pp. Universidad de Almería

[9] Bailey, B. J. 1985. Wind dependent control of greenhouse temperature. Acta Hort., 174: 381 ~ 386

[10] Baille, A., Leonardi, C., 2001. Influence of misting on temperature and heat storage of greenhouse grown tomato fruits during summer conditions. Acta Hort. 559, 271 ~ 278

[11] Baille M, Mermier M, Laury JC, Delmon D (1980) Le point sur les systemes d'ombrage sous serre. Internal Report M/801. INRA, Station de Bioclimatologie d' Avignon, Montfavet

[12] Baille M, Baille A, Delmon D (1994) Microclimate and transpiration of greenhouse rose crop. Agricultural and Forest Meteorology 71, 83 ~ 97

[13] Baille A, Kittas C, Katsoulas N (2001) Influence of whitening on greenhouse microclimate and crop energy partitioning. Agricultural and Forest Meteorology 107, 193 ~ 306

[14] Bakker, J. C., Adams, S. R., Boulard, T. and J. I. Montero, 2008. Innovative Technologies for an Efficient Use of Energy. Acta Hort. 801: 49 ~ 62

[15] Bakker, J. C., 1991. Analysis of humidity effects on growth and production of glasshouse fruit vegetables. Dissertation Agricultural University Wageningen, LU1449, The Netherlands, 155 pp

[16] Bartzanas, T., T. Boulard, and C. Kittas. 2002. Numerical simulation of airflow and temperature distribution in a tunnel greenhouse equipped with an insect – proof screen on the openings. Comp. Electr. Agric. 34: 207 ~ 221

[17] Beytes C., 2003. Ball RedBook, Volume 1: Greenhouses and Equipment (Ball Red Book). Ball Publishing; 17 Sub edition, 272 pp

[18] Bontsema, JU., Gieling, Th. H., Kornet, J., Rijpsma, E. and G. L. A. M. Swinkels, 2005. Effects of errors in measurements of light, temperature and CO_2 on energysaving in greenhouses (In Dutch). Report 510 Agrotechnology and Food Innovations, Wageningen, 145pp.

[19] Boodly, J. 1996. The Commercial Greenhouse. Delmar Cengage Learning; 2 edition 624

[20] Bot, G. P. A., van de Braak, N. J., Challa, H., Hemming, S., Rieswijk, Th., van Straten, G. and I. Verlodt, 2005. The Solar Greenhouse: State of the Art in Energy Saving and Sustainable Energy Supply. Acta Horticulturae 691: 501 ~ 508

[21] Boulard T, Baille A (1993) A simple greenhouse climate control model incorporating effects of ventilation and evaporative cooling. Agricultural and Forest Meteorology 65, 145 ~ 157

[22] Boulard, T., 2001. Solar energy for horticulture in France: hopes and reality. Workshop on Renewable Energy in Agriculture, November 19th, 2001: Castello Utveggio – Palermo, EUROPEAN COMMISSION, Energy and Transport, Energy Framework programme 1998 ~ 2002

[23] Bucklin, R. A., Jones, P. H., Barmby, B. A., McConnell D. B. and Henley, R. W. 2009. Greenhouse Heating Checklist. University of Florida, IFAS Extension, Publication CIR791

[24] Campen, J. B. and G. P. A. Bot, 2002. Dehumidification in greenhouses by condensation on finned pipes. Biosystems Engineering, 82(2): 177 ~ 185

[25] Campen, J. B., 2003. Improved temperature distribution by controllable side wall heating. Report Agrotechnology and Food Innovations B. V., Wageningen (in dutch), 132: 34

[26] Challa, H. and J. van de Vooren, 1980. A strategy for climate control in greenhouses in early winter production. Acta Hort., 106: 159 ~ 164

[27] Cockshull, K. E., Hand, D. W. and Langton, F. A. 1981. The effects of day and night temperature on flower initiation and development in chrysanthemum. Acta Hort., 125: 101 ~ 110

[28] Cohen S, Raveh E, Li Y, Grava A, Goldschmidh EE (2005) Physiological response of leaves, tree growth and fruit yield of grapefrui trees under reflective shading screens. Science Horticulturae 107, 15 ~ 35

[29] De Koning, A. N. M., 1988. The effect of different day/night temperature regimes on growt, development and yield of glasshouse tomatoes. The Journal of Horticultural Science, 63: 465 ~ 471

[30] De Pascale, S. and A. Maggio, 2004. Sustainable Protected Cultivation at Mediterranean Climate. Perspectives and Challanges. Acta Hort., 691: 29 ~ 42

[31] De Zwart, H. F., 2005. Evaluation of Roof Spraying as a Low Cost System for Sustainable Energy Collection. Acta Hort., 691: 597 ~ 603

[32] Demrati, H., T. Boulard, A. Bekkaoui, and L. Bouirden. 2001. Natural ventilation and microclimatic performance of a large – scale banana greenhouse. *Biosyst. Eng.* 80(3): 261 ~ 271

[33] Dieleman, A. and Kempkes, F., 2006. Energy screens in tomato: determining the optimal opening strategy. Acta Hort., 718: 599 ~ 606

[34] Elings, A, Kempkes, F. L. K, Kaarsemaker, R. C., Ruijs, M. N. A. van de Braak, N. J., and T. A. Dueck, 2005. The Energy Balance and Energy – Saving Measures in Greenhouse Tomato Cultivation. Acta Hort., 691: 67 ~ 74

[35] Enoch HZ (1984) Carbon dioxide uptake efficiency in relation to crop - intercepted solar radiation. Acta Horticulturae 162,137~147

[36] Garciá - Alonso, Y. , Espí, E. , Salmerón, A. , Fontecha, A. , González and J. López,2006. New Cool Plastic Films for Greenhouse Covering in Tropical and Subtropical Areas. Acta Hort. ,719: 137

[37] Gates, R. S. , J. L. Usry, J. A. Nienhaber, L. W. Turner, and T. C. Bridges. 1991. An optimal misting method for cooling livestock housing. Trans. ASAE 34(5): 2199~2206

[38] Giacomelli, G. , and W. Roberts. 1989. Try alternative methods of evaporative cooling. Acta Hort. 257: 29~30

[39] Hamer, P. J. C. , Bailey, B. J. , Ford M. G. and G. S. Virk,2006. Novel methods of heating and cooling greenhouses: a feasibility study. Acta Hort. ,719: 223~230

[40] Hand DW, Warren Wilson J, Hannah MA (1993) Light interception by a row crop of glasshouse pepper. Journal of Horticultural Science 68 ;695~730

[41] Healey KD, Rickert KG, Hammer GL, Bange MP (1998) Radiations use efficiency increases when the diffuse component of inside radiation is enhanced by shading. Australian Journal of Agricultural Research 49,665~672

[42] Hemming, S. 2005. EFTE: A high transmission covermaterial (in German). Gärtnerbörse 105 (2005)6

[43] Hemming, S. , Kempkes, F. , Van der Braak, N. , Dueck, T. and Marissen, N. ,2006c. Filtering natural light at the greenhouse covering - Better greenhouse climate and higher production by filtering out NIR? Acta Hort. ,711: 411~416

[44] Hoffmann, S. and D. Waaijenberg, 2002. Tropical and Subtropical Greenhouses - A Challenge for New Plastic Films. Acta Hort. ,578: 163~169

[45] Hurd, R. G. and Graves, C. J. ,1984. The influence of different temperature patterns having the same integral on the earliness and yield of tomatoes. Acta Hort. ,148: 547~554

[46] Jackson RD, Idso SB, Reginato RJ, Pinter PJ Jr. (1981) Canopy temperature as a crop water stress indicator. Water Resources Research 17,1133~1138

[47] Jun H, Imai K, Suzuki Y (1990) Effects of day temperature on gas exchange characteristics in tomato ecotypes. Scientia Horticulturae 42 ,321~327

[48] Katsoulas N, Baille A, Kittas C (2001) Effect of misting on transpiration and conductances of a greenhouse rose canopy. Agricultural and Forest Meteorology 106,133~247

[49] Katsoulas, N. , Kitta, E. , Kittas, C. , Tsirogiannis, I. L. and E. Stamati, 2006. Greenhouse Cooling by a Fog System: Effects on Microclimate and on Production and Quality of a Soilless Pepper Crop. Acta Hort. ,719: 455~461

[50] Katsoulas N, Kittas C, Tsirogiannis IL, Kitta, E, Savvas D (2007) Greenhouse Microclimate and Soilless Pepper Crop Production and Quality as Affected by a Fog Evaporative Cooling System. Transactions of the ASABE 50,1831~1840

[51] Kittas C, Baille A, Giaglaras P (1999) Influence of covering material and shading on the spectral distribution of light in greenhouses. Journal of Agricultural Engineering Research 73 ,341~351

[52] Kittas, C. , Katsoulas, N. , Baille, A. ,2001. Influence of greenhouse ventilation regime on microclimate and energy partitioning of a rose canopy during summer conditions. J. Agric. Eng. Res. 79 ,349~360

[53] Körner, O and H. Challa, 2003a. Process - based humidity control regime for greenhouse crops. Comp. Elect. Agric. 39: 173~192

[54] Körner, O. 2003. Crop based climate regimes for energy saving in greenhouse cultivation. PhD Thesis, Wageningen University

[55] Langton, F. A. and Hamer, P. J. C. 2003. Energy efficient production of high quality ornamental species. Final Report to Defra, project HH1330

[56] Li, S. , D. H. Willits, and C. A. Yunkel. 2006. Experimental study of a high - pressure fogging system in naturally ventilated greenhouses. Acta Hort. 719: 393~400

[57] Lorenzo P, Maroto C, Castilla N (1990) CO_2 in plastic greenhouse in Almeria (Spain). Acta Horticulturae 268,165~169

[58] Lorenzo, P. , García, M. L. , Sanchez - Guerrero, M. C. , Medrano, E. , Caparrós, I. , Giménez, M. , 2006. Influence of mobile shading on yield, crop transpiration and water use efficiency. Acta Hort. ,719: 471~478

[59] Miguel AF, Silva AM, Rosa R (1994) Solar irradiation inside a single span greenhouse with shading screens. Journal of Agricul-

tural Engineering Research 59,61~72

[60] Miguel AF, Van De Braak NJ, Bot GPA (1997) Analysis of the airflow characteristics of greenhouse screening materials. Journal of Agricultural Engineering Research 67,105~112

[61] Montero JI, Segal I (1993) Evaporative cooling of greenhouses by fogging combined with natural ventilation and shading. In: Fuchs M (Ed) Proceedings of International Workshop on Cooling Systems for Greenhouses, Agritech, Tel – Aviv, Israel, pp 89 ~99

[62] Montero, J. I. , A. Antón, C. Beil, and A. Franquet. 1990. Cooling of greenhouses with compressed air fogging nozzles. Acta Hort. 281: 199~209

[63] Montero, J. I. ,2006. Evaporative Cooling in Greenhouses: Effects on Microclimate, Water Use Efficiency and Plant Response. Acta Hort. ,719: 373~383

[64] Mortensen LM (1987) Review: CO_2 enrichment in greenhouses. Crop responses. Scientia Horticulturae 33,1~25

[65] Muñoz, P. , Antón A, Paranjpe A. , Ariño J. , Montero J. I. ,2007. Low Nitrogen Input Can Reduce Environmental Impact in the Mediterranean Region Without Reducing Greenhouse Tomato Yield. Agronomy for Sustainable Development

[66] Nederhoff EM (1994) Effects of CO_2 concentration on photosynthesis, transpiration and production of greenhouse fruit vegetable crops, PhD thesis. Wageningen, The Netherlands, 213

[67] Nelson, P. V. ,2002. Greenhouse Operation and Management. Prentice Hall; 6 edition,692

[68] Nisen A, Coutisse S (1981) Modern concept of greenhouse shading. Plasticulture 49,9~26

[69] Raveh E, Cohen S, Raz T, Grava A, Goldschmidt EE (2003) Increased growth of young citrus trees under reduced radiation load in a semi – arid climate. Journal Experimental Botany 54,365~373

[70] Rijsdijk, A. A. and Vogelezang. J. V. M. 2000. Temperature integration on a 24 – hour base: a more efficient climate control strategy. Acta Hort. ,519: 163~169

[71] Runkle, E. S. , Heins, R. D. , Jaster, P. and C. Thill, 2002. Environmental Conditions under an Experimental Near Infra – Red Reflecting Greenhouse Film. Acta Hort. ,578: 181~185

[72] Schapendonk AHCM, Gaastra P (1984) A simulation study on CO_2 concentration in protected cultivation. Scientia Horticulturae 23,117~229

[73] Sonneveld, P. J. and G. L. A. M. Swinkels,2005b. New Developments of Energy – saving Greenhouses with a High Light Transmittance. Acta Hort. 691: 589~595

[74] Sonneveld, P. J. , Swinkels, G. L. A. M. , Kempkes, F. , Campen, J. B. and G. P. A. Bot, 2006. Greenhouse with Integrated NIR Filter and a Solar Cooling System. Acta Hort. ,719: 123~130

[75] Tantau, H. J. 1998. Energy saving potential of greenhouse climate control. Mathematics and Computers in Simulation. 48: 93 ~101

[76] Toida, H. , Kozai, T. , Ohyama, K. , Handarto, 2006. Enhancing fog evaporation rate using an upward air stream to improve greenhouse cooling performance. Biosyst. Engin. 93,205~211

[77] Van de Braak, N. J. , Kempkes, F. L. K. , Bakker, J. C. and J. J. G. Breuer, 1998. Application of simulation models to optimize the control of thermal screens. Acta Hort. ,456: 391~398

[78] Vésine, E. , Grisey A. , Pommier F. , Chantry N. , Plasenti J. , Chassériaux, G. ,2007. Utilisation Rationelle de L' énergie dans les serres. Situation technico – économique en 2005 et leviers d' action actuels et futures. Rapport complet. Report Département DPIA, Direction DABEE, ADEME Angers, France. 134

[79] Willits DH (1999) Constraints and limitations in greenhouse cooling: Challenges for the next decade. Acta Horticulturae 534, 57~66

设施无土栽培叶菜中硝酸盐和维生素 C 的累积调控

刘文科[**]，杨其长，魏灵玲

(中国农业科学院农业环境与可持续发展研究所，中国农业科学院设施农业环境工程研究中心，北京 100081)

摘要：此文总结了设施无土栽培叶菜中硝酸盐和维生素 C 累积特征、产生原因及健康效应，综述了断氮和光调控设施无土栽培叶菜硝酸盐和维生素 C 累积的研究进展，提出了控制策略，并指出了中国应大力发展无土栽培实现优质高产设施蔬菜生产的必要性。

关键词：无土栽培；硝酸盐；维生素 C；品质控制；LED

Regulation of Nitrate and Vitamin C Accumulation of Protected Soill-cultivated Lettuce: a Review

Liu Wenke, Yang Qichang, Wei Lingling

(*Environment and Sustainable Development in Agriculture*, *Chinese Academy of Agricultural Sciences*, *Research Center for Protected Agriculture & Environment Engineering of CAAS*, *Beijing* 100081 *P. R. China*)

Abstract: This paper summarized the accumulation status of nitrate and vitamin C in soilless cultivated leafy vegetables under cover, and its major causes and healthy effects; also reviewed the advances in regulating nitrate and vitamin C accumulation in protected hydroponic leafy vegetables. Finally, the paper gave some control strategies of quality improvement for soilless cultivated leafy vegetables.

Key words: Soilless cultivation; Nitrate; Vitamin C; Quality control; LED

蔬菜是人们膳食中不可或缺的食物，为人体贡献了维生素、矿物质等多种有益物质，同时也提供了硝态氮、重金属等所含的有害物质。基于 AsA 和硝态氮在人体健康中公认的利弊作用，维生素 C（抗坏血酸，AsA）和硝态氮含量是衡量蔬菜品质优劣的重要指标，以增加 AsA 含量及降低硝态氮累积为目标的基因工程及农艺措施研究已成为国内外蔬菜品质调控方法研究的热点。AsA 是人体必需的抗氧化物质，可阻断人体内致癌物质亚硝胺的形成[1]，但因人体缺乏 AsA 合成的关键酶，无法自行合成，只能从食物中获取。据报道，人类膳食中 90% 以上的 AsA 来自蔬菜和水果[2]。然而，蔬菜对人体摄入硝态氮的贡献率也高达 80% 以上[3~4]。摄入的硝态氮可转化形成亚硝态氮，亚硝态氮可诱发高铁血红蛋白症，并可生成强致癌物质亚硝胺，诱发消化系统癌变[5~6]。鉴于人体健康的需要，富含 AsA、低硝态氮含量是人们追求的理想蔬菜品质。

---

[*] 中央级科研院所基本科研业务费项目（BSRF201004）；"十二五" 863 项目（2011AA03A1）资助。
[**] 作者简介：刘文科，男，博士，副研究员，硕士生导师，研究方向为设施蔬菜营养与品质生理。
Tel: 010-8106015，电子邮箱：liuwke@163.com。

1 设施叶菜硝酸盐和维生素 C 累积原因

硝态氮含量过高和 AsA 含量偏低是当前我国设施叶菜普遍存在的品质问题,在施氮过量和设施弱光胁迫条件下会更为严重,对公众健康造成危害。蔬菜是喜氮作物,由于设施栽培的氮肥用量远高于露地,加之设施内相对光强一般只有露地的 28% ~ 50%[7~8],在阴雨雪雾天气条件下易发生弱光胁迫,造成设施叶菜大量累积硝态氮而 AsA 含量降低。设施无土栽培是利用营养液进行蔬菜工厂化生产的栽培方式,国际上已普遍应用。由于高浓度硝态氮供应和设施弱光逆境,导致设施无土栽培叶菜生产中品质问题更为突出。为减少硝态氮摄入对人体健康的危害,欧盟制定了法规来限制设施叶菜中的硝态氮含量,FAO 和 WHO 等国际组织制定了硝态氮的日允许摄入量(ADI),中国规定了无公害蔬菜中硝态氮含量的限定值(GB18406—2001)。另外,为提高 AsA 的有效摄入,中国颁布的行业标准对绿色食品中各类蔬菜(叶菜、根菜等)中的 AsA 含量做了规定(NY/T743—745、NY/T748),而且中国营养学会推荐的成人 AsA 的每日摄入量为 100mg(中国居民膳食营养素参考摄入量,2006)。据调查,我国蔬菜硝态氮污染严重[9],通过蔬菜摄入硝态氮的量为 ADI 的两倍,远高于丹麦、英国和埃及等国家[10]。此外,设施叶菜 AsA 含量偏低,加之储运和烹饪损失,限制了 AsA 的有效摄入,难以达到推荐值。因此,以硝态氮含量过高和 AsA 含量偏低为特征的设施叶菜品质问题,对公众健康构成了重大危害。

2 设施叶菜硝酸盐和维生素 C 累积的断氮控制方法

设施叶菜硝态氮和 AsA 含量与蔬菜种类及品种、环境因子和农艺措施有关。在筛选栽培品种基础上,通过环境因子和农艺措施协同调控是削减设施叶菜硝态氮含量和增加 AsA 累积的有效途径。其中,以氮营养调控和光环境调控最为有效。与土壤栽培相比,设施无土栽培具有营养液和环境因子可控的优势。在过去 20 年里,国内外学者在设施无土栽培叶菜品质调控领域,尤其在采收前品质调控方面取得了喜人的突破。与栽培全过程调控方法不同,采收前调控方法是在采收前几天通过调整营养液成分和控制环境因子来提高蔬菜品质的方法,具有可操作性强、成本低等优势。研究表明,采收前断氮处理可减少水培叶菜硝态氮累积[11~17],增加 AsA 含量[11~12,17],是提高设施无土栽培叶菜品质的有效方法。断氮处理大致有三种方式:①供应无氮营养液;②供应水;③供应含渗调物质、钼酸铵的水溶液。试验表明,采收前三种断氮方式都能有效降低水培生菜的硝态氮含量,提高 AsA 含量[16~17]。比较而言,第三种方式下生菜硝态氮含量降低更快,其原因是渗调离子的渗调功能促进了硝态氮在叶片代谢库和储存库间的调配[17],而钼酸铵则增加了硝酸还原酶(NR)的活性[16]。

断氮处理提高水培叶菜品质的效果受以下因素的影响:①蔬菜种类及品种;②断氮处理方式;③处理前氮营养;④光照条件。硝态氮是易于运输的氮素储存形态,氮供应充足时奢侈吸收的硝态氮储存于液泡内,当氮亏缺时,蔬菜体内的硝态氮可在器官和亚细胞水平(液泡 - 细胞质)进行调配与再利用[14,18]。一般而言,叶菜可食部分包括叶片和叶柄,叶片和叶柄是累积硝态氮主要器官,只有同时降低叶片和叶柄中的硝态氮含量才能削减整株叶菜的硝态氮含量。然而,叶柄累积的大量硝态氮无法在叶柄中还原[19~20],需调配到叶片中进行。试验表明,断氮后小白菜叶片、叶柄和根系中的硝态氮含量稳步下降,发生了器官间的硝态氮调配。同时,叶片 NR 活性表现为降低—升高—降低的变化[14]。孙园园等(2008)[15]发现断氮后基质培菠菜叶片中硝态氮含量显著降低。人们利用硝态氮选择性微电极测定技术,发现断氮后叶菜叶片液泡和细胞质之间发生了硝态氮的调配,液泡硝态氮含量下降[14,21~22]。显然,断氮处理期间,伴随光合消耗,叶菜叶片中的硝态氮供给将经历过量—持平—缺乏三个阶段,迫使叶菜启动硝态氮空间调配过程。据报道,叶片和叶柄 NR 活性、代谢库大小决定着叶片和叶柄中硝态氮累积、同化和调配特征[19~20]。氮亏缺条件下硝态氮亚细胞水平的调配也是由细胞质内的 NR 活性决定的[23]。可是,NR 活性受底物浓度、光照条件和产物浓度等因素的影响。迄今,有关断氮条件下新老叶、叶片和叶柄中的硝态氮削减及调配特征方面的报道较少,涉及代谢关键酶活性变化及还原、同化产物累积规律的报道更少。所以,为达到整株削减硝态氮的目标,揭示断氮处理下叶菜体内硝态氮调配特征及代谢机理是非常必要的。

蔬菜中 AsA 的累积与其合成、再生循环和氧化分解代谢的活性密切相关。至今,关于采收前断氮处理对叶菜 AsA 代谢关键酶活性影响的报道较少,而有关新老叶、叶片与叶柄中 AsA 代谢对断氮处理的响应差异研究还是空白。与不断氮处理相比,三种断氮处理(无氮营养液、蒸馏水和 0.1mmol/L 氯化钾)提高了生菜新叶中的 AsA 含量,但对老叶中的 AsA 含量无影响[17],这种器官间响应差异是否与新叶对老叶的遮光有关,其代谢层面的原因有待揭示。孙园园等(2008)[15]发现,断氮处理后基质培菠菜叶片中 AsA 含量及半乳糖内酯脱氢酶(GalLDH)活性降低了,与 Mozafar(1996)[11]的菠菜断氮试验结果不一致,这可能与菠菜品种、氮营养、光照条件和栽培方式有关。氮营养对断氮生菜硝态氮和 AsA 累积影响的代谢机理尚需探明。通常,栽培氮营养决定着断氮时叶菜硝态氮和 AsA 的含量及植株代谢水平[15]。关于氮营养与叶菜硝态氮、AsA 累积的关系,孙园园等(2008)[15]的研究提供了重要线索。随着硝态氮供应水平的增加,菠菜叶片中硝态氮含量递增,而叶片 AsA 含量与 AsA 合成酶、再生循环还原酶活性呈先增后降的变化(拐点氮浓度为 10mmol/L),氮水平调控了 AsA 合成和再生循环代谢;另外,随铵硝比的提高,菠菜叶中硝态氮持续降低,而 AsA 含量不断升高,铵硝比变化调控了 AsA 的再生循环代谢。

3 设施叶菜中硝酸盐和维生素 C 累积的光调控方法

补光可克服设施弱光胁迫,促进叶菜体内硝态氮的还原同化与 AsA 合成,而且可协同采收前断氮处理确保品质调控的最佳效果。光通过光合作用为碳氮代谢提供原材料和能量,是影响蔬菜硝态氮与 AsA 累积和代谢的重要环境因子。光对蔬菜生长和品质形成的影响主要与光强、光谱分布和光照时间有关[28]。设施弱光胁迫时,蔬菜光合作用降低,光合产物和能量缺乏,产量和品质下降。众所周知,补光是克服设施弱光胁迫的有效方法。弱光胁迫下适宜补光可调控叶菜硝态氮和 AsA 的累积,这对无土栽培和土壤栽培均有效[29]。原因之一,光强提高可诱导增加硝态氮与 AsA 代谢关键酶的活性[30~34];原因之二,高光强可促进蔬菜光合作用,增加硝态氮与 AsA 代谢所需光合产物和能量的供给[35]。比如,有人通过协调光强与氮供给水平的关系,生产出低硝态氮的生菜[36]。研究发现,断氮处理后生菜叶片和叶柄硝态氮削减速率受光强控制,遮光处理抑制了叶片和叶柄中硝态氮的削减速率[37]。由此可知,适宜光强下叶片硝态氮还原同化速率要大于供给速率,从而启动亚细胞及器官水平的硝态氮调配过程,进而削减整株硝态氮的含量。适宜光强下,断氮将加速叶菜中硝态氮的削减进程;但弱光胁迫下,硝态氮还原同化速率很低,无法启动调配过程,致使断氮调控无法发挥作用。就 AsA 而言,光强调控叶菜 AsA 累积的机制要复杂些。研究表明,断氮对生菜新老叶中 AsA 累积的影响不同[17],这种器官间响应差异可能与新老叶所受光照强度大小有关。因此,弱光胁迫下,只有通过补光矫正才能使断氮调控获得更好的品质效益。

3.1 LED 补光优势分析

传统补光是以高压钠灯、金属卤素灯等为光源将设施光强提高到蔬菜光补偿点以上,延长光照时间,确保蔬菜产量及品质[28]。传统光源补光具有一定缺陷,主要是光谱分布固定、不可调控,无法针对蔬菜生理需求提供光谱,而且所含发热光谱成分多,耗能偏高。发光二级管(LED)为节能冷光源,可近距离照射植物,提供植物所需光谱,是设施补光的理想光源。据报道,植物光合作用主要以可见光区波长 610～720nm 的红橙光以及波长 400～510nm 的蓝紫光为吸收峰值区域。因此,开发这两个波段为主体的人工光源将会提高植物的光能利用效率,适宜植物的光合和生长[38~39]。近年来,随着 LED 技术的发展为实现这一目标提供了可能。LED 能够发出植物生长所需要的单色光,且便于组合,构成植物光合作用与形态建成所需要的光谱,实现有效补光。更重要的是,LED 为冷光源、具有节能环保、体积小、寿命长、易于分散与组合控制等诸多不同于其他电光源的特性[40],更适于设施栽培补光,应用潜力十分明显。

3.2 LED 补光对叶菜品质调控进展

研究证实,LED 红蓝光可调控无土栽培叶菜的硝态氮和 AsA 的累积[41~43],但相关代谢机理揭示较少。与荧光灯相比,LED 红蓝光栽培的生菜,其硝酸盐含量降低 15%～20%,糖含量提高[42]。Ohashi 等(2007)[41]采用 300μmol/(m²·s)的 LED 红蓝白光源进行叶菜栽培。与白光相比,红光降低了菠菜的硝态氮含量,蓝光或红蓝光处理增加了生菜和小松菜的 AsA 含量,也降低了生菜硝态氮含量。更为可喜的是,在不断氮条件下,采收前 LED 光照处理可降低生菜硝态氮含量,增加 AsA 的累积[43]。采收前用 500μmol/(m²·s)LED 红光照射 3d,生菜硝态氮含量降幅达 65%,AsA 含量增加[43]。在采收前用 150μmol/(m²·s)的 LED 红蓝光照射 3d,生菜硝态氮降低,AsA 和可溶性糖含量增加[44];并且发现,红蓝光组成显著影响生菜的品质变化。总之,光质调节可提高叶菜品质,但最佳光质及光质组合因植物种类和品种而异[38~39,41]。尽管 LED 光质,尤其是红蓝光对叶菜硝态氮和 AsA 累积影响的报道较多,但相关代谢机理的揭示较少。

4 设施叶菜硝酸盐和维生素 C 累积的营养液调控策略

采收前断氮与补光协同可实现无土栽培叶菜复合品质的大幅快速提高。众多研究表明,叶菜硝态氮含量削减和 AsA 含量增加常相伴发生[11~12,17,41~44],使得叶菜复合品质同步提高成为可能。断氮和补光是分别从营养和环境两个方面入手的调控方法,作用机制不同,具有协同调控硝态氮和 AsA 累积代谢的潜力。在设施蔬菜无土栽培中,以设施环境和营养液复合控制为核心理念,按照过程控制和采前控制两种思路,制定设施蔬菜硝酸盐累积控制的策略是实现低硝酸盐含量蔬菜生产的可行途径。通过栽培品种选择,控制光强和光质,结合功能型栽培液处理可解决设施无土栽培蔬菜硝酸盐累积和维生素 C 偏低的问题。

5 结语

蔬菜是高附加值的作物,大多数按鲜重出售,其价值取决于其品质状况(营养品质、卫生品质、外观品质和储藏品质),其中卫生品质和营养品质尤为重要。无土栽培将是未来设施农业的主要栽培形式,是工厂化栽培的必然选择。无土栽培介质杂质少,适宜的环境控制使病虫害发生频率较低,可避免重金属和农药的污染。在营养液循环利用技术普遍采用后,不仅可使水肥等资源得以高效利用,也使无土栽培的环境危害风险降至近乎为零。可以深信,不久的将来无土栽培将成为我国设施无公害蔬菜规模化生产的主体。在以化肥为特征的现代农业生产条件下,以硝酸盐累积和药残为主要特征的设施蔬菜卫生品质问题是世界性的问题,需要格外重视。从现在的研究进展看,在现有调控措施下无土栽培是生产低硝酸盐含量蔬菜的优选方法,值得我国发展和推广应用。特别是在设施土壤栽培存在许多当前无法解决的资源环境和连作障碍问题情况下,无土栽培技术的研发应用更显迫切。发达国家蔬菜无土栽培已经成为设施蔬菜栽培的主体,而我国仅有很少的栽培面积,差距明显。在我国大力发展设施蔬菜的同时,应加快无土栽培技术研究、发展和应用,以确保蔬菜硝酸盐品质和城乡居民的饮食安全,增强蔬菜产品的竞争力,促进我国由蔬菜大国向蔬菜强国迈进。

参考文献

[1] Tannenbaum S. R., Wishnok J. S., Leaf C. D. Inhibition of nitrosamine formation by ascorbic acid. American Journal of Clinical Nutrition, 1991, 53: 247~250

[2] Lee S. K., Kader A. A. Preharvest and postharvest factors influencing vitamin C content of horticultural crops. Postharvest Biology and Technology, 2000, 20: 207~220

[3] Gangolli S. D., van den Brandt P., Feron V. J., Janzowsky C., Koeman J. H., Speijers G. J. Nitrate, nitrite and N-nitroso compounds. European Journal of Pharmacology: Environmental Toxicology and Pharmacology, 1994, 292 (1): 1~38

[4] Eichholzer M., Gutzwiller F. Dietary nitrates, nitrites, and N-Nitroso compounds and cancer risk: a review of the epidemiologic evidence. Nutrition Review, 1998, 56 (4): 95~105

[5] Wolff I. A., Wasserman A. E. Nitrates, nitrites, and nitrosamines. Science, 1972, 177 (4043): 15~19

[6] Santamaria P. Nitrate in vegetables: toxicity, content, intake and EC regulation. Journal of the Science of Food and Agriculture, 2006, 86: 10~17

[7] 吴晓蕾,尚春明,张学东等. 番茄品种耐弱光性的综合评价. 华北农学报,1997,12 (2): 97~101

[8] 王克安,陈运起,张世德等. 冬暖大棚棚型结构温、光效应初探. 山东农业科学,1999,1: 15~17

[9] Zhou Z. Y., Wang M. J., Wang J. S. Nitrate and nitrite contamination in vegetables in China. Food Reviews International, 2000, 16 (1): 61~76

[10] Zhong W. K, Hu C. M., Wang M. J. Nitrate and nitrite in vegetables from north China: content and intake. Food Additives & Contaminants, 2002, 19 (12): 1125~1129

[11] Mozafar A. Decreasing the NO_3^- and increasing the vitamin C contents in spinach by a nitrogen deprivation method. Plant Foods for Human Nutrition, 1996, 49: 155~162

[12] 董晓英,李式军. 采前营养液处理对水培小白菜硝酸盐累积的影响. 植物营养与肥料学报,2003,9 (4): 447~451

[13] 罗健,程东山,林东教,刘士哲. 不同收获时期和控氮条件对水培小白菜硝酸盐含量的影响. 生态环境,2005,14 (4): 562~566

[14] 罗金葵. 小白菜硝酸盐累积机理研究. 南京农业大学硕士学位论文. 2006. 江苏·南京

[15] 孙园园. 氮素营养对菠菜体内抗坏血酸含量及其代谢的影响. 浙江大学硕士学位论文. 2008. 浙江·杭州

[16] Liu W. K., Yang Q. C. Short-term treatment with hydroponic solutions containing osmotic ions and ammonium molybdate decreased nitrate concentration in lettuce. Acta Agriculturae Scandinavica Section B-Soil & Plant Science (In Press)

[17] Liu W. K., Yang Q. C., Qiu Z. P. Spatiotemporal changes of nitrate and Vc contents in hydroponic lettuce treated with various nitrogen-free solutions. Journal of the Science of Food and Agriculture (Submitted)

[18] Echeverria E. D. Vesicle-mediated solute transport between the vacuole and the plasma membrane. Plant Physiology, 2000, 123: 1 217~1 226

[19] 刘忠, 王朝辉, 李生秀. 硝态氮难以在菠菜叶柄中还原的原因初探. 中国农业科学, 2006, 39 (11): 2 294~2 299

[20] 刘忠, 王朝辉, 李生秀. 菠菜叶片硝态氮还原对叶柄硝态氮含量的影响. 植物营养与肥料学报, 2007, 13 (2): 313~317

[21] Leij M. van der, Smith S. J., Miller A. J. Remobilisation of vacuolar stored nitrate in barley root cells. Planta, 1998, 205: 64~72

[22] 陈巍, 罗金葵, 尹晓明, 贾莉君, 张攀伟, 沈其荣. 硝酸盐在两个小白菜品种体内的分布及调配. 中国农业科学, 2005, 38 (11): 2 277~2 282

[23] Campbell W. H. Nitrate reductase structure, function and regulation: bridging the gap between biochemistry and physiology. Annual Review of Plant Physiology and Plant Molecular Biology, 1999, 50: 277~303

[24] Wheeler G. L., Jones M. A., Smirnoff N. The biosynthetic pathway of vitamin C in higher plants. Nature, 1998, 393: 365~369

[25] Smirnoff N. The function and metabolism of ascorbic acid in plants. Annals of Botany, 1996, 78: 661~669

[26] Agius F., Gonzalez-Lamothe R., Caballero J. L., Munöz-Blanco J., Botella M. A., Valpuesta V. Genetic engineering increased vitamin C levels in plants by overexpression of a D-galacturonic acid reductase. Nature Biotechnology, 2003, 21: 177~181

[27] Hemavathi, Upadhyaya C. P., Young K. E., Akula N., Kim H. soon, Heung J. J., Oh O. M., Aswath C. R., Chun S. C., Kim D. H., Park S. W. Over-expression of strawberry D-galacturonic acid reductase in potato leads to accumulation of vitamin C with enhanced abiotic stress tolerance. Plant Science, 2009, 177: 659~667

[28] 曹阳. 冬季温室补光对果菜类作物生长发育的影响 (综述). 河北农业科学, 2009, 13 (3): 10~12

[29] McCall D., Willumsen J. Effects of nitrogen availability and supplementary light on the nitrate content of soil-grown lettuce. The Journal of Horticultural Science & Biotechnology, 1999, 74 (4): 458~463

[30] Gautier H., Massot C., Stevens R., Sérino S., Génard M. Regulation of tomato fruit ascorbate content is more highly dependent on fruit irradiance than leaf irradiance. Annals of Botany, 2009, 103: 495~504

[31] Lillo C. Light regulation of nitrate reductase in green leaves of higher plants. Physiologia Plantarum, 1994, 62: 89~94

[32] Tamaoki M., Mukai F., Asai N., Nakajima N., Kubo A., Aono M., Saji H. Light-controlled expression of a gene encoding L-galactono-y-lactone dehydrogenase which affects ascorbate pool size in *Arabidopsis thaliana*. Plant Science, 2003, 164: 1 111~1 117

[33] Mishra N. P., Fatma T., Singhal G. S. Development of antioxidative defense system of wheat seedlings in response to high light. Physiology Plant, 1995, 95: 77~82

[34] Melzer J. M. Kleinhofs, A., Warner R. L. Nitrate reductase regulation: effects of nitrate and light on nitrate reductase mRNA accumulation. Molecular and General Genetics MGG, 1989, 217 (2~3): 341~346

[35] Blom-Zandstra M., Lampe J. E. M. The role of nitrate in the osmoregulation of lettuce (*Lactuca sativa* L.) grown at different light intensities. Journal of Experimental Botany, 1985, 36: 1 043~1 052

[36] Demšar J, Osvald J, Vodnik D. The effect of light-dependent application of nitrate on the growth of aeroponically grown lettuce (*Lactuca sativa* L.). Journal of the American Society for Horticultural Science, 2004, 129 (4): 570~575

[37] Liu W. K., Yang Q. C., Du L. F. Effects of short-term treatment with light intensity and hydroponic solutions before harvest on nitrate reduction in leaf and petiole of lettuce. Journal of Plant Nutrition (submitted after revision)

[38] Matsuda R., Ohashi K. K., Fujiwara K., Goto E., Kurata K. Photosynthetic characteristics of rice leaves grown under red light with or without supplemental blue light. Plant Cell Physiology, 2004, 45: 1 870~1 874

[39] Ohashi K. K., Matsuda R., Goto E., Fujiwara K., Kurata K. Growth of rice plants under red light with or without supplemental blue light. Soil Science and Plant Nutrition, 2006, 52: 444~452

[40] 崔瑾，徐志刚，邸秀茹. LED 在植物设施栽培中的应用和前景. 农业工程学报, 2008, 24 (8): 249~253

[41] Ohashi K. K., Takase M., Kon N., Fujiwara K., Kurata K. Effect of light quality on growth and vegetable quality in leaf lettuce, spinach and komatsuna. Environment Control in Biology, 2007, 45 (3): 189~198

[42] Urbonavi? iūt? A., Pinho P., Samuolien? G., Duchovskis P. Effect of short-wavelengh light on lettuce growth and nutritional quality. Sodininkyst? ir Dar? ininkyst?, 2007, 26 (1): 157~165

[43] Samuolien G., Urbonaviit A. Decrease in nitrate concentration in leafy vegetables under a solid-state illuminator. HortScience, 2009, 44: 1 857~1 860

[44] Zhou W. L. Liu W. K., Yang Q. C. Reducing nitrate concentration in lettuce by elongated lighting delivered by red and blue LEDs before harvest. Journal of Plant Nutrition, 2011 (In press)

[45] Gniazdowska-Skoczek H. Effect of light and nitrates on nitrate reductase activity and stability in seedling leaves of selected barley genotypes. Acta Physiologiae Plantarum, 1998, 20 (2): 155~160

[46] Smirnoff N. Ascorbate biosynthesis and function in photoprotection. Philosophical Transactions of the Royal Society of London Series B-Biological Sciences, 2000, 355: 1 455~1 464

设施园艺工程技术

日光温室热环境分析及设计方法研究*

马承伟**，徐凡，赵淑梅，李睿，刘洋

（中国农业大学 农业部设施农业工程重点开放实验室，北京 100083）

摘要：为了提高日光温室的设计与建造水平，提高其使用性能，需要根据科学理论，解决日光温室优化设计方法的问题。论文运用工程热物理的理论，分析了日光温室室内环境的形成机理，提出日光温室的优良热环境性能，归因于屋面透光率高、北墙增加太阳辐射接收量、屋面与墙体的优良保温性，以及墙体和地面的蓄热加温等四个方面作用，并提出日光温室优化设计的原则在于加强该四个方面的作用。分析了墙体设计和地面下沉等问题，提出墙体材料应外侧注重保温性，内侧注重蓄热性等观点。论文提出利用日光温室数学模型，模拟预测其环境性能，是日光温室优化设计的可行途径，并介绍了研发的日光温室热环境模拟预测软件。

关键词：日光温室；热环境；设计方法；模拟；软件

Solar Greenhouse Thermal Environment Analysis and Design Method Research

Ma Chengwei, Xu Fan, Zhao Shumei, Li Rui, Liu Yang

(*China Agricultural University*, *Key Laboratory of Agricultural Engineering in Structure and Environment*, *Ministry of Agriculture*)

Abstract: In order to improve the design and construction level of solar greenhouse, and advance its service performance, it is needed to solve the problem of solar greenhouse optimization design method based on scientific theory. The paper used the theory of Engineering Thermophysics, analyzed the formation mechanism of the solar greenhouse indoor environment, proposed that the solar greenhouse have excellent thermal environmental performance, because of the four factors below: the high roof transmittance, the increase of indoor solar radiation received by north wall, well heat insulation of the roof and wall, and heat storage and warming role of the wall and ground. And advised the principle of solar greenhouse optimization design is to enhance the four functions mentioned above. The paper also analyzed the wall design and ground subsidence, considered the wall material chosen should pay attention to the heat insulation outside, and heat storage inside. The paper proposed that using solar greenhouse mathematics model, simulating and predicting the environment performance, is a feasible way of solar greenhouse optimization design, and introduced the solar greenhouse thermal environment simulation and prediction software which researched and developed by authors.

Key words: Solar greenhouse; Thermal environment; Design method; Simulate; Software

---

* 基金项目：现代农业产业技术体系建设专项资金资助（Nycytx - 35 - gw24）。

** 作者简介：马承伟（1952— ），男，重庆人，博士，教授，博士生导师，中国农业工程学会高级会员，主要从事设施园艺环境工程研究。中国农业大学农业部设施农业工程重点开放实验室，北京 100083。E-mail：macwbs@cau.edu.cn。

日光温室是在中国得到广泛应用的一种主流园艺设施，目前面积已达到70万公顷以上，接近中国园艺设施面积的1/4。尽管日光温室建筑较为简陋，而且缺少完备的环境调控装备，但冬季室内的环境条件能够保证大多数园艺作物生长的基本要求。其最大的优点是，绝大多数日光温室冬季不需要加温，其节能性大大优于普通日光温室。在冬季，日光温室内部可以在不加温的情况下，维持高于室外21~25℃的气温[1]，夜间最低气温高于室外最低气温15~20℃。因此，在中国北方农村，在冬季夜间室外最低气温为－10℃左右的时期，日光温室仍能够在不依靠加温的情况下，夜间维持5~10℃的室内气温，可以进行大多数园艺作物的越冬生产。因此，日光温室的建设和运行费用极其低廉，生产成本很低，这是日光温室在中国获得大规模发展的主要原因。

但是，日光温室也普遍存在冬季夜晚温度过低等问题，遇寒潮、连续阴天等不利天气情况，室内仍难以保证适宜的条件，致使作物受到冻害，生产受到损失。据调查，目前还有相当多的日光温室，在冬季气温最低的时期，夜间室内气温会降低至5℃以下[2,3]，一些日光温室甚至出现1℃以下的低温[4,5]，作物受到冻害的情况还很普遍。

如何在保持日光温室建设和运行费用低廉的优点的情况下，进一步提高日光温室建设水平和性能，是中国设施园艺工程界非常关注的研究课题。由于日光温室环境与结构相关理论还未形成科学的体系，在其设计与新材料开发等方面，还缺乏系统的理论指导和科学的方法。因此目前日光温室的设计和建造还主要是依据有限的经验，普遍存在设计水平不高，性能潜力还未得到充分发挥的情况。尤其是一些性能欠佳的日光温室，其设计方案均存在若干不合理之处。

目前设施园艺环境工程界在日光温室合理设计的一些问题上，认识还相当模糊和混乱，甚至存在一些错误的看法。迫切需要运用科学的理论和可靠的方法，准确地掌握日光温室的技术原理，合理解决设计中诸如日光温室的总体方案、材料选择、结构和尺寸的确定等各种问题。

本文应用工程热物理的理论，通过对日光温室的室内环境形成机理的分析，探讨日光温室合理设计中的一些问题。并介绍笔者在日光温室优化设计的辅助工具软件研究开发方面的初步成果。

1 日光温室构造特征与热环境形成机理

从日光温室的构造特征，探明日光温室室内环境的形成机理和环境特点，对于解决日光温室的合理设计问题，具有重要的指导作用。

日光温室基本不用燃料加温，而主要在白昼依靠墙体和地面吸收并蓄积太阳热能，在夜间，墙面和地面成为加温的热源，将蓄积的热量释放回温室，在再加上其围护结构的良好保温性能，从而能够维持室内适于园艺作物生长的环境条件。这就是日光温室具有优良节能性能的基本技术原理，具体而言，从日光温室特有的建筑方案和构造特征的角度，可以分析归纳为四个方面。

1.1 屋面良好的日光透过特性

日光温室一律采用有利于日光透过的朝向南面的采光屋面，再加上对室内造成遮荫的骨架构件少，而且基本没有环境调控和管理设备遮荫的情况，因此日光透过率较高。

根据相关资料以及笔者对北京郊区等地温室的调查和测试,普通温室的平均日光透过率为 45%~60%,而日光温室的平均日光透过率可以达到 60%~80%[1,6],即日光温室比普通温室的平均日光透过率高 15% 以上。较高的日光透过率不仅使温室内的光照条件较为容易满足园艺作物生长的要求,还有利于使温室内进入较多的太阳热能。

1.2 北墙增加温室内太阳辐射接收量的作用

日光温室不透光的北墙具有截获和吸收太阳辐射热能的作用,与仅能在水平栽培面积上获得太阳辐射能的普通温室相比,日光温室内获得的太阳辐射热能大大增加。

如图 1 所示,日光温室北墙可以接受到室内地面以外部分的太阳辐射(相应于 H_w 的部分)。在日光正对温室屋面和北墙面的时刻,相对于室内地面所接受的太阳辐射(相应于 H_f 的部分),墙面增加接受的太阳辐射的比例近似为:

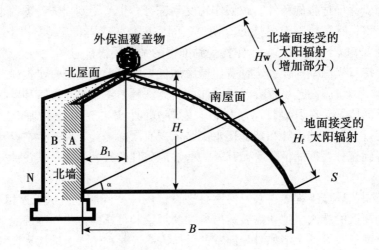

图 1　日光温室地面与墙面接受的太阳辐射

$$\beta = \frac{H_w}{H_f} = \frac{(H_r - B_1 \sin\alpha)\cos\alpha}{B\sin\alpha} \tag{1}$$

式中:B—跨度,m;B_1—北屋面水平投影宽度,m;H_r—脊高,m;α—太阳高度角,度。

例如在北京地区(北纬 40°),对于 $B=8m$,$B_1=1.4m$,$H_r=3.6m$ 的日光温室,在冬至正午时刻,太阳高度角为 $\alpha=26.5°$。可以计算得到 $\beta=0.746$。即由于温室北墙截获太阳辐射热能,温室内获得的太阳辐射热量增加了 74.6%。

需要说明的是,上述估计主要是按太阳直接辐射计算的,对于散射辐射,墙面接受的比例要低一些。此外,在正午前后的时刻,太阳高度与方位均在发生变化,不同时刻墙面和地面接受的太阳辐射比例有所变化。综合考虑各方面因素,更详细的计算分析表明,全天墙面增加接受的太阳总辐射约占地面接受太阳总辐射的 50%~60%。

在冬至以后,太阳高度角较大一些,墙面增加接受的太阳辐射比例比冬至日小。对于 11 月至次年 2 月这 4 个月的情况,考虑太阳辐射参数的变化,以及不同地区、不同日光温室建筑尺寸方面的差异,对多种日光温室内不同表面接受的太阳辐射分析表明,墙面增加接受的太阳总辐射一般占地面接受太阳总辐射的35%~60%。

由于北墙增加温室内太阳辐射接收量的作用,以及前述日光温室屋面透光率较高的特点,使日光温室内在冬季的白昼,吸收的太阳辐射热量比普通温室增加40%~70%,从而可以有效地提高温室内的空气温度,增加室内蓄积的热量。

1.3 墙体与屋面具有良好的保温性

与普通温室相比,日光温室的围护覆盖层的保温性高很多。对于普通温室,其覆盖层的传热系数[6~9],在采用单层覆盖时,一般为6~7W/(m^2·℃),在采用单层保温幕的情况下,可以降低至3~4W/(m^2·℃),即使采用双层保温幕,覆盖层传热系数也只能降低至2.5W/(m^2·℃)左右。

日光温室屋面在夜间覆盖草帘或用纤维以及泡沫塑料等材料制作的保温被,根据笔者的测试,其传热系数一般仅在1.6~2.0W/(m^2·℃)之间。同时,日光温室北面墙体具有很大厚度,一般土墙厚度可达1m以上,有的土墙平均厚度达到3~4m,其相应的传热系数约在0.3~1.0W/(m^2·℃)之间。砖墙厚度可达到0.6m以上,并往往在中间夹放保温材料,其传热系数约在0.6W/(m^2·℃)以下。综合考虑,日光温室屋面和墙面组成的全部围护覆盖的平均传热系数仅约为1.2~1.6W/(m^2·℃),比保温较好的普通温室还要低50%左右。

此外,日光温室外保温覆盖材料的覆盖方式可以保证覆盖非常严密,夜间日光温室的室内、外空气交换量极低,可以有效地减少冷空气渗入产生的热量损失。

1.4 北墙体蓄热和加温的作用

日光温室厚重的后墙可在白昼有效地蓄积所吸收的太阳热能,夜间成为室内的一个加温热源,将所蓄积的太阳热能缓慢地释放回温室内。

在北京地区(北纬40°)1月份的晴天正午时刻,日光温室北墙内表面的太阳辐射照度可以达到300W/m^2以上,其中1/2~2/3的热量传入墙体内部。夜间室内气温降低后,墙体内白昼蓄积的热量将逐渐释放回温室内,分析结果表明[10,11],在不同情况下,墙面放热的强度在8~60W/m^2之间,在一夜之内墙面释放的热量累计可以达到0.35~2.5MJ/m^2。仅该项热量就可以使室内气温提高4~8℃。

日光温室室内地面具有与北墙相同的作用,白昼吸收太阳热能,夜间释放热量到温室内,是日光温室夜间又一热量来源。两者相比较,墙面在夜间的单位面积放热强度高于地面,但由于墙面面积低于地面面积(一般墙面面积:地面面积=1/4~1/3),根据分析表明,日光温室冬季夜间室内获得的热量中,墙面放热量约占1/3,地面放热量约占2/3。

因此,在冬季夜间,日光温室北墙与地面成为日光温室室内的加温热源,是日光温室夜间在不用设备加温情况下,室内气温能够维持高于室外气温的重要原因。

2 日光温室设计中几个问题的分析

根据上述日光温室建筑的技术特点与室内热环境的形成机理,笔者认为,日光温室设计的原则,应是从温室的建筑方案、材料、构造和尺寸等方面,有目的地有效保证和强化上述四个方面的作用。即:(1)合理设计屋面,确保屋面具有较高透光率;(2)合理确定建筑的形体尺寸,提高北墙面在白昼接受的太阳辐射量;(3)尽量提高屋面和墙体的保温性;

（4）强化北墙和地面的蓄热作用。

以下仅就日光温室设计中，目前较为关注的两方面问题，进行分析和提出看法。

2.1 北墙体的合理设计

如前所述，日光温室北墙体具有保温和蓄热两方面的作用，其设计应强化保温和蓄热两方面的作用，涉及材料选择和合理厚度、构造等方面问题。目前普遍对北墙体蓄热方面的作用重视不够。

2.1.1 关于墙体材料

由于墙体两侧面分别面对温室内部和外部环境，其内、外不同部位材料发挥的作用有所不同。因此，普遍认识到，日光温室墙体应采用在不同部位使用不同建筑材料的复合构造墙体[12~16]。但对于墙体材料的选用和构造层次的划分，仍普遍地存在不同的看法。在工程实践中，通常较为注重墙体材料的保温性，而对其蓄热性不够重视，对不同部位墙体材料的选择有认识上的错误。

笔者认为，墙体从厚度方向可以大致分为内、外（图1中A与B）两个部分。外侧B部分的材料，其重点在保温，应该选用保温性良好的材料。内侧A部分的材料，其作用重点是蓄热，应该选用蓄热性能良好的材料，即比热容和密度较高的材料。同时，为了使墙面吸收的太阳热量容易向材料深层传递，在夜间深层材料蓄积的热量容易释放出来，所以，内侧材料应是导热良好，即保温性差的材料！在内侧采用保温性好、而蓄热性差的材料，反而是错误的做法。

2.1.2 关于墙体厚度

关于墙体的合理厚度，已有较多的研究，但目前仍存在不同的看法。生产中使用的日光温室墙体平均厚度从0.5~5m，差异很大。一般砖墙温室墙体厚度为0.5~0.8m。土墙日光温室的墙体则厚一些，通常在1m以上。尤其是，目前出现较多墙体截面呈梯形的日光温室（图2），其墙体很厚，底部厚度达到5~8m，由于占地较多，是否合理的问题，受到较多的关注和讨论。

从温室内环境的要求，日光温室墙体的厚度，应该能够满足保温和蓄热两方面的要求，这与所采用的材料有关。作者采用模拟的方法[17]，对不同厚度土墙的温室进行了模拟分析。模拟采用的温室建筑参数为：跨度10.0m，脊高4.0m，墙高3.0m，北屋面水平投影宽度1.4m，北屋面内表面倾角35.5°，温室覆盖PE薄膜，夜间覆盖草帘保温。其他条件按一般常见情况确定。

模拟根据北京地区（北纬40°）1月中旬的气象情况，按3个连续晴天（室外气温为-5~5℃）和3个连续阴天（室外气温为-10~-2℃）的天气变化周期考虑。墙体断面为矩形截面，厚度在0.8~6.0m的范围内。墙体材料（土壤）的密度1 350kg/m³，导热系数1.0W/(m·℃)，比热容1 500J/(kg·℃)。

模拟结果如图3与图4所示。由图可见，墙体厚度开始增加时，室内最低气温与夜间墙面热流量均很快增大，但墙体厚度增大到一定程度时，增加速率减缓，曲线较为明显的减缓约产生在2m厚度时。因此，可以认为2m左右为较为合理的厚度。

对于一般的砖墙温室，笔者按上述相同条件进行了模拟，墙厚为0.6m（250mm（砖）+100mm（聚苯乙烯泡沫塑料板）+250mm（砖））。模拟结果，其室内最低气温与夜

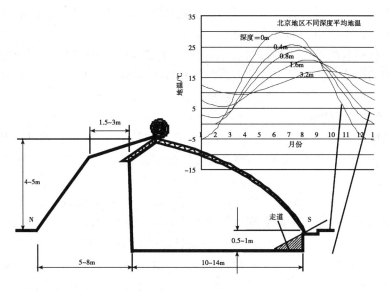

图 2 一种典型的土墙日光温室

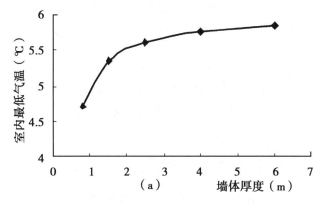

图 3 不同厚度墙体（土墙）与室内夜间最低气温

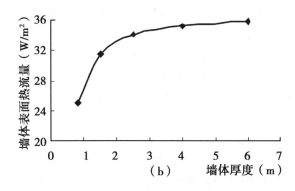

图 4 不同厚度墙体（土墙）与夜间墙面平均热流量

间墙面热流量均接近厚度为 1.5m 的土墙温室。说明材料的不同，尤其是其中加入了保温性良好的泡沫塑料，厚度可大大缩小。但尽管如此，该 0.6m 厚砖墙性能仍显得不够理想。在可能的情况下，仍以适当增加厚度为宜。

以上为一些初步的结果，是针对北京地区（北纬 40°）得出的。其他地区和气象条件下，以及不同的墙体材料、温室其他方面的不同具体条件，经济性和建筑技术等多方面不同情况下的合理墙体厚度问题，还有待开展更为深入和详细的研究。

2.2 温室内地面下沉的分析

在山东地区，日光温室普遍采用室内地面下沉的做法（图2），并已向其他地区推广。地面下沉的深度一般为 0.5～1.0m，但有逐渐向更深发展的趋势，目前有些温室地面下沉深度已超过 1.5m。

温室室内地面下沉，有利于室内地面的保温和地下土壤热量的蓄积。如图 2 所示，在冬季，自然地面以下的土壤温度随着深度的增加而增加。例如北京地区，1 月份地表（深度为 0m）的平均温度为 −5℃，而地下 0.8m 深度处土壤平均温度为 3℃，高于地面 8℃。温室内地面的下沉，使室内周边土壤与室外较高温度的土壤相邻接，可以有效减少周边向室外土壤的传热，从而可更好保持地下土壤蓄积的热量，提高地温和夜间室内地面的放热量，因此也有利于提高室内气温。

实验测试以及模拟分析的结果表明[18]，室内地面下沉对于提高日光温室内夜间最低气温有显著的效果，下沉深度为 0.5m 时提高 0.7℃，下沉 1.0m 时提高 1.1～1.3℃，1.5m 时提高 1.8～1.9℃。

但是，室内地面下沉的做法，将在温室南侧产生一阴影带（图2），目前山东地区的做法是将室内走道布置在南侧阴影带内，以避免影响植物的光照。但如下沉过多，阴影带将很宽，对室内光照的分布产生很大影响。此外，地面下沉还有需要防止室外雨水进入等问题。因此，室内地面下沉深度应有一定的限度，一般最深在 1m 左右为宜。

3 日光温室热环境模拟预测和优化设计的方法

目前日光温室的设计和建造还主要是依据有限的经验，为了提高日光温室设计建造水平，必须发展科学理论指导下的，依靠准确定量的计算进行优化设计的方法。

但日光温室作为一种特殊的建筑，其室内环境取决于复杂多变的外界条件、建筑体型、尺寸、墙体和屋面的材料与形式、地面以及栽培的植物、生产中的管理方式等多方面因素，如何优化设计，获得优良的性能，是涉及材料、建筑、气象、热工以及园艺等多专业领域的复杂问题，不能用简单的计算和分析准确地加以解决。

因此，通过科学严密的理论，采用准确的物理和数学模型，在模拟的基础上，结合一定的经验和分析计算的方法，掌握在各种条件下，不同日光温室设计方案的保温节能性能，以指导日光温室的设计，优化其建设方案，是一个有效的途径。

为此，笔者研究开发了日光温室热环境模拟预测软件（RGWSRHJ V1.1）[17]。该软件是根据工程热物理和温室环境调控等相关理论，构建日光温室内热环境的动态模型，并采用适当的数值算法，用 VC++ 计算机程序语言开发完成的工具软件。其目的是用于日光温室的研究和优化设计（图5）。

该软件依据的日光温室热环境动态模型在总体上采用了关系明确、便于升级和维护的模

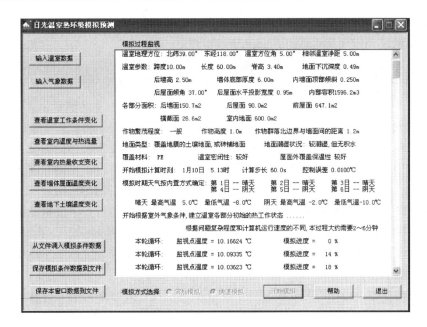

图5 日光温室热环境模拟预测软件

块化方案,由决定室内热环境的墙体、地下、覆盖层、通风和蒸发蒸腾5大模块,以及太阳辐射和室外气象条件等模块组成。通过分析各模块间的能量传递和平衡关系,预测日光温室室内热环境状态。

墙体与地下土壤的传热是日光温室热环境模型中较为核心的部分,属于非稳态传热的问题。采用了在日光温室横断面内的二维传热模型进行描述,模型求解的算法采用有限差分法。通过研究,在模型的边界条件、初始条件确定和方程组求解算法等方面采用了创新的方法,在保证模拟准确的同时,提高了运算的效率。

该软件具有较高的通用性,可以适用于不同地理位置和气象条件的各个地区,在日光温室类型方面,适用于常见的各种类型,包括不同截面形式、材料和尺寸墙体,不同地面下沉等各种情况。软件采用了方便用户使用的界面,使用者可以方便地通过对话框,输入气候条件、温室构造尺寸、材料参数等数据,并可方便地查看模拟运算的结果。输入的条件数据和模拟运行结果数据可以进行保存,以后可以方便地再次调用或查看。

软件模拟结果输出信息丰富。可以获得墙体、地下等部分任意点以及室内空气任意时刻的温度,以及各部位热流量等多种数据信息,便于对温室的热环境性能进行多方面的、综合性的分析。

运用该软件,可以根据给定的日光温室建设地点的地理位置、室外气象条件,采用的建筑材料和建筑、构造方案,以及不同种植条件和管理方式等情况下,较为准确地预测其室内热、湿环境全天随时间的变化情况,可对不同设计方案的日光温室性能进行预测评价、比较和优选。

4 结论与建议

本文对日光温室室内环境的形成机理进行了传热和热力学的分析,提出了日光温室具有比普通温室优良节能性能，主要归因于以下 4 个方面的原因。

（1）南向的采光屋面具有较好的日光透过特性。

（2）北墙面可以接受和吸收投射到室内地面以外的太阳辐射，使室内获得的太阳热能比普通温室大大增加。

（3）墙体和屋面具有优良的保温性能，有效地减少了夜间温室的热量损失。

（4）采用蓄热性能优良的材料建造的厚重的北墙，以及室内地下土壤具有蓄积太阳热能的作用，可在白昼有效地蓄积墙面和地面吸收的太阳热能，夜间释放回温室内。

基于上述日光温室室内环境的形成机理，提出了日光温室的设计原则，在于从温室的建筑方案、材料、构造和尺寸等方面，有效保证和强化上述四个方面的作用。

针对日光温室优化设计中的墙体设计与室内地面下沉的问题进行了分析。提出了墙体外侧应采用保温性良好的材料，内侧应采用导热和蓄热性良好材料的观点。采用模拟的方法，分析了墙体厚度对墙体热工性能的影响。根据实测与模拟的结果，分析了室内地面下沉对温室内环境的影响。

论文提出根据理论模型、采用模拟的方法预测温室设计方案热环境方面性能，是日光温室优化设计的有效方法。笔者介绍了开发的日光温室热环境模拟预测软件。由于日光温室中的热物理过程非常复杂，影响因素众多，又难于精确描述，因此其环境的准确模拟具有相当的难度。采用理论方法解决日光温室设计问题，该软件还是初步的尝试。其进一步的完善提高，还有待于对日光温室的环境形成机理、各种因素对室内环境的影响，运用工程科学的理论进行更加深入的研究。

参考文献

[1] 陈端生. 中国节能型日光温室建筑与环境研究进展. 农业工程学报，1994，10（1）：123~129
  Chen Duansheng. Advance of the Research on the Architecture and Environment of the Chinese Energy-Saving Sunlight Greenhouse. Transactions of the CSAE, 1994, 10 (1): 123~129. (in Chinese with English abstract)

[2] 刘乃玉，王春娜，王绍辉等. 日光温室环境参数的测定及分析. 北京农学院学报，2001，16（1）：74~79
  Liu Naiyu, Wangchunna, Wang Shaohui, et al. The Test and Study on Environment Parameter of Sunlight Greenhouse. Journal of Beijing Agricultural College, 2001, 16 (1): 74~79. (in Chinese with English abstract)

[3] 孟力力，杨其长，宋明军. 北京地区日光温室温光及蓄热性能的试验研究. 陕西农业科学，2008（4）：61~64

[4] 佟国红，李天来，王铁良等. 大跨度日光温室室内微气候环境测试分析. 华中农业大学学报，2004（增刊总第 35 期）：67~73
  Tong Guohong, Li Tianlai, Wang Tieliang, et al. Experiment Research on Microclimate Environment in a Large-scale Sunlight Greenhouse. Journal of Huazhong Agricultural University, 2004 (Sup. Sum. 35): 67~73. (in Chinese with English abstract)

[5] 徐凡，刘洋，马承伟. 天津地区典型日光温室使用现状调查. 北方园艺，2010（15）：19~24
  Xu Fan, Liu Yang, Ma Chengwei. Investigation of the Using Status of Typical Greenhouse in Tianjin. Northern Horticulture, 2010 (15): 19~24. (in Chinese with English abstract)

[6] 马承伟. 农业生物环境工程. 北京：中国农业出版社，2005

[7] 周长吉. 温室工程设计手册. 北京：中国农业出版社，2007

[8] 日本施設園芸協会. 施設園芸ハンドブック（第五版）. 東京：園芸情報センター－，2003

[9] ANSI/ASAE EP406.4 JAN03 Heating, Ventilating and Cooling Greenhouses, ASAE Engineering Practice, 2003

[10] 马承伟，卜云龙，籍秀红等. 日光温室墙体夜间放热量计算与保温蓄热性评价方法的研究. 上海交通大学学报，

2008, 26 (5): 411~415

Ma Chengwei, Bu Yunlong, Ji Xiuhong, et al. Method for Calculation of Heat Release at Night and Evaluation for Performance of Heat Preservation of Wall in Solar Greenhouse. Journal of Shanghai Jiaotong University, 2008, 26 (5): 411~415. (in Chinese with English abstract)

[11] 马承伟, 陆海, 李睿等. 日光温室墙体传热的一维差分模型与数值模拟. 农业工程学报, 2010, 26 (6): 231~237

Ma Chengwei, Lu Hai, Li Rui, et al. One-dimensional finite difference model and numerical simulation for heat transfer of wall in Chinese solar greenhouse. Transactions of the CSAE, 2010, 26 (6): 231~237. (in Chinese with English abstract)

[12] 陈端生, 郑海山, 刘步洲. 日光温室气象环境综合研究—I. 墙体、覆盖物热效应研究初报. 农业工程学报, 1990, 6 (2): 77~81

Chen Duansheng, Zheng Haishan, Liu Buzhou. Comprehensive Study on the Meteorological Environment of the Sunlight Greenhouse. 1. Preliminary study on the thermal effect of the wall body and covering materials. Transactions of the CSAE, 1990, 6 (2): 77~81. (in Chinese with English abstract)

[13] 佟国红, 王铁良, 白义奎等. 日光温室节能墙体的选择. 可再生能源, 1990 (4) (总第110期): 14~16

Tong Guohong, Wang Tieliang, Bai Yikui, et al. Selection of energy efficiency wall in solar greenhouse. Renewable Energy, 2003 (4) (110 Issue in All): 14~16. (in Chinese with English abstract)

[14] 郭慧卿, 李振海, 张振武等. 日光温室北墙构造与室内温度环境的关系. 沈阳农业大学学报, 1995, 26 (2): 193~199

Guo Huiqing, Li Zhenhai, Zhang Zhenwu, et al. The Relationship Between the North Wall Construction and Interior Temperature Environment in Solar Greenhouse. Journal of Shenyang Agricultural University, 1995, 26 (2): 193~199. (in Chinese with English abstract)

[15] 李小芳, 陈青云. 墙体材料及其组合对日光温室墙体保温性能的影响. 中国生态农业学报, 2006, 14 (4): 185~189

Li Xiaofang, Chen Qingyun. Effects of different wall materials on the performance of heat preservation of wall of sunlight greenhouse. Chinese Journal of Eco-Agriculture, 2006, 14 (4): 185~189. (in Chinese with English abstract)

[16] 佟国红, David M. Christopher. 墙体材料对日光温室温度环境影响的CFD模拟. 农业工程学报, 2009, 25 (3): 153~157

Tong Guohong, David M Christopher. Simulation of temperature variations for various wall materials in Chinese solar greenhouses using computational fluid dynamics. Transactions of the CSAE, 2009, 25 (3): 153–157. (in Chinese with English abstract)

[17] 马承伟, 韩静静, 李睿. 日光温室热环境模拟预测软件研究开发. 北方园艺, 2010 (15): 69~75

Ma Chengwei, Han Jingjing, Li Rui. Research and Development of Software for Thermal Environmental Simulation and Prediction in Solar Greenhouse. Northern Horticulture, 2010 (15): 69~75. (in Chinese with English abstract)

[18] 李睿. 下沉型日光温室保温蓄热性能的测试与研究. 北京: 中国农业大学, 2010

Li Rui. Research and Test on Heat Preservation and Storage Capability of Sunken Solar Greenhouse. Beijing: China Agricultural University, 2010

空气—空气热泵技术在温室环境综合控制中的应用

仝宇欣[*]

(千叶大学环境健康科学研究中心,日本千叶县柏市柏叶 6-2-1,277-0882)

摘要：近年来,随着市场对设施农业产品需求的不断增加与温室技术的不断进步,设施农业发展迅速。目前,如何开发一种高效可持续发展的生产系统,在提高产量与质量的同时提高经济效益是设施农业发展面临的重大课题之一。利用多功能空气-空气热泵技术对温室内环境进行综合控制被认为是解决上述问题的一种有效措施。这种热泵可用于温室内冬季加温、夏季夜晚降温、夏季白天降温并进行 CO_2 施肥、除湿与增湿以调控温室内水气压亏缺、同时可增加温室内空气循环。文中详细阐述了运用这种热泵技术对温室内环境进行调控时出现的一系列问题及其相应的解决措施。在东京附近进行的实验结果表明,利用这种热泵进行温室加温与降温时的性能系数平均值分别可达到 4 和 10。文中也对这种热泵技术在中国温室内应用的可行性做了分析。结果表明,在室外最低温度高于 -5℃ 的地方应用这种技术可能会收到较好的效果。综上所述,空气—空气热泵技术可被认为是一种节能、减排和经济高效的温室环境控制技术。并且随着热泵技术的不断进步,这项技术将会在温室环境控制中得到更广泛的应用。

关键词：空气—空气热泵；性能系数；节能

Application of the Air-air Household Heat Pump Technologies for Integrated Greenhouse Environment Control

Tong Yuxin

(*Center for Environment, Health and Field Sciences, Chiba University, Kashiwa-no-ha* 6-2-1, *Kashiwa, Chiba* 277-0882, *Japan*)

Abstract: In recent years, horticultural production using greenhouses has increased greatly in many countries due to the increase in demands of horticultural products all year around and the advancing greenhouse technologies. Nowadays, the challenges for greenhouse industry are how to develop sustainable production systems with high yield and quality and to make the investments economically feasible. An integrated greenhouse environment control system employing the air-air household heat pumps (HHPs) with multi-functions can help address these challenges. The HHPs can be used for heating in winter, nighttime cooling in summer, daytime cooling together with CO_2 enrichment, dehumidification and humidification to control water vapor deficit, as well as increase in air circulation. The problems associated with the integrated greenhouse environment control system using the HHPs and their possible solutions are discussed in this article. The experiments conducted near Tokyo using the HHPs show that the coefficient of performance (COP) is 4 for heating and 10 for cooling on average. The possibility of using the HHP technologies in greenhouses at different locations in China is also analyzed. Considerable potentials of using the HHPs are expected at locations where the minimum outside air temperature is higher than -5℃. The analysis indicates that

---

[*] 1 Corresponding author. Postdoctoral Researcher, Chiba University. Tel：81-080-3588-7163.
　　E-mail：yxtong 07@ hotmail. com.

the integrated greenhouse environment control system with the HHPs, especially with the improving HHPs technology, can be used widely as an energy-saving and carbon emission-reducing technology with high input resource utilization efficiencies, and offer economic benefits.

Key words: Air-air heat pump; Coefficient of performance; Energy saving

1 Integrated greenhouse environment control

During the last twenty years, impressive yield increase in greenhouse crop production has been achieved in the Netherland due to the progress in greenhouse engineering and improvement in crop cultivars. However, the actual productivities of the crops in many countries are still around 25% of the potential productivities due to diseases, pest insects and the unfavorable environment (Ikeda, 2010). The unfavorable environment has become the primary reason for low yield. Thus, recently, the challenges for greenhouse industry are to increase yield and improve quality by controlling the environmental factors at the optimum levels with minimum resources (such as energy, water etc.) and environmental pollution (such as CO_2 emission, etc.).

Development of an integrated greenhouse environment control system, which can create the most appropriate environment for maximization of plant growth with minimum input of energy, water and pesticide and atmospheric pollutants, can be of substantial importance. A stepwise increase in dry mass, leaf area index or net photosynthetic rate by integrated environment control has been schematically illustrated by Kozai, et al., (2009). The authors classified the environmental factors into five levels, depending on the factors that limit growth and production: Level 1: temperature; Level 2: level 1 + air circulation; Level 3: level 2 + CO_2 concentration; Level 4: level 3 + relative humidity; Level 5: level 4 + nutrition and water uptake.

The greenhouse environment control is a complicated procedure since many variables are inter-related. Many researchers have worked on greenhouse environment control using various control units and strategies. However, those are often difficult to implement in greenhouses because various control units are necessary resulting in high investment costs. As shown in Figure 1, heat pumps, especially air-air household heat pumps (called HHPs hereafter), which can be used for controlling various environmental factors, have been improved remarkably due to competition in the market and recognized as energy efficient equipments (Kozai, 2009). Thus, considerable potentials exist in developing an integrated greenhouse environment control system with heat pumps. The objectives of this paper are to discuss the advantages of application of the integrated greenhouse environment control system with the HHPs, and the problems associated with this system and their possible solutions.

2 Advantages of application of the HHPs for integrated greenhouse environment control

2.1 The HHPs used for greenhouse heating

Comparing with combustion-based heating systems (such as the oil heater), the advantages of using the electricity driven HHPs for greenhouse heating include: (1) Reductions in primary energy

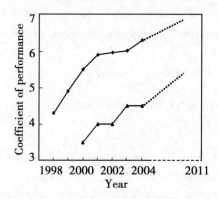

**Figure 1 Coefficient of performance of the heat pumps is being improved
(modified data from Kozai, 2009)**

consumption and costs, because the heat pump operates by transferring and upgrading heat. Tong *et al.* (2010) reported that the primary energy consumption was reduced by 25% ~60% by using the HHPs instead of oil heater; (2) Reductions in the emissions of CO_2 and atmospheric pollutants (NO_x; particulate matter less than 2.5 μm in diameter) at the global level (Russoa, *et al.*, 2009), because the electricity driven heat pump generates no pollutants at the local level. A reduction of about 55% ~70% in CO_2 emission by using the HHPs instead of oil heater has been reported by Tong, *et al.* (2010); (3) Achievement of spatially uniform distribution of the air temperature and water vapor pressure deficit (VPD) by using the fans installed in the heat pumps due to the improvement in air circulation; and 4) Enhancement of photosynthesis, transpiration, and uptake of water and nutrients by using the fans installed in the heat pumps due to the decrease in leaf boundary layer resistance (Kozai, *et al.*, 2009). In addition, the advantages of using the multiple HHPs (two or more HHPs are used in one greenhouse) are the the HHPs operate independently and defrost asynchronously, affected by the spatial air temperature around them. Thus, even if a single HHP breaks down, the others can mostly compensate the heat capacity and maintain the inside air temperature for plant growth, and even if one of the HHPs enters the defrost cycle, the others can continue working in heating cycle to maintain the system COP and inside air temperature (Tong, *et al.*, 2010).

2.2 The HHPs used for greenhouse cooling

It is essential to develop a suitable greenhouse cooling method with minimum energy consumption for crop production in summer in the most regions and/or all year round in tropical regions to meet the increasing demands for high value horticultural products. The advantages of using the electricity driven HHPs for greenhouse cooling can be summarized as: (1) Reduction in primary energy consumption due to high COP, especially when the air temperature outside the greenhouse is lower than that inside. A high COP of 8 ~9 has been reported by Ohyama, *et al.* (2002) when the HHPs were used for cooling a closed crop production system; (2) Decrease in leaf temperature; (3) The same as 3 & 4 in section 2.1; (4) Enhancement of the water utilization efficiency through recycling the condensed water which is collected by the inside heat exchanger of the heat pumps; (5) Minimum diseases spread, the pest insect can be well prevented in a closed greenhouse; and (6) Prolonged the greenhouse ventilators closed period, a closed greenhouse allows for better environment control, such as control of CO_2 concentration, VPD and so on, compared with a semi-closed greenhouse. Except the advantages mentioned above, the advantages of using the electricity driven HHPs for greenhouse cooling together with the CO_2 enrichment system during the daytime also include: (1) Extended the period for keeping the CO_2 concentration inside the greenhouse at a higher level than that outside; (2) Uniformly enriched the CO_2 to the greenhouse by using the fans of the heat pumps; and (3) Enhancement of the utilization efficiencies of both the enriched CO_2 and solar energy, especially when the high CO_2 concentration (such as 800 ~ 1 500ppm) can be achieved under relatively high solar radiation (such as 500 W · m^{-2} or higher).

2.3 The HHPs used for controlling the relative humidity/VPD inside greenhouses

The control of the relative humidity/VPD has often been ignored in greenhouses because it is difficult to conduct by using the conventional environment control systems. In the integrated greenhouse environment control system with the HHPs, it is available and necessary to control the relative humidity/VPD inside greenhouses at the optimum level for plant growth to improve the rates of photosynthesis and transpiration and limit the disease spread. The HHPs can be used for dehumidification by condensing the water vapor of the greenhouse air at the evaporators and humidification by supplying liquid water to the condensers to humidify the greenhouse air (Kozai, et al., 2009).

3 Problems associated with the HHPs for greenhouse environment control and the possible solutions

3.1 Decrease in the COP for heating of the air source heat pumps by the defrost cycle

One disadvantage of the air source heat pump while heating is its defrost cycle. If the outside air temperature falls near or below its freezing point, moisture in the air passing over the outside heat exchangers will frost on it. The frost accumulation decreases the efficiency of the heat exchangers by reducing its ability to absorb heat from the ambient air and transfer to the refrigerant. In another words, the COP decreases as the frost accumulates (Byun, et al., 2008). To minimize the decrease in the COP, the frost must be removed. This can be carried out in a number of ways, such as: (1) reverse-cycle defrosting; (2) electricity heat defrosting; (3) compressor shut down defrosting; and (4) hot gas defrosting (Hewitt & Huang, 2008). The defrost cycle should be long enough to melt the frost and short enough to be energy-efficient. Unnecessary defrost cycles reduce the seasonal performance of the HHP. Therefore, the demand-based defrosting method is generally efficient since the defrost cycle is started only when it is required. The defrost cycles can be prevented or decreased by the methods which can increase the surface air temperature of the outside heat exchangers and/or decrease the absolute humidity around the outside heat exchangers. One possible method to decrease the defrost cycles is to improve the design of the heat exchanger, such as increasing its surface area, increasing the wind speed passing it, doing its surface coating and so on, due to the possible decrease in the amount of condensate accumulation on it (Hayashi, 2009). In addition, the defrost cycles can be prevented if the heat source of the HHPs is changed from the ambient air to the geothermal or groundwater.

3.2 The COP of the HHPs affected by the number of installed and/or operating heat pumps

The heating and cooling capacity of the HHP is usually in the range of 2.2 ~ 6.3 kW (Kozai, et al., 2010). Thus, the multiple HHPs are usually needed to control a greenhouse environment successfully. For achieving the maximum COP, the number of installed and/or operating HHPs should be decided by analyzing the heating/cooling load of the greenhouses. The maximum COP can be obtained if the heating/cooling load of a greenhouse is in a range of 60% ~ 80% of the total heating/cooling capacity of the operating HHPs (Kozai, et al., 2009). If the heating/cooling load of the greenhouse is lower than 60% of the total heating/cooling capacity of the operating HHPs, the COP will decrease with decreasing heating/cooling load of the greenhouse. Software should be designed / programed to control the optimum number of the operating HHPs with changing heating/cooling load of the greenhouse.

3.3 Optimum CO_2 concentration control in greenhouses with the HHPs during the daytime

It is essential to keep the CO_2 concentration at the optimum level for photosynthetic growth of plants in greenhouses, since the low CO_2 concentration can significantly reduce the photosynthesis of plants. To make the investments economically feasible, CO_2 enrichment for keeping its concentration higher than that outside should be conducted only when the ventilators are closed to minimize the amount of CO_2 leak out. However, when the HHPs are used for greenhouses cooling at daytime on a hot day, the air temperature inside the greenhouses is difficult to keep at the set-point under high solar radiation and/or high outside air temperature. Thus, the ventilators of the greenhouses should be opened. During the ventilators open period, a hybrid cooling system with the HHPs for air circulation and an evaporative cooling system (such as fogging) can be employed in the greenhouses, the CO_2 enrichment can be done by keeping its concentration at the same level as that outside.

4 The possibility of using the HHPs for greenhouse environment control in China

The area of the greenhouse in China has been significantly increased in the recent years. However, the yield and the quality of the greenhouse produces are still low, and the yield in China is about one-sixth of that in the Netherland, mainly due to less control of greenhouse environment (Li, 2009). Thus, it is imperative to control the greenhouse environment at the optimum level by using the recently improved technology. The COPs of 4 for heating and 10 for cooling on average have been achieved by employing the HHPs (Tong, 2010). Thus, Considerable potentials are expected in employing the HHPs for greenhouse environment control in greenhouses at the locations in China where the minimum outside air temperature is higher than −5℃, especially in the Chinese solar greenhouses in which less number of the HHPs is needed due to the relatively low heating load (Figure 2), although the further research is needed.

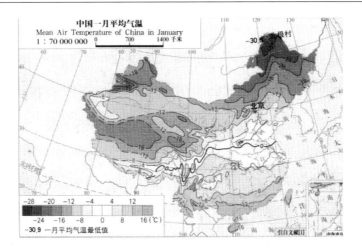

Figure 2 Considerable potentials are expected in employing the household heat pumps for greenhouse environment control in China where the minimum outside air temperature is higher than −5℃

References

[1] Ikeda, T. The optimal environment for agriculture production. Japan-China International Workshop of Horticulture 2010. Japan, Tokyo

[2] Kozai, T., Ohyama, K., Tong, Y., Tongbai, P., Nishioka, N. Integrative environmental control using heat pumps for reductions in energy consumption and CO_2 gas emission, humidity control and air circulation. Acta Hort. In press

[3] Kozai, T. Solar assisted plant factory. Tokyo, Japan: Ohmsha, 2009. 58 (In Japanese)

[4] Tong, Y., Kozai, T., Nishioka, N., Ohyama, K. Greenhouse Heating Using Heat Pumps with a High Coefficient of Performance (COP). Biosystems Engineering. 2010, 106 (4): 405~411

[5] Russoa, S. L., Cesare, B., Massimo, V. C. Low - enthalpy geothermal energy: An opportunity to meet increasing energy needs and reduce CO_2 and atmospheric pollutant emissions in Piemonte. Geothermics. 2009, 38 (2): 254~262

[6] Tong, Y., Kozai, T., Nishioka, N., Ohyama, K. Greenhouse Heating Using Heat Pumps for Reductions in Primary-energy Consumption and CO_2 Emission. Annual Conference of Environment and Biosystems Engineering in Japan. Kyoto, September 8~10, 2010

[7] Ohyama, K., Kozai, T., Kubota, C., Chun, C., Hasegawa, T., Yokoi, S., Nishimura, M. Coefficient of performance for cooling of a home-use air conditioner installed in a closed-type transplant production system. Journal of Society of High Technology in Agriculture. 2002, 14 (3): 141~146 (In Japanese)

[8] Byun, J., Lee, J., Jeon, C. Frost retardation of an air - source heat pump by the hot gas bypass method. International journal of refrigeration. 2008, 31: 328~334

[9] Hewitt, N., Huang, M. Defrost cycle performance for a circular shape evaporator air source heat pump. International journal of refrigeration. 2008, 31: 444~452

[10] Hayashi, M. Effective Use of a Heat Pump in a Greenhouse. Tokyo, Japan: Nougyou Denka Kyoukai, 2009. 35 (In Japanese)

[11] Kozai, T., Tong Y., Nishioka, N., Ohyama, K. Greenhouse environment control with home-use air conditioner. Agricultural Electrification. 2010, 63 (6): 2~8

[12] Li T. The problems in Chinese solar greenhouse development and the solutions. The1st High-level International Forum on Protected Horticulture. Shouguang, China. 2009 (In Chinese)

[13] Tong, Y. Integrated greenhouse environment control using heat pumps with high coefficient of performance. Library of Chiba University: Chiba University, 2011

山地日光温室性能分析*

邹志荣**，张 勇***

（西北农林科技大学，陕西 杨凌 712100）

摘要：西北地区山地日光温室建设面积大，种类繁多，但是相对缺乏系统的理论研究，因此在实践中存在建设混乱的问题。本文针对西北地区山地有代表和推广几种类型，通过对其性能的研究，结果表明，五型温室性能较好，山地温室其次，甘泉温室和普通温室较差。

关键词：温室结构；太阳能利用；温室性能

Performance Analysis of Hillside Solar Greenhouse

Zou Zhirong, Zhang Yong

(*Northwest University of Farming and Forestry Technology*, Yangling, 712100, *P. R. China*)

Abstract: Large areas of Hillside Solar Greenhouse was built in Northwest China, several types of which was developed. However, The absence of theoretical studies make it chaotic in Hillside Solar Greenhouse construction. In this paper, several types of greenhouse in mountainous region of Northwest China was introduced, Performance of the greenhouses were analysed. The result shows that Temperature and Light Capability of Wuxing type greenhouse is better than other greenhouse types.

Key words: Greenhouse structure, Solar energy utilization, Temperature and Light Capability of greenhouse

山地温室是近几年中国山区设施农业发展的新类型，其性能与生产效果明显。但温室性能仍然不清楚。所以，研究山地温室性能与效果成为研究的热点。

1 材料与方法

1.1 试剂及仪器

PDR 数据记录仪。

IN-TMP（内置温度传感器）

测量范围：$-30 \sim 60$℃

分辨力：0.1℃

准确度：±0.5℃（$-10 \sim 60$℃）；±1℃（其他）

---

* 基金项目："十一五"国家科技支撑计划"西北旱作农业新农村建设关键技术集成与示范"(2008BAD96B08 - 3)。

** 作者：邹志荣（1956— ），男，陕西延安人，教授，博士，博士生导师，主要从事设施农业研究。

*** 作者：张勇（1977— ），男，陕西榆林人，讲师，博士，主要从事现代农业园区规划设计和温室建筑结构研究。

IN-TMP（外置温度传感器）

测量范围：-50~120℃

分辨力：0.1℃

准确度：±0.5℃（-10~70℃）；±1℃（其他）

IN-RH（相对湿度）

测量范围：0~99%RH

分辨力：1%RH

准确度：±3%（20~70℃）±5%（其他）

IN-PR（内置大气压力）

测量范围：10~1 100hPa

分辨力：0.1 hPa

准确度：±1.5 hPa

lx（光照度）

测量范围：0~200 000lx（可定制）

分辨力：1lx

准确度：±7%

重复测试：±7%

温度特性：±0.5%/℃

1.2 试验温室

测试温室坐北朝南，分别位于甘泉现代农业示范园、安塞现代农业示范园、安塞井居山地温室基地内。以普通温室为对照，如表1所示。

表1 温室结构标准（m）

类型	跨度	高度	长度	墙体厚度
五型棚	11.5	5.5	100	3.25
甘泉棚	9.0	3.9	60	1.1
山地棚	11.5	5.5	80	无限
普通棚	8.0	3.0	60	0.5

1.3 测试时间

2011年1月24日至2月19日。

2 结果与分析

2.1 温室室内光照分析

通过光照分析可以得出，2011.1~2月示范园五型温室45 382lx、甘泉温室23 880.8lx、山地温室的采光性41 673lx，普通类型的温室性能36 161.4lx。同一时间下室外的光照强度为74 702lx。（图1）

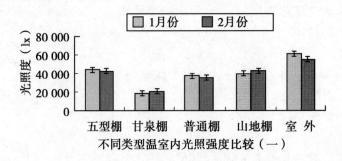

不同类型温室内光照强度比较（一）

图1 普通温室室内光照强度图—全月对比

2.1.1 温室全月光照分析
2.1.2 试验温室单日光照分析

在不同的管理条件下，同一天里（1月30日），五型温室光照强度是64 456.7lx、山地温室光照强度是5 929 804lx；普通温室是55 021.7lx、甘泉温室是23 079.2lx。

2.1.3 典型天气条件下不同类型温室光照分析

表1 典型天气条件下不同类型温室光照情况（lx）

	雪天	阴天	晴天
五型棚	7 682.5	37 515.8	66 077.6
甘泉棚	2 015.3	18 927.9	32 845.8
普通棚	10 491.0	32 004.9	52 076.4
山地棚	13 778.4	46 548.4	56 646.3
室 外	21 247.9	57 160.5	68 461.6

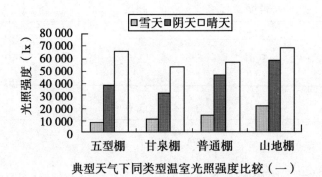

典型天气下同类型温室光照强度比较（一）

图2 三种温室光照强度对比

通过光照分析可以得出，阴雪天山地温室的采光性最好。示范园五型温室其次，甘泉温室最差。在不同的管理条件下，同一天里，五型温室和山地温室的采光性能最高、普通类型的温室性能最差（表1和图2-图3）。

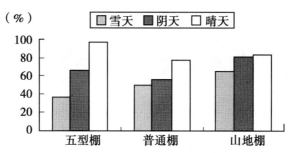

典型天气下不同类型温室光照强度比较（二）

注：假定当天室外光照强度为100%

图3 三种温室光照强度对比

2.2 试验温室温度分析

2.2.1 温室月平均温度分析

表2 温室月平均温度（℃）

平均温度	五型棚	甘泉棚	普通棚	山地棚	室外
1月	15.5	11.5	11.5	12.9	-9.5
2月	17.9	14.5	13.9	16.4	-1.9

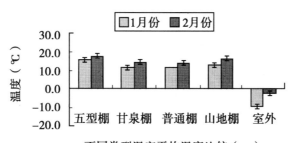

不同类型温室平均温度比较（一）

图4 三种温室室内温度对比

通过温度分析可以得出，五型温室热性能最好，山地温室其次，甘泉温室和普通温室热性能接近（表2和图4－图6）。

2.2.2 温室单日温度分析

通过温度分析（图7）可以得出，山地温室中午能达到的温度最高，但是有过热的缺陷，温度超过了42℃。早晨的温度示范园五型温室较高，能达到12~15℃。同一时间，山地温室早晨的温度能达到10℃。示范园普通温室性能最差，只能达到7℃。

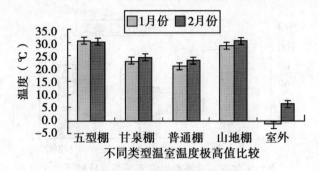

图5　三种温室室内温度极高值对比

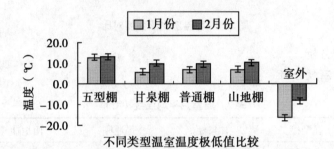

图6　三种温室室内温度极低值对比

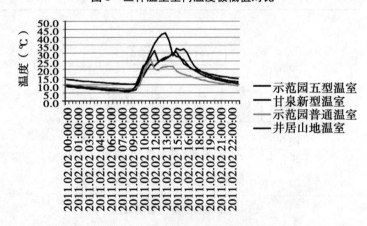

图7　三种温室室内温度对比

2.2.3　温室典型天气温度分析

表3　典型天气平均温度（℃）

	雪天	阴天	晴天
五型棚	11.9	17.4	19.4
甘泉棚	7.7	14.2	16.4
普通棚	9.1	14.6	14.7
山地棚	13.8	12.6	17.9
室　外	-8.6	-3.9	0.6

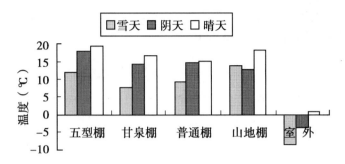

典型天气下不同类型温室平均室温比较

图8 典型天气下三种温室室内温度对比

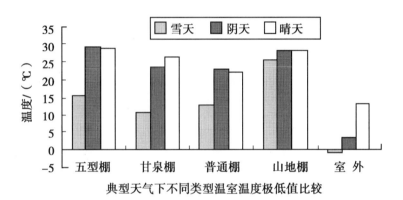

典型天气下不同类型温室温度极低值比较

图9 典型天气条件下三种温室室内温度极值对比

表4 典型天气最低温度（℃）

	雪天	阴天	晴天
五型棚	11	12.9	12.8
甘泉棚	6.1	9.2	8.9
普通棚	7.9	9.2	9
山地棚	8.6	9.8	10.1
室　外	−13.2	−11.1	−6.8

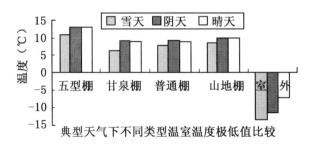

典型天气下不同类型温室温度极低值比较

图10 典型天气条件下三种温室室内温度极值对比

通过温度分析可以得出，典型天气五型温室温度较高，山地温室温度其次，甘泉和普通温室温度接近（图8至图10和表3至表4）。

3　结论

综合分析，山区示范园五型温室温光性能最优，山地温室次之，甘泉示范园温室和普通温室性能较差。

Designing a Greenhouse Plant: Novel Approaches to Improve Resource Use Efficiency in Controlled Environments

A. Maggio, S. De Pascale and G. Barbieri

(Department of Agricultural Engineering and Agronomy University of Naples Federico II-Via Università, 10080055 Portici - Naples Italy)

Abstract: Greenhouse cultivation is among the most advanced technological systems in agricultural productions since it allows to optimally adjust the Environment to the actual plant needs and vice versa. Much progress in greenhouse technology has been achieved by developing high-tech covering materials with an improved light transmittance, designing efficient cultivation units and water/nutrients delivery systems (hydroponics), improving the control of environmental parameters through sophisticated software. In contrast, the development of a *greenhouse plant* with characteristics that have been properly tailored for this environment has rarely been the focus of a specific research. Such specificity is required by the peculiar modifications, both metabolic and morphological, that plants undergo in the greenhouse environment. Although greenhouse production is generally associated to high technological-input systems, advances in plant biotechnology have mainly been addressed to generate field-rather than greenhouse-high-tech crops. In addition, the definition of biophysical models for predicting resource fluxes in a confined/controlled environment could greatly benefit of the possibility of changing specific plant traits to test and validate such models. This approach could generate an unprecedented *feed-back/feed-forward* research system able to identify optimal plant/environment interactions for an efficient resource use. Here we provide a few examples on how the information generated via model analysis can be tested and, more importantly, improved by implementing biotechnological tools.

Key words: Biophysical models; Biotechnology; Stress tolerance

Introduction

Successful agricultural productions rely on optimal management of both technical inputs and environmental variables. In open field, the Environment is a critical yield and quality determinant since it affects the economic sustainability and geographical distribution of most crop species. In contrast, environmental control in protected agriculture allows to improve yield by adjusting critical growth parameters to the actual plant needs. This fundamental difference in terms of plant-environment interactions has never been specifically addressed to define whether and how a greenhouse plant should differ from a field plant. Conceivably, plants that would efficiently use the available resources may have different morphological, physiological and metabolic traits that would make it specific to a given cultivation system. Nevertheless, the development of a technologically improved *greenhouse plant* that actually responds to such requirements lags behind the development of most advanced greenhouse technologies. Consequently, while innovative and efficient water and nutrient delivery systems are available, we do not actually have a plant that is specifically designed for an efficient use of these resources. Reasonably, a greenhouse plant should likely have specific leaf reflection properties, shoot vs. root developmental patterns, water and nutrient uptake systems as well as other specific features that have remained so far unexplored. Clearly, a better understanding of the complex plant/greenhouse environment interactions would be helpful to define a greenhouse plant able to improve the resource use efficiency and to reduce the environmental impact of the production process. This could be critical to fully exploit the potential of the entire greenhouse system. Based on our experimental results and published literature, in the following sections, we will present a few examples and research approaches that may provide useful insights for *designing* a greenhouse plant.

Materials and methods

All the experiments were carried out at the University of Naples Federico II experimental field located in Portici, Italy (14° 20' E; 40° 49' N; 20 m a. s. l.) in a greenhouse covered with a 60% cut-off screen (Experiment 1 and 2) or without screen (Experiment 3).

Experiment 1

To assess plant response to increasing salinity in terms of biomass distribution and morphological modifications, seeds of cherry tomato (Licata F1-COIS 94) were germinated in Styrofoam flats containing a mixture of sand and peat moss (1 : 1) and subsequently transferred, at the stage of 2 fully expanded leaves, in 15L buckets filled with aerated Hoagland solution (EC = 2.5dS/m; pH = 6.0), which was replaced weekly. Salinization was accomplished two weeks after transplanting by adding equal increments of NaCl:CaCl (2 : 1 molar basis) to reach, over a 6-day period, eight different EC levels: 2.5 (non-salinized control); 4.2; 6.0; 7.8; 9.6; 11.4; 13.2; 15.0dS/m. The EC of the nonsalinized control treatment (2.5dS/m) was equal to the EC of the standard nutrient solution (Maggio *et al.*, 2007). Plants were arranged in a randomised block design with three replicates. Each salinity treatment consisted of 45 single plant buckets (15 buckets per replicate). At harvest, fresh weight and dry weight were measured separately on leaves, stems fruits and roots, after drying them at 60℃. Data were analyzed by ANOVA and means were compared by the least significant difference (LSD) test.

Experiment 2

The objective of this experiment was to understand whether we could modify plant morphological traits through a functional control of nutritional requirements. Seeds of *Microtom* tomato were germinated in plastic flats containing a mixture of sand and peat moss (1 : 1) and subsequently transferred, after one week, into 10 plastic containers (60cm length, 40cm width, 15cm height) that were randomized on the greenhouse bench. Five of these containers were filled with aerated Hoagland solution and 5 with aerated Hoagland solution minus phosphorus. Each container hosted 10 plants placed onto floating Styrofoam panels to a final density of 40 plant/m^2. After one week of cultivation at the +/ − P regime, the nutrient solution was replaced and all the plants were grown until flowering using regular Hoagland solution. At flowering, plants were harvested. Leaf area, leaf number, shoot and root fresh and dry weights were measured on two plants per replication. Data were analyzed by ANOVA.

Experiment 3

In this experiment we added proline to the irrigation water to understand whether and how osmolytes treatment could improve salinity tolerance (Maggio et al., 2002). Seeds of *Raphanus sativus* L. cv. Saxa2 were germinated in a climatic chamber (24°C-U. R. 80%; 12h photoperiod, 200μmol/m^2 PAR). Seven days after germination, the young seedlings were transferred into plastic containers filled with a mix of perlite (50%) 2 and peat-moss (50%) at a final plant density of 120 plant/m^2. Plants were irrigated three times per day with a nutrient solution (De Pascale et al., 2001). Ten days after transplanting, 3 salinity treatments were imposed (L1, L2, L3) by adding 50, 100 or 150mmol/L NaCl to the nutrient solution corresponding to an EC of 6.8, 11.3, 15.6 dS/m, respectively. A non-salinized control (L0, EC = 2.4dS/m) was also included. After 7 and 14 days after beginning of salinization, a 8.5mmol/L L-proline solution was added to half of the plants exposed to different salinization levels. Plants were harvest 42 days after transplanting. Before harvest, the stomatal conductance was measured on 5 plants per treatment on fully expanded leaves. On the same plants it was measured fresh and dry weight upon dehydration at 60°C. Na^+ and Chlorides were determined using an Auto Analyzer Bran Luebbe.

Results and discussion

Morphological Modifications and Functional Adaptation to Stressful Environments

Plant response to salinity is generally described in terms of relative yield as a continuous function of root zone salinity, expressed as electrical conductivity of the solution in contact with the roots (ECe) (Maas and Hoffman, 1977). The relative plant biomass, i.e. the percentage of the maximum dry mass obtained under non-stressed conditions, progressively decreased after 2.5dS/m (Figure 1). However, in contrast to most published literature, the relative yield response to increasing EC after the tolerance threshold (2.5dS/m) revealed two linear regions with a sharp slope change at approximately 9dS/m. Specifically, a 6% plant dry mass decrement per dS/m was observed until approximately 9dS/m, whereas after this EC value the yield decrease was only 1.4% per dS/m (Figure 1). The plant dry weight decay at increasing EC corresponded to different shoot and root responses (Figure2). The shoot dry weight gradually decreased until approximately 10dS/m. In contrast, the root dry weight was virtually unaffected within the same salinity range. After 10dS/m, we observed a sharp increase of the root dry mass which remarkably increasd the root-shoot ratio. The bilinear response after the specific salinity tolerance threshold has been reported by other authors (Dalton et al., 1997). Nevertheless, the physiological significance of this *slope change* has never been functionally analyzed in terms of salinity tolerance. Several authors have suggested that morphological root and shoot traits can play an important role in plant stress adaptation. Dalton et al. (2000), for example, have indicated that a reduced root relatively to the shoot development may increase the tomato salinity tolerance by delaying the onset of a critical level of ion accumulation/toxicity into the shoot. A systematic study on the contribution of these traits to salinity tolerance in specific cultural contexts has never been pursued, however. In addition to grafting techniques that may involve uncontrollable physiological disturbances, root habits and characteristics could be modified via genetic engineering since genetic determinants that affect root length, lateral root development and root hair formation (Muday et al., 1995; Hochholdinger et al., 2001) have been thoroughly characterized. This hypothesis can be also tested vs. the large collections of root mutants available in model species and, to a less extent, in tomato or other species. In this respect, the possibility of integrating the output of biophysical models with biotechnological tools should be further considered.

Controlled Nutritional Deficiency to Improve Root vs. Shoot Growth

It is possible to change the relative root-to-shoot growth by modulating the plant nutritional status in specific phenological stages. Biddinger et al. (1998) have demonstrated that phosphorus starvation increases the root: shoot biomass ratio, however a prolonged phosphorus starvation will also decrease CO_2 assimilation and stomatal conductance. Based on these findings we attempted to impose a transient root starvation to promote a rapid root development. One week phosphorus starvation, followed by regular phosphorus fertilization caused a 30% increment of the root biomass without significantly affecting the shoot development (Figure 3). Although margins for practical applications of these principles need to be further established, modulating the fertilization regime to generate specific plant modifications may be considered as a strategy to activate important plant functions while minimizing fertilizers overdosage. Stimulating a rapid root development may contribute to avoid temporary water stresses that may occur in substrates with poor water retention properties, whereas supplemental foliar applications of a specific element may compensate possible nutrient deficiencies. Changes in the relative hydraulic resistance in response to nitrate availability has also been reported (Gloser et al., 2007). Genes encoding for specific proteins responsible for ion transport have also been identified and could be over-expressed in model plants to define novel strategies to improve nutrient use efficiency. All together these results indicate that nutritional control is another important tool that should be considered to improve water and nutrients use efficiency in greenhouse plants.

Osmoprotectant Treatments to Control Plant Growth and Ion Partitioning in Saline Environment

Plant treatment with organic molecules that may act as compatible solutes has been proved to induce a partial stress protection (Makela et al., 1996). The exact mechanism of action of these osmolytes is not univocal, however, and it may depend on complex interactions with other environmental variables (Maggio et al., 2002). Exposure of radish plants to a 15mmol/L proline solution reduced the stomatal resistance at any salinity level tested (Table 1) and increased the relative dry matter accumulation (data not shown). In addition to ameliorate the general plant water status, the proline treatment had also an effect on Na^+ and Chlorides distribution within the plant (Figure 4). Despite their higher transpiration rate (Table 1), proline treated plants had reduced leaf Na^+ and Chlorides concentrations, which may have contributed to delay the appearance of toxicity symptoms.

Preliminary results indicate that these ions may preferentially accumulate into the hypocotyls which, in this species, seemed to serve as detoxification sinks (Figure 4). The molecular components of most biosynthetic pathways of plant compatible solutes have been determined and targeted, via genetic engineering, to different plant tissues and organs to enhance the accumulation of specific metabolites (Maggio et al., 2001). This is another example of possible plant performance control through a targeted application and/or overproduction of specific metabolites.

Conclusions

There are many under-explored approaches that should be considered in agricultural productions to identify innovative strategies to both improve resource use efficiency and reduce the environmental impact of agricultural practices. Greenhouse agriculture is an excellent scenario where such innovations could be implemented because it allows to have the best control on plant-environment interactions. Precision cultivation protocols can be derived from a profound knowledge of the functional biology of plant adaptation to environmental stresses. In this respect, biotechnology and genetic engineering should be envisioned not only as advanced techniques to generate plants with improved performances but also as powerful tools to understand what should be *potentiated* in the plant to have a more efficient greenhouse production process.

Tables

Table 1. Daily variations of stomatal resistance (RS) in response to salinity and proline treatments. ** indicate significant differences at $P < 0.01$, n.s. = not significant differences.

(s/cm)

	RS (10:00am)	RS (noon)	RS (4:00pm)
-Pro			
L0	0,64	1,35	0,90
L1	0,88	1,68	0,82
L2	0,92	2,03	0,97
L3	1,07	2,62	1,52
Mean	0,88	1,92	1,05
+Pro			
L0	0,29	0,50	0,74
L1	0,60	0,97	0,65
L2	0,83	1,21	0,95
L3	0,85	1,56	1,44
Mean	0,64	1,06	0,94
Proline effect	**	**	**
EC effect	**	**	**
Interaction (P x EC)	n.s.	n.s.	n.s.

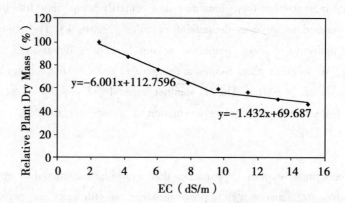

Figure 1   Tomato relative plant biomass in response to increasing EC of the nutrient solution. Values are means ± S. E. ($n = 9$)

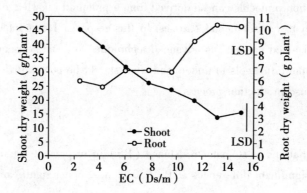

Figure 2   Tomato relative shoot and root dry weights in response to increasing EC of the nutrient solution. Values are means ± S. E. ($n = 9$)

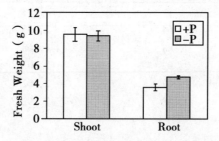

Figure 3   Tomato shoot and root fresh weights in response to phosphorus deficiency. Values are means ± S. E. ($n = 10$)

Literature Cited

[1] Biddinger, E. J., Liu, C., Joly, R. J. and Raghotama, K. G. 1998. Physiologically and molecular responses of aeroponically grown tomato plants to phosphorus deficiency. J. Am. Soc. Hort. Sci. 123: 330~333

[2] Gloser, V., Zwieniecki, M. A., Orians, C. M. and Holbrook, N. M. 2007. Dynamic changes in root hydraulic properties in response to nitrate availability. Journal Exp. Bot. 58: 2 409~2 415

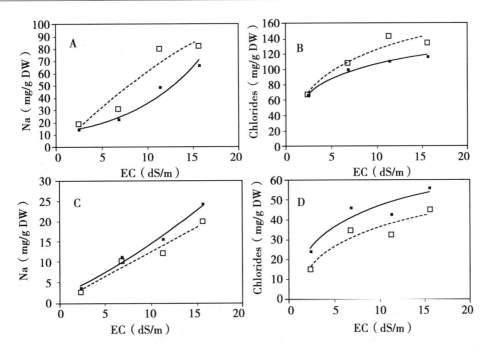

Figure 4 Leaf (top panels) and hypocotyls (bottom panels) Na and Chlorides contents of *Raphanus sativus* L. in response to proline treatment (black square = +proline; open square = -proline; DW = Dry Weight)

[3] Dalton, F. N., Maggio, A. and Piccinni, G. 1997. Effect of root temperature on plant response functions for tomato: comparison of static and dynamic salinity stress indices. Plant Soil 192: 307~319

[4] Dalton, F. N., Maggio, A. and Piccinni, G. 2000. Simulation of shoot chloride accumulation: separation of physical and biochemical processes governing plant salt tolerance. Plant Soil 219: 1~11

[5] De Pascale, S., Maggio, A., Fogliano, V., Ambrosino, P. and Ritieni, A. 2001. Irrigation with saline water improves carotenoids content and antioxidant activity of tomato (*Lycopersicon lycopersicum* L.). Journal of Hort. Sci. and Biotechnology 76: 447~453

[6] Hochholdinger, F., Park, W. J. and Feix, G. H. 2001. Cooperative action of SLR1 and SLR2 is required for lateral root-specific cell elongation in maize. Plant Physiology 125: 1 529~1 539

[7] Maas, E. V. and Hoffman, G. J. 1977. Crop salt tolerance. Am. Soc. Civil Eng. J. Irrig. Drain. Div. 103: 115~134

[8] Maggio, A., Raimondi, G., Martino, A. and De Pascale, S. 2007. Salt stress response in tomato beyond the salinity tolerance threshold. Environ. Exp. Bot. 59: 276~282

[9] Maggio, A., Miyazaki, S., Veronese, P., Fujita, T., Ibeas, J. I., Damsz, B., Narasimhan, M. L., Hasegawa, P. M., Joly, R. J. and Bressan, R. A. 2002. Does Proline Accumulation Play an Active Role in Stress-induced Growth Reduction? ThePlant Journal 31: 699~712

[10] Maggio, A., Joly, R. J., Hasegawa, P. M. and Bressan, R. A. 2002. Can the quest for drought tolerant crops avoid *Arabidopsis* any longer? p. 99~129. In: Crop Production Under Saline Environments: Global and Integrative Perspectives. S. S. Goyal, S. K. Sharma, D. W. Rains (eds.), Journal of Crop Production Vol. 7, n. 1~2

[11] Makela, P., Peltonen-Sainio, P., Jokinen, K., Pehu, E., Setala, H., Hinkkanen, R. and Somersalo, S. 1996. Uptake and translocation of foliar-applied glycinebetaine in crop plants. Plant Science 121: 221~230

[12] Muday, G. K., Lomax, T. L. and Rayle, D. L. 1995. Characterization of the growth and auxin physiology of roots of the tomato mutant *diageotropica*. Planta 195: 548~553

下沉式机打土墙结构的日光温室性能与适应性分析*

丁小明**，周长吉***，魏晓明

（农业部规划设计研究院 设施农业研究所，北京 100125）

摘要：下沉式机打土墙日光温室建造速度快、建设成本低、室内空间大、蓄热保温性能好、冬季室内环境温度高、湿度小，近年来在山东、新疆、陕西、山西、宁夏、内蒙古等地得到迅速推广。但由于对其结构和性能的理论研究滞后，各地在建设和应用中的效果褒贬不一，对其未来发展的争论不断。为了正确评价和应用这种形式温室，在实地调研的基础上，结合文献分析，从温室结构、使用性能、土地利用、建造要求等方面全面分析了其优缺点，并提出了未来推广应用的适应性条件，可供各地建设日光温室参考。

关键词：日光温室；土墙；下沉式；半地下；结构；性能；适应性

Performance and Adaptability of Solar Greenhouse with Sinking Floor and Adobe Wall

Ding Xiaoming, Zhou Changji, Wei Xiaoming

(*Institute of Facility Agriculture, Chinese Academy of Agricultural Engineering, Beijing 100125 P. R. China*)

Abstract: Solar greenhouses with adobe wall and sinking indoor floor have such advantages as easily built; less cost; huge inner space; better heat preservation. They also have higher temperature and lower humidity in winter. Such type of greenhouses was constructed rapidly in Shandong, Xinjiang, Shanxi, Shaanxi, Ningxia and Inner Mongolia province in recent years. But less research were devoted to structure and performance of the greenhouses. So its effect in different areas is in good ones and in bad ones. Future of such type of greenhouses to extend over the northern China has different ideas. In order to give such type of greenhouses a scientific evaluation, this paper discussed the structure, performance, land use capability, construction requirement and so on, according to recent field research and scientific literature Merit and demerit of the greenhouses were also analysed. District adaptability was concluded. It shall be helpful for northern area in China willing to build such type of greenhouses.

Key word: Solar greenhouse; Adobe wall; Sinking floor; Structure; Performance; Adaptability

为进一步落实"三农"政策，加快农业产业结构调整的步伐，增强现代农业的示范和引领作用，国家和各级政府部门近年来对设施农业从政策、资金、科研支撑、人才培训等方

---

* 基金项目：公益性行业（农业）科研专项（200903009）。
** 第一作者：丁小明（1976— ），男，山东泰安人，高级工程师，主要从事设施农业装备研究。E-mail:dingxmcn@qq.com。
*** 通讯作者：周长吉（1964— ），男，甘肃武威人，研究员，博士，主要从事温室工程研究、设计和标准化工作。北京市朝阳区麦子店街41号 农业部规划设计研究院。100125，E-mail：zhoucj@facaae.com。

面上给予了高度重视和重点倾斜，各地相继出台了设施农业的跨越式发展规划。在此背景下，中国特色的节能型日光温室在我国北方地区迅速发展，并成为了设施农业发展的主导形式。为适应这种快速发展的需要，同时也为了进一步提高日光温室的温光性能，一种以"寿光五代"为代表的下沉式机打土墙结构日光温室迅速普及，成为了各地争先推广和效仿的新型日光温室形式。这种温室在建设现场采用挖掘机就地掘土，用压路机分层压实，即形成温室墙体，较传统的砖墙结构日光温室，既节约了建筑材料，又加快了建设速度，并且由于取土使温室地面下降自然形成了温室的"半地下式"形式，同时墙体厚度也加大，使温室的保温蓄热性能也随之提高。因此，它很好地适应了国内当前设施农业快速发展的需要，自山东寿光首先试验成功后，近年在山东[1]、河北、山西[2~3]、西北[4~7]、东北地区[8]迅速得到推广。但由于不同地区土质条件、地下水位、管理水平等因素的差异，各地在推广应用这种形式的温室中表现出不同的使用效果，有的地区使用效果显著，有些地区则出现了很多问题，如甘肃兰州红古区普通日光温室种植的番茄果实灰霉病平均发病率为12%，而下挖0.6m的厚土墙日光温室番茄果实灰霉病平均发病率为34%，明显高于普通日光温室[9]。此外，诸如温室内成了水窖、墙体在雨水浸泡下倒塌、温室遇大风大雪整体倒塌等问题也时有发生。我国幅员辽阔，土地资源、地理纬度、地下水位特征、降水量、土壤质地等实际情况差距很大，而下挖式机打土墙日光温室的结构和性能与这些因素息息相关，各地在推广应用这种形式的温室中应因地制宜、科学分析，循序渐进地在改进中推广。本文从这种温室的结构特点出发，分析了其建设和应用的各种优缺点，可供温室建设者在具体应用中参考。

1 温室的结构特点

下沉式机打土墙日光温室，其结构的基本特征就是土墙结构、地面下沉、墙体下部宽、上部窄，如图1所示。表1为不同地区推广应用这种温室采用的结构总体尺寸。由表1可见，各地下沉式机打土墙日光温室与传统日光温室相比，温室跨度加大、脊高提高、后屋面缩短，墙体厚度加厚，后屋面仰角基本保持在45~50°，总体尺寸基本保持在室内净跨10~12m、下沉深度0.5~1.5m、墙体厚度3~6m，通常温室长度在80m以上，温室总体上呈现大型化的发展趋势。

表1 不同地区推广下沉式机打土墙结构日光温室的结构总体尺寸（m）

地区	跨度	脊高	后屋面水平投影	地面下沉深度	墙体底宽	墙体顶宽	墙体高度	后屋面仰角
山东Ⅳ型	10	4.2~4.3	0.8	0.3~0.4	3.5~4.5	1.0~1.5	3.0~3.2	45~50°
山东Ⅴ型	11	4.6	1.0	0.6~1.0	4.0~5.0	1.5~2.0	3.2~3.5	45~50°
山西省临县[2]	11	4.5	/	0.5~1.2	6.0	2.0	3.5	45°
青海[6]	10	4.2	/	1.4~1.5	3.6	2.3	3.7	40~43°

注：①脊高从室内地面算起；②跨度为室内净跨。

2 优点

（1）建设速度快 土墙结构日光温室，墙体的施工方法一般有三种：一是人工干打垒；二是机械干打垒；三是下挖式机打墙。

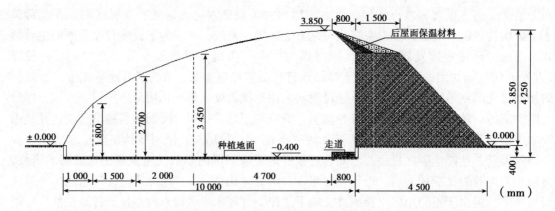

图 1　典型的下沉式机打土墙结构日光温室（图中尺寸为山东Ⅳ型温室尺寸）

早期的土墙结构日光温室多采用人工干打垒的方法建造，用夯土模装土，层层夯实。由于人工取土、人工夯实，工效低，墙体表面粗糙，接缝不严密，墙体密度不均匀，劳动强度大，保温性相对较差。

后来采用机械干打垒，采用机械取土、机械夯实，从春季到秋季可以实现全天候连续打墙，较人工干打垒工效有了显著的提高[10]。

下挖式机打土墙的施工则动用挖掘机、链轨式推土机等机械化作业机械，工效进一步提高。如山东Ⅴ型（寿光型）土墙下沉式日光温室的施工程序为（图1）：先测量放线画出8m宽墙体地基；然后用链轨拖拉机压实墙体地基；再用挖掘机从温室地面取土，摊平在墙上，每上40cm土为一层，用链轨拖拉机压，共压8层，建成地面以上3m高墙，并使室内地面下挖0.7m，达到室内墙高3.7m，顶宽3m的墙体；切削后墙成顶宽2m，并使墙体从墙基到墙顶向北倾斜1.2m左右。最后盖护墙膜，以防遇雨塌墙。这种温室建造速度快，仅墙体作业，一个作业机组每天可以完成2个100延米的日光温室。

（2）建设投资低　机打土墙每延米造价80元～100元，仅为高2.5m、墙体厚度580mm的砖墙墙体（240mm砖+100mm土夹层+240mm砖）造价的1/10～1/8，成本较低。温室总体造价，包括塑料薄膜、草苫和电动卷帘机在内，每667m² 造价大约在5万元左右，造价大约为传统砖墙结构日光温室的50%[11]。

（3）蓄热性能好、保温能力强　压实土壤的蓄热系数要高于普通建筑材料，温室墙体的厚度越大，其白天蓄热越多，夜间放热越多，尤其是凌晨以后保温作用越强，室内温度下降缓慢。另外，下沉一定深度后，温室的保温比提高，还可有效利用地热资源。试验表明，下沉1m温室较不下沉日光温室，平均室内气温提高1.3℃，土壤10cm深处平均地温提高1.1℃。而下沉1.5m温室比不下沉温室平均气温提高1.7℃，15cm深处地温平均提高1.6℃[12]。

（4）室内空间大、病虫害较轻　下沉式机打土墙日光温室室内空间大，温度和湿度的变化较为缓慢，室内空气均匀性好，通风状况良好，室内相对湿度较低，不利于病虫害的发生。

3 缺点

（1）占地面积大、土地利用率低 日光温室是单体太阳能建筑，为了不影响温室的正常越冬生产，在光照时间最短的冬季12～1月必须保证温室至少有4～6h的全地面光照时间，温室之间必须留有足够的空间，此外，在温室的两端还必须留出足够的操作空间。虽然下沉式机打土墙日光温室的跨度增加到12m左右，在跨度方向上土地利用率有所提高[13]，但对于跨度为12～14m的日光温室，跨度方向需要占地25m左右，对100m长温室，长度方向则需要占地120m，整个温室共需占地面积3 000m²，土地利用率仅为40%[14]。

（2）土壤破坏大、恢复周期慢 温室建造利用机械化作业，熟土层几乎全部用于修筑墙体，新建和改建温室因墙体用土量过大，造成熟土层基本用完，生土层冷板硬，矿质元素大量减少，有机质损失殆尽，土壤微生物和土壤酶恢复需较长时间，对于新建温室必须首先通过培肥地力和恢复土壤结构，然后才能协调好水、温、气、热的关系，为作物的健壮生长提供基本条件。长远看，如果今后日光温室不再使用，要恢复土地原貌，则又需要恢复地力，建立生态，增加投入。

（3）后屋面短、夜间保温性差 下沉式机打土墙结构日光温室后屋面水平投影大多在0.5～1.0m之间，前屋面采光面大，透光率高，从而使白天温室内升温快。但是较传统的温室设计后屋面明显缩短[15]，造成温室夜间散热面积大，晚间保温性差。在气候较温暖的河南、山东和河北南部地区，由于高厚墙体白天蓄积大量的热量，晚上大量放热，抵消了温室后屋面短的负面影响，从而可以保证温室内的温度，但纬度较高、气候寒冷的西北、东北地区，后屋面短的缺点将逐渐凸显。

（4）温室间距小、换土困难 为了保证合理的采光，温室的合理间距应在10m以上，但在很多地方一方面受限于目前的土地政策，土地的流转困难，为了最大限度地利用土地，温室栋与栋之间的距离大多在4～5m之间，在山东苍山有些地方甚至不到2m。但温室的土壤环境与高度集约化蔬菜生产方式，土壤的理化、生物学性质，耕种几年后会出现诸如连作障碍、土壤盐分积聚、土壤过湿等问题。时间一长，会出现蔬菜质量下降、品质变劣、病虫害加重、植株早衰等问题[16～17]。为了改良温室的土壤，比较好的办法就是彻底换土。而目前的土地资源管理非常严格，不允许随便取土。但土壤是多种病源物越冬越夏的场所，遇到适宜的条件，继续侵染危害。所以，把耕作层的土壤全部换掉，既消灭了初侵染源，又改造了室内土壤结构。而下沉式机打土墙日光温室由于占地面积较大，在土地资源有限制的地区建造间距普遍较小，今后温室的换土或重建将会出现困难。

（5）室内立柱多、不利机械化作业 温室由于跨度大，虽然也存在无立柱的钢架温室，但该类型日光温室高度高，空间大，造价昂贵。为了降低建造成本，多采用4排立柱的方式。室内交错排列的立柱虽然提高了温室的承载能力，同时也给作业机械化带来了不便。

（6）场区排水困难 温室室内下挖后，地势较周边降低，尤其是出现雨雪天气后，温室屋面集水和场区汇水全部汇集在温室前屋面处而排不出去，甚至出现温室前沿被冲塌的危险。

4 适应性分析

（1）下挖深度与地理纬度的关系 建设地的纬度会影响采光前屋面角的选取，同样也会影响下挖深度。为了减少下沉深度对室内采光性能的影响，张峰[18]在保证室内高效采光率的情况下提出了确定室内适宜下沉深度的方法，即用某时刻太阳的具体位置来计算此时刻下沉壁面在室内所产生的阴影区域面积，然后以室内采光率为约束条件来确定下沉深度。以室内总阴影率不大于30%为临界确定了不同地区的下挖深度（表2）。而《园艺作物节能栽培工程技术集成与产业化示范》课题则对大跨度（10～14m）日光温室的下沉深度进行优化研究[19]，提出山东省（北纬34°25′～38°23′）日光温室的合理采光时段屋面角为30.4～34.6°，下挖壁面在室内产生的阴影率不宜超过15%，跨度为10～12m的日光温室的适宜下挖深度为0.8～1.1m，与张峰[18]近似。

表2 各地区日光温室适宜下沉深度

地区	纬度	经度	适宜下沉深度（m）	地区	纬度	经度	适宜下沉深度（m）
江苏无锡	31°34′	120°18′	1.1～1.2	河北承德	40°59′	117°57′	0.7～0.8
陕西宝鸡	34°39′	107°09′	0.9～1.0	吉林长春	43°54′	125°19′	0.6～0.7
新疆和田	37°13′	79°55′	0.8～0.9	山东寿光	36°53′	118°44′	0.9～1.0

（2）下挖深度与当地地下水埋深的关系 温室的下挖深度必须考虑当地的地下水埋深。地下水埋深指地下水位与地面高程之差。我国各地的地下水埋深差距很大，同时由于降雨量随着季节变化而不同。2011年3月的《地下水动态月报》[20]显示，松辽平原大部分地区地下水埋深小于10m，其中辽宁省平原区大部分地区地下水埋深小于5m。在黄淮海平原，北京市、河南省平原区大部分地区地下水埋深分别为10～50m、1～5m，而天津市、河北省、山东省则为1～10m，江苏省和安徽省淮河平原区地下水埋深一般小于5m。与2010年3月份相比河北省东部、山东省西北部和天津市南部地下水埋深减少水位上升。表3是2010年8月初的《地下水动态月报》，宁夏低至1.34m。可见宁夏平原地区不适合建设下沉式日光温室，而其他地区也并不是全都适合，如甘肃河西走廊有的地方最浅地下水埋深才1.07m。

表3 盆地和平原2010年8月初地下水埋深

省（区）	平原	地下水埋深（m）		
		平均	最浅	最深
山西	大同盆地	7.10	1.63	27.80
内蒙古	呼包平原	8.08	0.53	23.05
陕西	关中平原	21.95	2.51	108.17
甘肃	河西走廊	9.75	1.07	63.75
宁夏	银川和卫宁平原	1.34	0.07	3.51
青海	湟水河谷平原和柴达木	4.39	0.41	17.04
新疆	吐鲁番盆地	28.39	4.87	101.20
山东	青岛	3.23	/	/
	潍坊	9.24	/	/
	临沂	3.72	/	/
	淄博	13.38	/	/

注：山东数据来自文献[21]，其他数据则来自文献[22]。

（3）土壤质地对墙体质量的影响　机打土墙就地取土，而且土壤还要进行蔬菜等作物的种植，对于用于蔬菜种植的温室首先要满足《无公害食品　蔬菜产地环境条件》[23]的要求。为了满足机打土墙的要求，要求土质忌过黏、过酸、过碱，土壤的酸碱度在 pH 为 6.5～7.5。同时强调土壤耕作层不宜过浅，至少在 40cm 以上。

所以在考虑机打土墙的时候，要考虑我国的土壤质地分布。中国的土壤分为三类 11 级，分别是沙土（粗沙土、细沙土、面沙土、沙粉土）、壤土（粉土、粉壤土、黏壤土）、黏土（沙黏土、粉黏土、壤黏土、黏土）。其中壤土兼有沙、黏土的优点，是最理想的一种土壤质地。此外，为了得到好的土墙质量，在机打土墙时要判断土壤的湿密度、干密度和含水率在最佳值（表4）。

表4　土壤的最优含水量和最大干密度[24]

土壤类别	最优含水量（%）	最大干密度（g/cm³）
沙土	8～12	1.80～1.88
粉土	16～22	1.61～1.80
黏土	19～23	1.58～1.70
粉黏土	12～15	1.85～1.95

（4）当地降水量与墙体耐久性的关系　土墙保温性能好，有较大的强度，有一定的耐久性，可以就地取材，成本低，但缺点是耐水性差，有时容易出现裂缝，若遇较大降雨，温室后屋面流下来的雨水侵入墙体会造成局部坍塌。中国北方的降雨量差距较大，东北较多，西北（除陕西）较少，华北居于两者之间[22]。所以建设机打土墙日光温室对于降雨量较多的地方，应采取必要的防雨措施。

（5）当地气温与下沉深度的关系　下沉式机打土墙日光温室主要目的还在于提高温室的蓄热和保温性能，提高温室内的温度，所以对于冬季气温较高的河南省、安徽省等地区可不采用该形式的温室，而对于冬季温度较低的东北、西北地区，采用下沉式则有利于保温和增加温室日间的蓄热能力。

（6）土地资源对温室建设的制约　机打土墙要使用大量的土壤，所以应考虑当地的土地资源状况，对于土地资源紧张的地区，应有节制的发展，而在土地相对丰富的地方，在满足地下水埋深、土壤质地等条件的情况下，可以适当发展。

5　对策建议

（1）因地制宜设计、科学规范施工　日光温室的结构、建造，受当地地理条件、自然环境和气候影响甚大，各地建造日光温室一定要因地制宜，必须经过科学的论证、设计，经过小面积试用后再进行示范、推广，要根据自身的特点合理的确定施工规范，不可盲目模仿。2010年2月的大雪造成寿光30%的日光温室变形，5%的倒塌[25]。所以，下沉式机打土墙要根据纬度、地下水位、土壤质地、风雪荷载等情况合理选择，以免大面积建设后造成不必要的损失。

很多研究表明，温室墙体的厚度越大，其白天蓄热越多，夜间放热越多，但随着厚度的逐步增大，墙体厚度增加其边际增温效果明显下降。杨建军[26]给出了合理的墙体厚度值，认为陕西省杨凌地区为1.0m，甘肃省白银地区为1.3m，宁夏银川地区为1.5m，新疆塔城地区为1.4m，超过这个厚度，其增温效果不明显。

墙体高度也是影响温室性能的一个重要因素，不同地区的日光温室后墙高度不同，山东寿光式的高后墙、短后坡温室与辽宁海城式的矮后墙长后坡温室，以及介于其间的河北省永年式温室等，性能有很大差异。墙体作为日光温室重要围护结构之一，对温室内的热环境有直接影响，且后墙面积或总表面积的比值每增加0.1，温室室内夜间温度可提高0.5℃[27]。但是墙体高度提高后，其建造费用也将增加，所以在实际生产中还要考虑其经济性。

综上可见，作为下沉式机打土墙日光温室高达4~5m的后墙是否造成浪费，应该作进一步的研究，避免盲目引进，造成浪费。此外，大跨度引起的温室骨架荷载的变化在西北、东北等地大风大雪的负荷下能否经受考验也应当加以规范和验证。

（2）发挥资源特点、科学合理发展　下沉式机打土墙日光温室要使用大量的土壤，而且要占很大的面积。中国的土地资源各省市分布不均（图2），而且受土地政策的影响，土地的流转难易差别很大，所以对于土地资源紧张的地区，如北京市、山东省、河南省、河北省等地要有所限制的发展，尤其要限制在基本农田上建设大量的机打土墙结构日光温室，而土地资源相对丰富的内蒙古自治区、新疆维吾尔自治区、甘肃省、辽宁省西部等地可根据各自的气候和水资源状况在科学设计和规范施工的前提下合理发展。

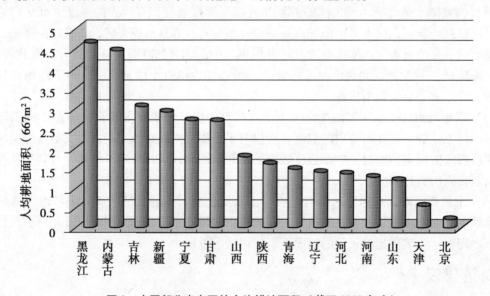

图2　中国部分省市区的人均耕地面积（截至2009年底）

注：图中数据来自2009年各地统计公报。

（3）加强科学研究、解决设计和施工中的问题　下沉式机打土墙日光温室虽然已有研究，但其结构、性能和地区适应性仍然缺少系统的理论研究，目前仍然存在着盲目追求高度、跨度、长度和下挖深度的问题。近年来，设施农业在西北、东北地区大面积推广，下沉式机打土墙日光温室也在大量建设，各地在如何根据自己的气候等条件设计合理的结构，仍然缺少研究。此外，温室的施工存在的如土壤质地的选择、土壤最佳含水率和压实系数的确定以及墙体的高度和厚度的合理比值等问题仍然需要科学地突破。

参考文献

[1] DB37T 391-2004 山东ⅠⅡⅢⅣⅤ型日光温室（冬暖大棚）建造技术规范
[2] 马卫华. 改良型冬暖半地下式抗风雪日光节能温室结构及建造技术. 农业技术与装备, 2010 (11): 44~45
[3] 王海啸, 李杰林. 黄土高原地区坑式日光温室保温性能试验研究. 山西农业大学学报, 2006, 5 (6): 11~13
[4] 岳玲, 凌广通, 张振军等. 赤峰市日光温室厚墙体的建造技术. 内蒙古农业科技, 2005 (1): 44, 46
[5] 田雨. 8m跨度竹木拱架机制土墙半地下式日光温室建造技术. 内蒙古农业科技, 2009 (4): 117
[6] 靳伟, 林顺花, 薛寒青等. 青海半地下式日光节能温室结构及建造. 青海农林科技, 2010 (4): 13~14, 40
[7] 王华民, 杨孟, 李维山等. 吐鲁番市下挖式日光温室建筑技术. 农村科技, 2009 (5): 11~12
[8] 佟国红, 罗新兰, 刘文合等. 半地下式日光温室太阳能利用分析. 沈阳农业大学学报, 2010, 41 (3): 321~326
[9] 宋学栋. 半地下式日光温室番茄灰霉病预防技术. 中国蔬菜, 2010 (3): 26~27
[10] 王华民, 杨孟, 葛同江. 日光温室墙体机械干打垒技术. 农村科技, 2009 (1): 12
[11] 李培之, 于丽艳, 王志和. 浅析寿光市高效节能型日光温室管理技术. 温室园艺, 2010 (9): 30~32
[12] 李睿. 下沉型日光温室保温蓄热性能的测试与研究. 中国农业大学. 2010
[13] 周长吉, 刘晨霞. 提高日光温室土地利用率的方法评析. 中国果菜, 2009 (5): 16~20
[14] 韩太利, 魏家鹏. 寿光新型日光温室的结构特点与推广应用. 中国蔬菜, 2010 (13): 7~9
[15] JB/T 10594-2006 日光温室和塑料大棚结构与性能要求
[16] 胡留杰, 曾希柏, 白玲玉等. 山东寿光设施菜地土壤砷含量及形态. 应用生态学报, 2011, 22 (1): 201~205
[17] 曾希柏, 白玲玉, 苏世鸣等. 山东寿光不同种植年限设施土壤的酸化与盐渍化. 生态学报, 2010, 30 (7): 1 853~1 859
[18] 张峰, 张林华, 刘珊. 日光节能温室下沉深度对其采光性能的影响. 山东建筑大学学报, 2008, 23 (6): 473~474, 477
[19] 园艺作物节能栽培工程技术集成与产业化示范［国家科技支撑计划课题（2008BADA6B05）］. 科技计划成果, 2010, 17: 9
[20] 水利部. 地下水动态月报（2011年3月）[EB/OL]. [2011-03-30]. http://www.mwr.gov.cn/zwzc/hygb/dxsdtyb/201103/t20110330_257789.html
[21] 山东省水利厅. 山东省平原区地下水动态（2009年1月）[EB/OL]. [2009-02-16]. http://www.sdwr.gov.cn/sdsl/pub/cms/1/2092/2123/489/15837.html
[22] 水利部. 地下水动态月报（2010年8月）[EB/OL]. [2010-09-02]. http://www.hydroinfo.gov.cn/gb/dxsdtyb/201009/t20100902_235306.html
[23] NY 5010-2002 无公害食品 蔬菜产地环境条件
[24] 李学智. 房屋建筑工程见证取样和送检指南. 北京: 化学工业出版社, 2008
[25] 魏家鹏, 于贤昌. 近期雪灾对寿光蔬菜日光温室的影响. 中国蔬菜, 2010 (9): 7~8
[26] 杨建军, 邹志荣, 张智等. 日光温室土墙厚度及其保温性的优化. 农业工程学报, 2009, 25 (8): 180~185
[27] 温祥珍, 梁海燕, 李亚灵等. 墙体高度对日光温室内夜间气温的影响. 中国生态农业学报, 2009, 17 (5): 980~983

雪灾中钢骨架结构日光温室倒塌原因及对策*

白义奎[1,2,3]**，李天来[1,3]***，王铁良[2,3]，刘文合[2,3]

［1. 沈阳农业大学园艺学院，沈阳 110866；2. 沈阳农业大学水利学院，沈阳 110866；
3. 设施园艺省部共建教育部重点实验室（沈阳农业大学）沈阳 110866］

摘 要：日光温室作为设施农业生产的主要载体，得到了广泛的应用。但是，我国目前还缺少通用的、强制性的日光温室设计、制造和施工标准，不合理的设计、施工存在较大的安全隐患，并对其耐久性能产生影响。本文详尽地分析了钢骨架日光温室在雪灾中倒塌、被破坏的原因，提出了相应的改进建议，为日光温室设计、建造、施工提供依据。
关键词：日光温室；钢骨架；设计；建筑结构参数；安全性；耐久性

Countermeasures and Causation About Collapse of Steel Skeleton structure Solar Greenhouse in Snow Disaster

Bai Yikui[1,2,3], Li Tianlai[1,3], Wang Tieliang[2,3], Liu Wenhe[2,3]

［1. College of Horticultural, Shenyang Agricultural University, Shenyang 110866, P. R. China
2. College of Water Conservancy, Shenyang Agricultural University, Shenyang 110866, P. R. China
3. Key Laboratory of Protected Horticulture (SYAU), Ministry of Education, Shenyang 110866, P. R. China］

Abstract: Solar greenhouse has widely application as an important establishment in protected agriculture. However, in China, there is a lack of all-purpose and compellent standard on design, fabrication and construction of solar greenhouse at present. Incorrect design and construction about solar greenhouse has hidden trouble on safety, and influent its durability. This thesis puts forward detailed analysis of the cause of collapsing and destroying of steel skeleton structure solar greenhouse, and put forward appropriate recommendations for improvement. It will provide the basis for design, fabrication and construction of solar greenhouse.
Key words: Solar greenhouse; Steel skeleton; Design; Parameter of construction; Safety; Durability

日光温室作为发展设施农业的主要载体，在我国得到了广泛的应用，全国设施总面积已达 310 万 hm²。辽宁省是日光温室发源地，也是我国设施蔬菜重点发展区域。截至 2009 年底，辽宁省设施农业占地面积达到 43.55 万 hm²（653.2 万亩），其中日光温室面积达到

---

* 基金项目：中国博士后科学基金特别资助项目（项目编号：200801395）；"十一五"国家科技支撑计划重点项目（项目编号：2008BADA6B05）。
** 白义奎，沈阳农业大学水利学院，教授，博士生导师，沈阳农业大学园艺学院园艺学博士后，110866 沈阳市东陵路 120 号，E-mail：baiyikui@163.com。
　　研究方向：设施环境工程。
*** 李天来，沈阳农业大学园艺学院，教授，博士生导师，110866 沈阳市东陵路 120 号，E-mail：ltl@126.com。

27.22万hm^2（408.3万亩），辽宁日光温室蔬菜的面积和产量均列国内第一位，成为设施农业生产的支柱产业。然而从日光温室结构设计、建造水平来看，与发展水平还存在一定的差距，尤其是传统的竹木结构温室占60%以上，即便是钢骨架温室，也还存在缺乏科学、合理的规范化设计，结构不合理、施工质量差、管理水平低等问题。然而我国目前还缺少通用的、强制性的日光温室设计、制造和施工标准，致使一些温室产品与国外产品在质量上还存在较大的差距。随着近几年来自然灾害频发，设施园艺生产也遭受了不同程度的损失。2007年3月辽宁地区发生百年一遇的极端天气，沈阳地区降雪（水）51mm，鞍山地区降雪（水）76.8mm，大连地区暴风雨、雪，降水量达50mm以上，伴有偏北大风，陆地风力7级，阵风9~12级。过大的风、雪荷载严重超出了温室大棚设计标准（一般日光温室设计标准为10~15年一遇的风雪荷载），造成了重大的损失。然而通过调查发现，在同一温室和大棚生产基地，相同结构、相同材料建造的温室，有的倒塌，有的却安然无恙。发生上述事故除灾害性天气因素外，不合理的设计、施工及维护管理等也是致灾的主要原因。本研究通过分析钢骨架日光温室在雪灾中倒塌、破坏的原因，提出了相应的改进建议，为日光温室设计、建造、施工提供依据。

1 钢结构日光温室破坏原因

日光温室钢骨架通常采用平面桁架结构，这种结构在平面内具有较强的承载能力和抵抗变形的能力。根据其受力特点，其在竖向荷载的作用下，在支座处会产生水平推力作用，因此要求支座处必须具有稳固的水平支撑；作为平面桁架结构，也要求其受力时必须保证竖向在一个平面内；桁架平面内支撑较多而平面外支撑较少，可能造成平面外结构失稳而发生破坏，因此必须设置足够的纵向系杆结构来加强平面外的支撑；钢桁架结构长期处于高温、高湿环境，一定程度上降低了结构的耐久性能；另外，缺乏科学合理的设计，在材料的选用、施工工艺等方面存在诸多问题，也是造成温室破坏的主要因素。

1.1 设计问题

（1）设计荷载取值偏小，甚至漏选实际存在的荷载项目。如草帘集中作用、半跨雪荷载作用、卷放帘时产生的集中荷载的影响、作物吊重等附加荷载。同时由于温室建筑的特殊性，尤其在温室前坡未知会产生较大的局部积雪荷载。

（2）农业生产建筑的特殊性，及操作管理的特殊性，要考虑风、雪荷载组合，同时应考虑检修（清扫积雪）荷载的存在。

（3）纵向系杆设置位置不当或过少。结构受力分析及最不利杆件的选取，直接影响系杆位置的设置，部分受压较大杆件缺少侧向（平面外）支撑，是造成失稳破坏的主要原因。

（4）缺少必要的结构分析，对温室骨架结构受力不了解，凭经验建设。如采用单钢管结构作为前坡骨架，其承受竖向荷载能力有限，很容易发生受弯破坏。

1.2 结构不合理

温室骨架的整体稳定性差是钢骨架结构温室倒塌、破坏最主要的原因。

（1）倒塌温室一般缺少南底角的地梁和后墙的圈梁，或者梁的强度不够。骨架与前底角基础和后坡、后墙的连接不牢固，缺少必要的水平支撑；尤其是土墙温室，有的温室骨架直接放在土墙顶面，遇水后很容易失去支撑作用；或有的虽然设置支撑，也只是简单支撑在摄制的水泥柱顶面，缺乏可靠的连接，稍有变形支撑点很容易从柱顶脱落。另外，骨架前底角处不设基础、或简单设置砖块支撑也是温室倒塌破坏的主要原因之一。

（2）纵向系杆与骨架连接不牢固。有的纵向系杆和骨架仅采用一点焊接，几乎起不到拉结作用；有的纵向系杆与骨架之间虽然设置斜拉支撑，但只有一侧，使得单片骨架的稳定性降低，发生侧向变形而失去应有的支撑作用。

（3）纵向系杆没有固定到山墙上或固定不牢。有些温室的拉杆仅简单地插入土墙或搭在砖墙上。有些虽也固定在山墙上，但紧靠山墙没有骨架，导致骨架没有形成一个稳固整体，受压后骨架整体侧摆变形而失稳破坏。

（4）骨架腹杆数量不够，有些两根腹杆没有连上，甚至腹杆平行设置而形成几何可变体系，大大降低了骨架承载能力。

（5）温室高跨比不当，脊高低、跨度大。有些日光温室高跨比较小，屋面坡度小，结构配置不合理，不仅影响日光温室的采光，同时也大大降低了温室的承载能力。

1.3 材料不合格

（1）桁架的上弦钢管和下弦钢筋不达标准的现象普遍。大多数农户为了省钱往往采用小钢厂生产的非国标钢管、钢筋等劣质钢材焊接骨架，钢管壁厚仅有 1.5~1.8mm（国标钢管壁厚 2.75mm）。

（2）拉杆材料过于单薄。使用钢筋、钢丝绳做拉杆代替 $\phi21.25 \times 2.75$ 钢管的现象普遍，有的甚至使用 8 号或 10 号铁线。这些材料的抗压能力较差，特别是钢丝绳康拉能力很强，但几乎不能承受压力，稳定性较差，一旦一侧失去张拉力，整个温室将失去纵向支撑。

（3）桁架腹杆和斜拉使用的钢筋过细，焊缝质量不容易保证，影响骨架承载能力。另外，在材料选择上，有些使用 16~18mm 钢筋代替钢管作为温室骨架上弦，不仅浪费材料，也使骨架平面外稳定性降低。

1.4 施工质量低劣

（1）焊接质量差。为了节省投资，农民往往自行焊接温室骨架，焊接质量良莠不齐。焊缝长度不够、单面焊等，存在大量虚焊，以及焊材与母材不匹配等。

（2）骨架制作时不设胎具，直接在地面焊接，很难保证骨架的平整度；安装缺乏专业的施工队伍及质量保证措施，安装的骨架间距不一、竖向平整度差等，直接影响骨架结构的承载能力。

（3）温室年久失修，墙体裂缝，骨架脱焊现象普遍。钢架防锈技术低且不注意保护，骨架锈蚀严重。

1.5 其他原因

（1）有的保温被、草苫子、棉被等外保温覆盖材料吸水过多，以及外覆盖材料表面积雪、冻结后卷帘时产生较大的附加荷载，加大了温室骨架的负荷，使温室发生破坏。

（2）栽培黄瓜、甜瓜、番茄等作物的温室，由于作物吊重加大了温室桁架的负荷，温室倒塌的几率增加了。

（3）极端天气的影响。目前一般农业生产建筑，如日光温室设计标准基本按 10~15 年一遇风雪荷载标准设计，荷载取值偏低。尤其近几年气候恶化，恶劣气候出现几率增大，过大的风雪荷载（如 2007 年辽宁地区降雪达到百年一遇）也是造成温室倒塌的一个不可忽视的原因。

2 对策

温室设计建造时，除在材料选择上必须保证外，在施工质量、结构构造做法等方面应注意以下几点。

2.1 钢骨架结构材料选择

钢骨架结构材料是保证承载能力的最主要因素，必须按结构设计要求确定。通常采用 Q235（钢管）或 HPB235（钢筋）强度级别即可。不同跨度温室骨架材料截面选用见表 1。

表 1 不同跨度温室骨架材料选用
Table 1 Steel skeleton material in different span greenhouse

材料	骨架构造			
跨度（m）	6.0~7.0	7.5~8.0	8.5~9.5	10.0~12.0
上弦（mm）	φ21.25×2.75	φ21.25×2.75	φ26.75×2.75	φ33.50×3.25
下弦（mm）	φ12	φ14	φ16	φ18
腹杆（mm）	φ8	φ8	φ10	φ12
纵向系杆（mm）	φ21.25×2.75	φ21.25×2.75	φ21.25×2.75	φ26.25×2.75
纵向系杆数量（道）	5	6~7	7~8	8~10

＊说明：温室骨架间距 0.85m；φ21.25×2.75（4 分管，D15）；φ26.75×2.75（6 分管，D20）；φ33.50×3.25（1 英寸管，D25）。

2.2 制作施工

（1）钢平面桁架的施工。制作骨架的胎具要严格保持水平，准确放骨架形状线，然后施焊。各节点焊缝长度不小于 25mm，且双面满焊，若为镀锌钢材，则焊完后，应除去焊垢，再刷防锈漆和银粉各 2 道。

（2）骨架安装就位必须准确，并与后墙体垂直。温室两山墙处应各放一榀骨架。骨架的系杆伸入山墙不少于 120mm，并应保证可靠连接，必要时刻加焊横向钢筋以加强锚固，或者采用必要的拉结锚固在地下。纵向系杆与骨架焊接必须得到保证。应采用三角形斜拉构造，以保证系杆与骨架下弦、上弦的连接。系杆沿温室纵向必须连续，当系杆长度不足时，系杆的接长必须采用搭接焊，且搭接长度不小于 100mm。

（3）为保证温室的整体稳定及侧向变形，在温室的两个端部的两榀桁架间，应设置水平支撑。水平支撑在每两道纵向系杆之间设置，可采用 φ12 钢筋焊接成"×"形。

（4）加强两铰拱式结构支座的稳定。两铰拱式结构要求在平面内不能发生水平位移（移动），否则很容易发生破坏。建造及施工要求必须有稳固的支撑，前底脚要设置基础及地圈梁，基础埋深一般不应小于 500mm，地圈梁尺寸不应小于 120 mm×120mm；后坡在骨架支撑位置也要设置圈梁，圈梁尺寸不应小于 60 mm×240mm。必要时可在前底脚地圈梁及后坡骨架支撑位置设置的圈梁上设置预埋件，与骨架焊接在一起。

3 结论与建议

钢平面桁架结构日光温室一般均采用无柱设计，具有温室骨架承载能力大、耐久性能好，温室内部作业空间加大，同时有利于机械化生产等优点，得到了广泛的应用。但如果温室骨架材料选择不当、温室结构设计不合理以及存在的施工质量差等问题，也会造成严重的结构失效问题，造成不必要的损失。另外，正常使用以及正常的后期维护对保证结构耐久性能也是十分重要的。

参考文献

[1] GB 50017—2003 钢结构设计规范. 中国建筑工业出版社. 2003
[2] JGJ 81—2002 建筑钢结构焊接技术规程. 中国建筑工业出版社. 2002
[3] GB 50009—2001 建筑结构载荷规范. 中国建筑工业出版社. 2002
[4] 喻志武. 节能日光温室的结构及其建造. 新疆农业科技，1994，(6)：20~22
[5] 冯广和. 建造温室要注意的几个问题. 西北园艺，2003，(7)：5~6
[6] 徐群. 建筑材料在温室结构中的应用. 天津农林科技，1995，(3)：33~34
[7] 张立芸，徐刚毅，马承伟，兰淼. 日光温室新型墙体结构性能分析. 沈阳农业大学学报，2006，37(3)：459~462
[8] 丛祥安，杨晓波，吴建虎. 日光温室的钢骨架制作技术. 现代农业，2006，(11)：15~16
[9] 周启武，胡斌，彭正昶. 日光温室钢骨架的承载能力推荐及建造中应考虑的问题. 机械研究与应用，2003，(16) 增刊：100，103
[10] 白义奎，明月. 影响日光温室钢骨架结构安全及耐久性能因素分析. 房材与应用，2005，33(5)：14~15

日光温室浅层土壤水媒蓄放热增温效果[*]

方慧[1,2]**，杨其长[1,2]***，梁浩[3]，王烁[1,2]

(1. 中国农业科学院农业环境与可持续发展研究所，北京 100081；2. 农业部农业环境与气候变化重点开放实验室，北京 100081；3. 中国农业大学水利与土木工程学院，北京 100083)

摘要：墙体与地面蓄热是日光温室主要储热模式，研究表明最冷月份墙体夜晚所释放的热量约占温室得热量的35%~40%[1]，但过厚的墙体设计以及大量的建筑材料的使用造成土地资源浪费、成本上升，能否将墙体热量转移到土壤，是事关墙体能否轻简化的重要一点。本试验以太阳能为热源，以水为蓄热介质，以温室浅层土壤为蓄热体，白天通过水的循环将热量收集并储存到温室浅层土壤中，夜间通过土壤的自然放热将热量释放到温室中，提高温室夜间温度。结果表明：此蓄热方法增加了温室的蓄热量，在盖上保温被以后，试验温室与对照温室温差开始增加，平均气温差为4.0℃。实验温室土壤60cm以上区域温度一直高于对照温室，0cm处夜间平均温差为3℃，30cm处夜间平均温差为3℃，60cm处夜间平均温差为5℃。因此，此方法既提高了空气温度，也提高了作物根部土壤温度。

关键词：日光温室；浅层土壤；水媒蓄放热；增温

Rising temperature experiment on heat release and storage with water in solar greenhouse[*]

Fang Hui[1,2], Yang Qichang[1,2]***, Liang Hao[1,2], Wang Shuo[1,2]

(1. Institute of Environment and Sustainable in Agriculture, Chinese Academy of Agricultural Sciences, Beijing 100081, China; 2. Key Lab. For Agro-Environment & Climate Change, Ministry of Agriculture, Beijing 100081, P.R. China)

Abstract: Wall and ground play an important role in heat storage in greenhouse. Studies showed that the amount of heat stored in wall account only 35% ~40% of the total stored energy in the coldest month. But the wall was designed too thick, and the use of a large number of construction materials can cause the waste of land resources and the rising cost. Whether the heat stored in the wall can be transferred into the soil is a key to simplify the wall. In this study, the solar energy is used as heat source, the water is used as heat storage medium and the shallow soil in greenhouse is used for heat storage. The heat was collected and stored in shallow soil through water circulation during the day, and the greenhouse temperature increased at night due to the soil heat releasing. The results showed that: this method increased the heat storage of greenhouse. The temperature difference began to increase after covered the insulation. The average air temperature in experiment greenhouse is 4.0 ℃ higher than in normal greenhouse. The soil

---

[*] 基金项目：北京市科委重点项目（D111100000811001）；国家自然科学基金（31071833）。

** 作者简介：方慧（1983— ），女，主要从事设施农业环境工程方面的研究。中国农业科学院农业环境与可持续发展研究所，北京100081。E-mail：fh2002124@163.com。

*** 通讯作者：杨其长（1963— ），男，博士，研究员，博士生导师，主要从事设施园艺环境工程研究。中国农业科学院农业环境与可持续发展研究所，北京100081。E-mail：yangq@ieda.org.cn。

temperature at 0cm, 30cm, 60cm in the experiment greenhouse is 3.0℃, 3.0℃, 5.0℃ higher than in the normal greenhouse respectively. Therefore, this method not only increased the air temperature, but also increased the temperature of soil around which crops grow.

Key Words：Solar greenhouse；Shallow soil；Heat release and storage with water；Temperature-increasing

日光温室中的能量主要通过太阳能获得。白天，太阳辐射到温室后温室内温度比较高，在晴天最高温能达到35℃以上，部分以热量的形式存储在后墙、山墙和土壤中；夜晚，当室内气温下降时，墙体和土壤中蓄积的热量又源源不断地向温室供应。由于日光温室的蓄热和保温能力有限，在温室后半夜温度比较低，难以满足作物生长需求[2~7]。

为增加温室夜间温度，王顺生等在后墙内侧安装太阳辐射集热调温装置，在冬季晴天能将集热器中的水温提高并储存在保温水箱中，夜间利用该热水加热地温，能将地温提高3.2~3.8℃[8]。H. Ali-Hussaini 等在温室内建造了蓄热水池，水池深5~20cm，水池内表面涂黑，水池中装满一定浓度的盐水，外面覆上透明塑料膜，利用蓄热水池蓄集的热量加热温室，起到较好的加热效果[9]。I. Grmiadellis 等提出了在温室中用蓄热介质收集太阳能为温室加热，并验证了其可行性[10]。山本雄二郎提出了温室中采用地下热交换的设想，并在1967年建造了第一个地中热交换温室进行试验，效果良好，初步确定了其可行性[11]。高仓直和山川健一提出了较为完整的地中热交换温室的数学模型[12]。马承伟对单栋塑料大棚地中热交换进行了研究，一般白昼系统运行时，室内空气向管道贮热后，气温可降低6~8℃，空气焓值降低13~20kJ/kg，夜间室内空气通过管道被土壤加温，气温升高达4~5.5℃，空气焓值增加7~10kJ/kg[13]。张峰等利用卵石作为蓄热体，应用于南疆地区的日光温室中，取得了蓄热增温的效果[14]。

本文以太阳辐射产生的热量为能源，水为蓄热介质，以温室浅层土壤为蓄热体，白天将温室中的热量通过水收集并传递到温室浅层土壤中储存起来，增加温室蓄热量；夜间通过温室浅层土壤热量的传导与释放，增加温室空气温度。

1 试验条件与测试方案

1.1 试验温室概况

试验于2010年1月3日至2月7日在北京市大兴区实验基地日光温室内进行。试验温室为东西走向，坐北向南，温室方位角为0°，后墙高2.5m，脊高3.2m，跨度8m，长75m，后墙为50cm砖墙，后墙外表面用5cm厚的珍珠岩做外保温处理。用双层塑料薄膜做两处隔断将温室内部空间平分成25m长的三个部分。选取东西两段25m长的日光温室作为试验温室和对照温室。试验期间温室内没有种植作物。在进行试验之前先测定了试验温室与对照温室基础温度，如图1所示。在晴天早上揭开保温被时，试验温室与对照温室温度波动较大，但在夜间试验温室与对照温室温度温差小，本试验是在夜间进行，故对试验影响不大。

1.2 蓄热装置设计

日光温室热量采集与释放原理是白天通过集热器将后墙上的热量转移并储存到温室浅层土壤中，以提高土壤温度，增加温室蓄热量；夜间通过后墙和土壤的自然放热，以提高温室夜间温度。通过此方法实现温室白天的热量在时间上的转移。整套装置包括后墙集热器、连

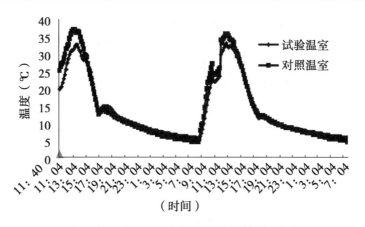

图 1 试验前试验温室与对照温室跨中气温对比

Figure 1 Temperature comparison between experiment greenhouse and normal greenhouse before experiment

接管路、循环水泵、地下蓄热器。图2为蓄放热装置示意图。

集热器：后墙内表面平行安装16组并联的换热管，换热管连接在分集水器上，每组换热管的长度是150m，呈U型布置，换热管的规格为ϕ19mm，材质为PE塑料管。16组换热管呈32排，每排间距为5cm均匀布置在后墙表面。换热管内装满蓄热介质水，白天蓄热介质水吸收太阳的辐射能，并通过水的循环将热量转移到浅层土壤中。

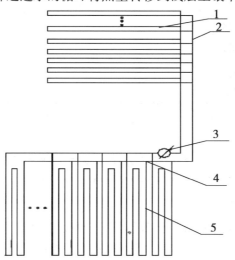

1. 集放热器 2. 分集水器 3. 水泵 4. 分集水器 5. 浅层土壤换热装置

图 2 蓄放热装置示意图

Figure 2 Schematic of heat storing and releasing heating equipment

蓄热器：蓄热系统采用的是浅层地表土壤蓄热模式，即在浅层地埋设换热管，通过换热管内水的流动与土壤换热[15]。该系统由48组换热管组成，换热管采用规格为32mm的PE管，换热管安装在距离地表以下60cm处，每组换热管的长度是28m，呈U型布置，地面共有换热管192排，每排换热管长7m。换热管均匀布置在从后墙到距离后墙7m的范围内，每排换热管的间距为35cm。48组换热管连接在分集水器上，该分集水器与后墙集热器上的分集水器串联接通。白天，集热器中的热量通过水的循环转移到地下换热管中，再通过水管和土壤的传热，将热量转移到浅层土壤中并储存起来。

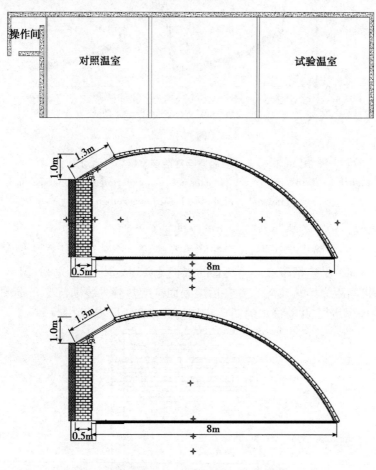

图3　试验温室与对照温室测温位点布置
Figure 3　Layout of temperature measuring points in experiment greenhouse and normal greenhouse

1.3　试验测试方案

测定晴天试验温室与对照温室相比蓄热效果，测定集热器集热效率。试验时间是1月29日至1月30日，29日天气晴好，上午8：30揭开保温被同时开启循环水泵进行蓄热，下午16：00关闭循环水泵，17：00盖上保温被。每间隔10min记录各测点的温度（图3）。

后墙墙体温度的测量：试验区域距地面1.0m的墙体处，在后墙上利用电钻打孔埋入传感器来测量墙体分层温度，打孔深度分别是墙体表面、25cm、50cm和55cm处，在集热管进水口和出水口布置两个测点。

土壤分层温度测量：在试验温室与对照温室的中部布置土壤温度的测点，实验温室中土壤测温位点深度分别是0cm、10cm、20cm、30cm、50cm和60cm处，对照温室土壤测温位点深度分别是0cm、30cm和60cm。

空气温度测量：在实验温室中部距后墙1m、4m和6m处设置三个测温位点，在对照温室中部距后墙4m处设置一个测点，测温位点距地面高度为1.0m。

2 试验结果分析

2.1 温度变化

2.1.1 试验温室南北方向温度影响效果

后墙和浅层地表内的换热管布置均匀，在温室东西方向，热量分布均匀。在温室的南北方向除地面向空气中均匀放热外，后墙蓄积的热量在夜间也向温室放热，放热的方向是从北到南，这种放热的形式可能会造成温室南北方向上的温度分布不均，本文重点分析了温室南北方向上的温度变化。从图4可以看出南北方向白天温度分布不均，距离后墙6m的地方温度稍低，主要是受前屋面塑料膜传热影响，距离后墙1m与距离后墙6m处的最大温差是上午的11:30为4.9℃；在下午的17:00盖上保温被以后，南北方向上的温差不大，温度比较均匀（传统温室内是靠近后墙高），温差一直在0.7℃的范围内，说明夜间后墙的放热对南北方向的温度梯度影响较小。

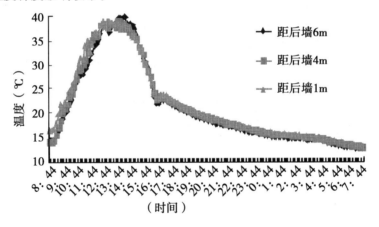

图4 温室南北方向气温变化曲线
Figure 4 Temperature curves in south – north direction

2.1.2 跨中温度影响效果

如图5所示，在掀开保温被后到下午14:00前，温室内温度主要受太阳辐射的影响，温差不明显在9:00~14:00时间范围内两温室的平均温差为0℃；通过土壤的蓄热和传热的影响，在14:00以后到下午17:00时盖上保温被之前，两温室内开始有温差，平均温差为3.8℃；在盖上保温被以后，温室内温度主要受蓄热量的影响，这时温室内温差加大，在17:00~8:00时间范围内两温室的平均温差为4.2℃。晴天太阳辐射大，白天通过循环水蓄积的热量较多，夜间通过土壤和后墙的放热也能提高并维持温室夜间温度。试验期间室外最低温为-8.9℃，出现在早晨7:00；实验温室最低温为13.2℃，出现在早晨揭保温被前8:00，而对照温室最低温为8.7℃，出现在早上8:00揭开保温被之前，此时试验温室与对照温室温差为4.5℃。

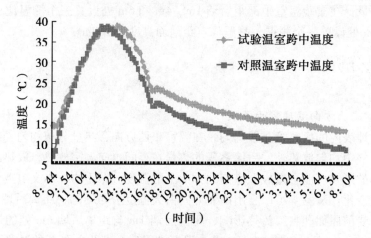

图5 试验温室与对照温室跨中气温对比
Figure 5 Temperature comparison between experiment greenhouse and normal greenhouse

2.1.3 土壤温度影响效果

如图6所示。日光温室内土壤温度日变化昼高夜低,土壤各层(0cm、10cm、30cm、50cm和60cm)的最高温度依次为42.7℃、23.7℃、16.9℃、15.7℃和20.9℃。最低温度依次为13.9℃、13.8℃、12.4℃、10.2℃和11.3℃。平均温度分别为22.8℃、19.2℃、15.3℃、13.9℃和15.0℃。随着深度的增加土壤温度降低,到60cm处受换热管换热的影响,土壤温度开始升高,而传统温室中土壤温度随深度下降,下降到11℃左右,浅层土壤平均温度就基本保持不变[17]。

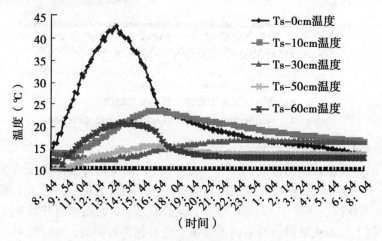

图6 土壤分层温度曲线
Figure 6 Temperature curves of soil layers

一天内,不同土层地温的变化趋势差异较大。地表0cm和10cm处温度主要受太阳辐射影响,白天温度高,晚上温度低,波动较大;土壤50cm和60cm处温度主要受循环水的影响,白天蓄热,温度高,夜晚放热,温度低;土壤30cm处白天温度低,夜晚温度高。利用水的循环将后墙的热量转移到温室的浅层土壤中明显增加了土壤的蓄热量。

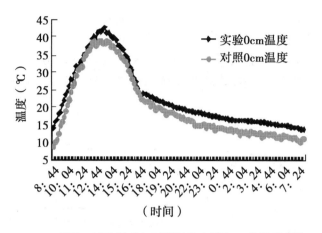

图 7 试验温室与对照温室土壤 0cm 处温度对比

Figure 7　Soil temperature comparison at 0cm between experiment greenhouse and normal greenhouse experiment greenhouse and normal greenhouse

如图 7 所示，地表温度白天主要受太阳辐射影响，平均温差为 2.6℃；盖上保温被以后，由于实验温室的蓄热量比较大，所以温差较大，平均温差为 3.0℃，而且盖上保温被以后，实验温室温度下降趋势缓于对照温室。

土壤 30cm 处为作物生长的根部温度，一般作物根部温度适宜时，可适当降低温室空气温度。从图 8 可看出实验温室 30cm 处土壤温度明显高于对照温室 30cm 处的土壤温度。两温室 30cm 处土壤温度白天低，夜间高，主要是因为热量传递在时间上的延后。实验温室最低温出现在中午 12：50 左右，为 12.4℃，最高温出现在夜间 0：00 左右，为 16.9℃，平均温度为 15.3℃。对照温室最低温出现在中午 14：40 左右，为 9.7℃，最高温出现在夜间 0：00 左右，为 13.9℃，平均温度为 12.3℃。试验温室比对照温室平均高 3.0℃。

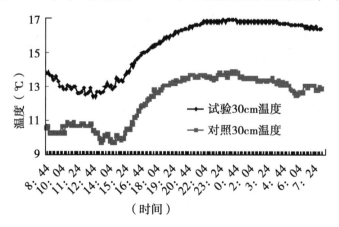

图 8 试验温室与对照温室土壤 30cm 处温度对比

Figure 8　Soil temperature comparison at 30cm between experiment greenhouse and normal greenhouse

试验温室中白天开启循环水泵，将后墙上的热量转移到 60cm 深度的土壤中，所以土壤的温度开始慢慢升高，下午太阳辐射降低，后墙温度也开始慢慢降低，后墙的集热管收集的热量也开始减少，则转移到土壤中的热量也减少，随着转移热量的减少和高温土壤与周围低温土壤之间热量的传递，60cm 土壤的温度开始慢慢降低到 13℃ 以后维持稳定。实验温室 60cm 处土壤最高温度为 20.9℃，最低温度为 13℃，平均温度为 15.0℃。对照温室 60cm 处土壤温度白天低，最低温是在 15：00 左右为 8.0℃；夜间高，最高温室是 11.5℃，在揭开保温被之前 8：00 左右，平均温度为 10.0℃。两温室平均温差为 5℃（图9）。

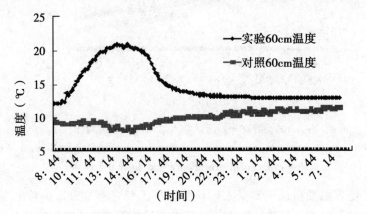

图9 实验温室与对照温室土壤 60cm 处温度对比

Figure 9 Soil temperature comparison at 60cm between experiment greenhouse and normal greenhouse

表1 为试验温室白天蓄热夜间放热时土壤分层平均温度和对照温室自然蓄放热时土壤分层平均温度。试验温室白天地表通过太阳的辐射直接蓄热，60cm 处土壤通过水的循环将热量收集并储存蓄积，两种蓄热方法都增加了土壤温度，使试验温室 0cm 和 60cm 温度均高于 30cm 处。对照温室白天只通过太阳辐射蓄热，温室从 0cm 到 60cm 土壤温度逐渐降低。白天增加了试验温室的蓄热量，夜间试验温室各层土壤温度均高于对照温室。

表1 试验温室与对照温室土壤分层平均温度

Table 1 Average temperature of soil layers in experiment greenhouse and normal greenhouse respectively

	试验温室（℃）		对照温室（℃）	
	白天蓄热	夜间放热	白天蓄热	夜间放热
0cm	31.5	17.4	28.9	14.4
30cm	13.4	16.5	10.6	13.3
60cm	17.7	13.3	8.9	10.6

2.2 集热效率分析

集热器的集热性能由集热效率决定，集热效率可通过水获得的热量与照射在集热器上的辐射量的比值求得[16~18]。

$$\eta_c = \frac{Q_u}{A_c I_c} = \frac{m_c c_p (T_{out} - T_{in})}{A_c I_c} \times 1\,000$$

式中：Q_u——瞬时有用集热量，kJ；

η_c——瞬时集热效率；

A_c——有效集热面积，m²；

I_c——投射至集热器采光面积上的太阳辐射照度，W/m²；

m_c——质量流量，kg/s；

c_p——水的定压质量比热，4.1867kJ/kg.℃；

T_{in}——集热器进口温度，℃；

T_{out}——集热器出口温度，℃；图10为集热器进出水温度与室内太阳辐射照度。从集热器进出水口温度可以看出在上午9：00以前和下午15：30以后集热效果不明显，所以晴天最佳集热时间是9：00~15：00。根据公式可以算出集热器的瞬时集热效率，如图11所示。集热器的平均集热效率31.1%。

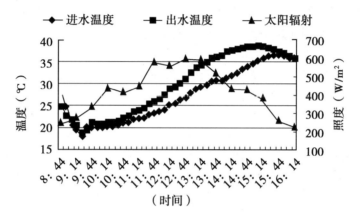

图10 集热器进出水温度与室内太阳辐射照度

Figure 10 Water temperature in and out pipes and Solar radiation in greenhouse

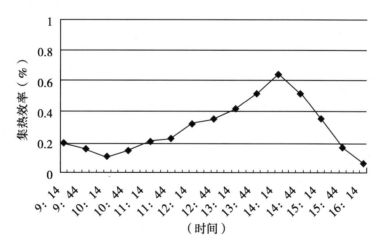

图11 集热器瞬时集热效率分析

Figure 11 The instantaneous efficiency analysis on heat collection

3 结论与讨论

(1) 本试验以太阳能为能源,以日光温室后墙集放热器中的水为蓄热介质,以日光温室中的浅层土壤为蓄热体,白天通过水的循环将温室中的热量转移并储存到浅层土壤中,以提高温室蓄积的热量,下午盖上保温被后,浅层土壤蓄积的热量又源源不断的释放到温室中,以提高温室低温时段的温度,蓄热效果明显。

(2) 通过白天的蓄热,温室浅层土壤的温度升高明显,60cm 深度土壤温度平均升高 5℃,30cm 处土壤温度平均升高 3℃。由于土壤蓄热量增加,温室夜间空气温度也上升,与对照相比温提高加 4.0℃。

(3) 温室集热器的平均集热效率为 31.1%,集热效率偏低。在今后实验中需改进集热器集热材料,如采用热阻小的材料,增加材料的透光率等,以进一步提高集热器的集热效率。

(4) 在阴天或是雨雪天气,该装置应用效果不明显。在今后试验中可增加辅助加热设备,以满足增个冬季作物生长对温度要求。

参考文献

[1] 亢树华,房思强,戴雅东等. 节能型日光温室墙体材料及结构的研究. 中国蔬菜. 1992 (6):1~5
[2] 杨仁全,马承伟,刘水丽等. 日光温室墙体保温蓄热性能模拟分析. 上海交通大学学报,2008,26 (5) 449~453
[3] Bai Yikui, Liu Wenhe, Wang Tieliang, et al. Experimental research on environment and heat preservation effect of solar greenhouse: type Liaoshen I. 农业工程学报,2003,19 (5):191~196
[4] 李小芳,陈青云. 日光温室山墙对室内太阳直接辐射得热量的影响. 农业工程学报 2004,20 (5):241~245
[5] 张立芸,徐刚毅,马承伟等. 日光温室新型墙体结构性能分析. 沈阳农业大学学报,2006,37 (3):459~462
[6] 马承伟,卜云龙,籍秀红等. 日光温室墙体夜间放热量计算与保温蓄热性评价方法的研究. 上海交通大学学报,2008,26 (5) 411~415
[7] 王晓东,马彩雯,吴乐天等. 日光温室墙体特性及性能优化研究. 新疆农业科学,2009,46 (5):1 016~1 021
[8] 王顺生. 日光温室内置式太阳能集热调温装置的研究. 北京:中国农业大学,2006
[9] H. Ali-Hussaini, K. O. Suen. Using shallow solar ponds as a heating source for greenhouse in cold climates. Energy Convers. 1998,9 (13):369~1 376
[10] Ì. Grmiadellis, E. Traka - Mavrona. Heating greenhouses with solar energy-new trends and developments CIHEAM - Options Mediterraneennes vol. 31:119~134
[11] 山本雄二郎. 地中—空气热交换的应用例(日). 农业气象,1996,2 (2):77~79
[12] 高仓直,山川健一. 地中热交换设计(日),农业气象,198 1,37 (3):187~196
[13] 马承伟,黄之栋,穆丽君. 连栋温室地中热交换系统贮热加温的试验. 农业工程学报,1999,15 (2):160~164
[14] 张峰,张林华,刘文波等. 带地下卵石床蓄热装置的日光温室增温实验研究. 可再生能源,2009,27 (6):7~9
[15] 马春生. 主动式温室太阳能地下蓄热系统的研究. 山西:山西农业大学,2003
[16] 王奉钦. 太阳能集热器辅助提高日光温室地温的应用研究. 北京:中国农业大学,2004
[17] 赵玉兰,张红,张战栋等. CPC 热管式真空管集热器的集热效率分析. 太阳能学报 2007,28 (9):1 022~1 025
[18] 孟力力. 基于 VB 和 MATLAB 的日光温室热环境模型构建与结构优化. 北京:中国农业科学院,2008

华北地区几种日光温室墙体热流量测定与分析[*]

胡 彬[**]，马承伟[***]，王双瑜，阳 萍，徐 凡，曹晏飞

（中国农业大学农业部设施农业生物环境工程重点开放实验室，北京 100083）

摘要：为了研究日光温室墙体的放热与吸热情况，测定了华北地区几种温室墙体内表面热流量。热流量数据按每日和测试全期进行整理，统计分析了每个测点的日平均和最大蓄热流量及出现时间、持续时间，日平均和最大放热流量及出现时间和持续时间，并统计全部点的平均值和最大值，以此来进行分析墙体热流的变化规律、温室的保温性能。统计数据表明，大跨度、厚土墙、下沉温室的墙体具有较高蓄热和放热流量；墙体蓄热时间主要在太阳辐射较好的9：00~15：00；温室墙体在蓄热时间段的蓄热流量平均为110~150W/m²左右，在放热时间段的放热流量平均为20~50W/m²，通过简单估算得出根据日光温室的类型不同，墙体在夜间的放热可以使温室室内提高2.08~10℃。

关键词：日光温室；墙体；热流量；测定

Determination and Analysis of heat Absorption and Release of Wall in Several Kinds of Solar Greenhouses in Northern Region of China

Hu Bin[**], Ma Chengwei[***], Wang Shuangyu, Yang Ping, Xu Fan, Cao Yanfei

(*Key Laboratory of Agricultural Bio-Environmental Engineering, Ministry of Agriculture, China Agricultural University, Beijing* 100083, *P. R. China*)

Abstract: The study observed heat release of wall of greenhouse, and measured heat flux of several kinds of walls in northern region of China. The data of heat flux is analyzed in everyday and in the entire time of the tests. The outcomes of data analysis are the average and maximum regeneration current capacity of every measure point, and its epoch and duration in everyday. At the time, the average and the maximum exothermic current capacity of every measure point and its epoch and duration in everyday were also included. Moreover, the mean and maximum value of all measure points had been calculated. The outcomes of our tests can be used to reveal the rules of heat flow of the wall and evaluate the function of heat preservation of the greenhouse. The study can conclude that a sunken greenhouse with a great span, a thick soil wall, has the superiority in heat preservation, and the time of wall's heat absorption is about 9 o'clock to 15 o'clock. The wall of greenhouse can absorb about 110 ~ 150W/m² equally in aborting time, and can release about average 20 ~ 50W/m² in the releasing time. At last, it can predict that the temperature of the greenhouse could be improved about 2.08 ~ 10℃ by the energy released from the walls in different kinds of solar greenhouses.

Key words: Solar greenhouse; Wall; Heat flow; Determination

日光温室以其特殊的围护结构，能够充分利用太阳能、最大限度地保存室内热量从而有

---

[*] 基金项目：现代农业产业技术体系建设专项资金资助（Nycytx-35-gw24）。
[**] 作者简介：胡彬（1988— ），女，湖北黄冈人，在读硕士研究生。E-mail：hubincau@ yahuoo. cn。
[***] 通讯作者：马承伟（1952— ），男，重庆人，教授，博士生导师。E-mail：macwbs@ cau. edu. cn。

效地节省能源，基本实现越冬生产而在我国发展迅速。墙体作为日光温室的重要围护结构，日光温室的主要蓄热体[1]，对日光温室内热环境有着直接的影响。陈端生[2]通过测定分析认为，日光温室最理想的墙体结构是：内侧由吸热、蓄热较好材料组成蓄热层，外侧由导热、放热较差材料组成保温层，中间设隔热层。

对于日光温室的墙体的温度，国内学者做了较多的理论研究与实验设计[2~4]，郦伟等[5]就日光温室温度分布曾建立其热环境模型。但是目前对于传热的理论研究和实验研究还不是很成熟。佟国红等[6]采用频率响应法，李小芳等[7]采用了反映系数法，孟力力、杨其长等[8]采用离散化的分层计算传热的方法，佟国红等[9]采用CFD软件Fluent等方法，马承伟等[10]采用了一维差分模型与数值模拟的方法，王谦等[11]用传热学中平板表面对流换热和物体内部热传导传热计算方法，分别模拟和分析了日光温室墙体的传热的方法，为日光温室内传热的理论研究分析和实际建造提供了宝贵的经验。但是有时实际情况中日光温室的情况复杂，空气、墙体等的传热系数难以测定，通过温度计算出的热流与实际有一定的差别。

掌握墙体中热流的状况及其变化规律，对于日光温室墙体的蓄热保温性能分析评价、设计与建造有着重要的意义。李建设等[12]，与白青[13]对日光温室的墙体热流量进行了测定，选取的温室为宁夏地区的土质墙体日光温室，对土质日光温室墙体昼夜放热关系，及与太阳辐射的关系进行了研究，对于日光温室内热环境的研究提供了参考。但是由于实验条件、时间、地域等的局限，要全面掌握日光温室墙体中热流的状况及其变化规律，还需要进行更多、更全面和系统的观测和研究分析。

为此，本实验选取了华北地区的北京市（通州）、天津市（西青区杨柳青）、山东省（寿光两处）等地4栋典型日光温室，在冬季气温最低的时期，对其墙体内表面热通量进行了测定。

1 实验测试内容、方法与仪器

2010年12月底至2011年2月底，对华北地区北京市、天津市、山东省等地4处典型日光温室进行了墙体内表面热流量的测定。

仪器为北京世纪建通技术开发有限公司JTNT-C多通道热流温度检测仪，该仪器为16通道热流、32通道温度多通道测试仪。测试时各热流板采用黄油将紧贴于内墙面，仪器自动记录各时刻的墙体内表面热流量，测试时间间隔均为10min。

测试结果数据主要分为蓄热与放热两个时间段来进行统计。热流量数据的整理主要按每日和测试全期，统计分析每个测点的日平均和最大蓄热流量及出现时间、持续时间，日平均和最大放热流量及出现时间和持续时间，并统计全部点的平均值和最大值。

2 实验温室测点布置与测试结果

2.1 一号试验温室的测试

试验温室位于北京市通州区北京市富通环境工程公司生产基地。温室长50m，净跨7.2m，后墙高2.15m，脊高3m，东西山墙及北墙为从内到外240mm红砖+10cm聚苯板+240mm红砖，后屋面主要材料为聚苯板。温室前屋面夜间覆盖保温被保温。测试期间，日光温室内种植甘蓝、油麦菜等叶菜类蔬菜。温室平均5、6天温室灌水一次，测试剖面未灌水。

测试点设在靠近温室中部的位置，为由西向东21m处，共布置3个测点，分别位于高度0.5m（1号测点）、1.0m（2号测点）、1.5m处（3号测点）。

测试时间：2010年12月26日13点30分至2011年1月3日11点10分。测试全期的统计数据如表1所示。

表1　一号温室测试全期墙体热流统计数据

测点	平均蓄热流量（W/m²）	持续时间（h）	平均放热流量（W/m²）	持续时间（h）
1	70.10	5.5	13.99	17
2	208.50		37.01	
3	62.74		12.66	
平均	113.78		21.22	

从测试结果可以看出：

（1）该温室墙体在5.5h的蓄热时间段能够吸收平均110W/m²左右的热流，在17h的放热时间段能够释放平均20W/m²左右的热流。

（2）在天气晴好的情况下，温室大概在9:10左右到下午15:00左右时间段内蓄热，在15:30~9:00时间段内放热；阴天由于晚揭帘和早盖帘，使蓄热时间减少，放热时间加长。

2.2　二号试验温室的测试

试验温室位于天津市西青区杨柳青菜博园内。温室长60m，净跨7.3m，后墙高度为2.55m，脊高3.4m。东西山墙及北墙为从内到外240mm红砖+10cm聚苯板+240mm红砖。后屋面由聚苯板、三合土、毛毡等组成。温室前屋面夜间覆盖保温被保温。测试期间，日光温室种植芹菜。测试期间温室内未灌水。

测试点设在靠近温室中部的位置，为由西向东25.5m处。共布置4个测点，分别位于高度0.5m、1.5m、2.0m和2.5m（依次为1~4号测点）处。

测试时间：2011年1月4日至2011年1月13日。测试全期的统计数据如表2所示。

表2　二号温室测试全期墙体热流统计数据

测点	平均蓄热流量（W/m²）	持续时间（h）	平均放热流量（W/m²）	持续时间（h）
1	118.56	6	23.20	16
2	95.36		14.47	
3	98.84		23.30	
4	42.55		8.68	
平均	88.83		17.41	

从测试结果中大致可以得出：

（1）在天气晴好的情况下，温室大概在 8:50 左右到下午 15:00 左右时间段内蓄热，在 15:30～8:00 左右放热；阴天由于晚揭帘和早盖帘，使蓄热时间减少，放热时间加长。

（2）数据显示该温室墙体在 6h 的蓄热时间段能够吸收大约为 $130W/m^2$ 左右的热流，在 16h 的放热时间段能够达到大约 $30W/m^2$ 左右的热流。

2.3 三号试验温室的测试

试验温室位于山东省寿光市。温室坐北朝南，东西延长，长度 69m，净跨 12.5m，后墙剖面为梯形，材质为土，高度 3m，上宽 2.2m，下宽 6m，脊高 3.9m，下沉 1.6m。后屋面从内到外的材料为土壤、棉毡和草帘。测试期间，日光温室种植黄瓜。温室温度达到 28℃ 以上进行通风。

测试点设在靠近温室中部的位置，为由西向东 34.5m 处，共布置 3 个测点，分别位于高度 1.0m、1.5m 和 2.0m（依次为 1～3 号测点）处。

测试时间：2011 年 1 月 14 日至 2011 年 1 月 22 日。测试全期的统计数据如表 3 所示。

表 3　三号温室测试全期墙体热流统计数据

测点	平均蓄热流量（W/m^2）	持续时间（h）	平均放热流量（W/m^2）	持续时间（h）
1	103.67	5.5	29.37	17.5
2	241.30		79.22	
3	105.94		32.03	
平均	150.31		46.88	

从测试结果中大致可以得出：

（1）在天气晴好的情况下，温室大概在 9：00～15：00 时间段内蓄热，在 15：30 到翌日早上 8：30 左右放热；天气状况不好的时候，由于晚揭帘和早盖帘，使蓄热时间减少，放热时间加长。

（2）数据显示该温室墙体在 5.5h 的蓄热时间段能够吸收大约 $150W/m^2$ 左右的热流，在 17.5h 的放热时间段能够达到大约 $47W/m^2$ 左右的热流。

2.4 四号试验温室的测试

试验温室位于山东省寿光市。温室坐北朝南，东西延长，长度为 88m，净跨 11m，后墙为梯形，后墙高 2.8m，上宽 2m，下宽 5m，脊高 3.5m 下沉 1.0m，后屋面从内到外的材料为土壤、棉毡、草帘。测试期间，室内种植辣椒等。温室温度达到 28℃ 以上进行通风。

测试点设在了靠近温室中部的位置，为由西向东 30m 处。共布置 5 个测点，分别位于高度 1.0m、1.5m 和 2.0m（依次为 1～3 号测点）处。

测试时间：2011 年 1 月 22 日至 2011 年 2 月 21 日。测试全期的统计数据如表 4 所示。

表4 四号温室测试全期墙体热流统计数据

测点	平均蓄热流量（W/m²）	持续时间（h）	平均放热流量（W/m²）	持续时间（h）
1	71.51		19.92	
2	235.11	6	66.59	17
3	67.08		17.06	
平均	124.57		34.52	

从测试结果中大致可以得出：

①在天气晴好的情况下，温室大概在9:00~15:00时间段内蓄热，在15:30到翌日8:30左右放热；天气状况不好的时候，由于晚揭帘和早盖帘，使蓄热时间减少，放热时间加长。

②数据显示该温室墙体在蓄热时间段能够吸收大约为124W/m²左右的热流，在放热时间段能够达到大约35W/m²左右的热流。

3 分析与讨论

3.1 晴阴天墙体热流量日变化

为了更清楚地了解日光温室墙体蓄热和放热过程，需要掌握其在不同天气情况下，全天内表面热流量的变化情况。以下为从测试数据中摘取的具有代表性的天气下墙体内表面热流量日变化。

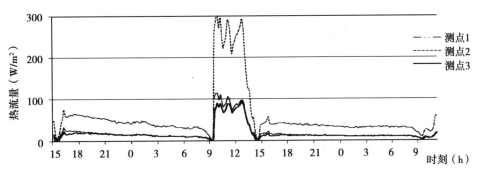

图1 一号温室晴天墙体热流量日变化（1月2日）

从测试结果可以看出

（1）墙体蓄热时间主要在太阳辐射较好的9:00~15:00时间段。

（2）在放热时间段的开始一个小时内，放热热流量会达到一个顶峰值，并且释放出较多的热量，之后放热量会逐渐趋于平缓。如果能够采取一定的措施使热量不在开始的短时间内流失太多，是否可以存贮到夜间外界气温降低时再释放，以提高温室的越冬能力。

（3）从各温室的数据来看，均为墙体上的测点2的峰值较高，主要原因为测点2处于墙体高度的中部，一般在太阳直射的位置，能够接收到较多的太阳辐射。

（4）从图3可以看出，四号温室在阴天太阳辐射很微弱的环境下，全天未揭帘，几乎没有接收太阳辐射，因此，墙体夜间和白昼均处于放热的状态，平均热流还能达到

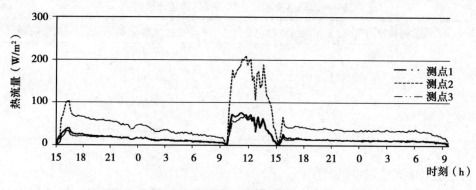

图 2　一号温室阴天墙体热流量日变化（12 月 29 日）

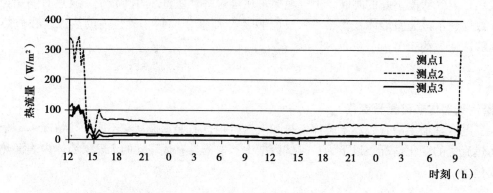

图 3　四号温室遇恶劣天气整天不揭帘墙体热流日变化（1 月 28 日）

31.18W/m^2，和一号、二号温室的夜间平均热流接近。说明墙体具有较强的蓄热和持续放热的能力，阴天能够在白天蓄热很少甚至完全未蓄热的情况下，释放在前些时期天气晴好时蓄积、未能全部释放的热量。

3.2　温室蓄热保温性能评价

对温室内温度的数据按照墙体蓄热与放热时期进行统计分析，测试全期的统计数据如表 5 所示。

表 5　四个温室测试全期温度统计数据（℃）

时期	统计项目	一号温室	二号温室	三号温室	四号温室
墙体蓄热时期	室内平均温度	16.18	19.04	25.2	25.98
	室外平均温度	-2.37	-2.39	-1.36	3.15
	室外温差	18.55	21.43	26.56	22.84
墙体放热时期	室内平均温度	5.29	6.06	14.8	15.9
	室外平均温度	-5.51	-7.04	-7.62	-4.61
	室外温差	10.8	13.1	22.42	20.51

从测试结果来看，三号温室与四号温室无论是在放热时间段平均热流量还是夜间温室内

最低温度、夜间平均室内外温差,都要优于一号、二号温室。一号、二号温室为砖墙加聚苯板,三号、四号温室为厚土墙、下沉、大跨度,所以,看来下沉、大跨度、厚土墙日光温室在一定的程度上比红砖加聚苯板的墙体确实有优势,能够提供较高的夜间温室内温度。

从四个温室的统计数据来看,无论是蓄热时间段还是放热时间段,平均热流量都与室外温差有着一定的正相关性,同时夜间室外温差大温室,在白天的室外温差也较大。主要原因在于温室的保温性较好,在白天的时候吸收热量,贮存在温室中,达到小环境中的热流交换,比外界温度要高。在夜间的时候,保温性较好的温室能够将白天蓄热的热量释放出来,少受外界低温的影响。

通过查阅三号温室与四号温室的室内太阳辐射测定的数据,可以判断当日揭帘与盖帘的时间,从数据中更可以看出放热时间比温室盖上保温被的时间要长,往前提到半小时左右,往后推迟半小时左右。主要原因为揭帘后半小时内,室外温度和太阳辐射还不够强,墙体表面仍然需要放热,在盖帘前半小时内由于此时室外的温度和太阳辐射已经不太强了,墙体表面由于通过吸热温度较高,此时已经开始放热了。

通过对墙体内表面布置的测点的结果可以看出,墙体内表面中中部的测点的蓄热流量是其上下部位的 2~3 倍,主要原因为墙体中部处于太阳光可直射的部位,同时墙体下部由于潮湿和与温室地面相连,墙体上部与后屋面相连都会对墙体吸放热有影响。

在三号温室和四号温室通过对每日的热流量进行分析时,发现连阴天开始的几天内夜间的平均热流量比连晴天开始的几天内夜间的平均热流量要大。这是因为连阴天开始的阶段墙体有前些晴好天气所蓄积的热量,而连晴天开始的阶段墙体过往蓄积的热量基本在前些时段的阴天释放完毕。

4 结论

从两个月中对华北地区几种日光温室的墙体内表面热流量及室内外温度的测定,可以得出以下结论:

(1) 温室墙体在蓄热时间段的蓄热流量平均为 110~150W/m² 左右,在放热时间段的放热流量平均为 20~50W/m²。

(2) 根据日光温室的类型不同,墙体在夜间的放热可以使温室室内提高 2.08~10℃。

(3) 下沉、大跨度、厚土墙日光温室在一定的程度上比红砖加聚苯板的墙体确实有优势,能够提供较高的夜间温室内温度。

(4) 天气晴好时墙体蓄热时间主要在太阳辐射较好的 9:00~15:00 时间段,天气状况不好的时候,由于晚揭帘和早盖帘,使蓄热时间减少,放热时间加长。

(5) 放热时间比温室盖上保温被的时间要长,往前提到半小时左右,往后推迟半小时左右。

(6) 墙体内表面中部测点的蓄热流量是其上下部位的 2~3 倍。

(7) 连阴天开始的几天内夜间的平均热流量比连晴天开始的几天内夜间的平均热流量要大。

参考文献

[1] 陈端生. 中国节能型日光温室建筑与环境研究进展. 农业工程学报,1994,10(1):123~129

[2] 陈端生. 日光温室气象环境综合研究 I 墙体、覆盖物热效应研究初报. 农业工程学报, 1990, 6 (2): 77~81
[3] 白青, 张亚红, 刘佳梅. 日光温室土质墙体内温度与室内气温的测定分析. 西北农林学报, 2009, 18 (6): 332~337
[4] 亢树华, 戴雅东, 房思强等. 日光温室优型结构的研究. 农业工程学报, 1996 (增), 30~35
[5] 郦伟, 董仁杰, 汤楚宙等. 日光温室的热环境理论模型. 农业工程学报, 1997, 13 (2): 161~163
[6] 佟国红等. 日光温室墙体传热及节能分析. 农业系统科学与综合研究, 2003, 5 (19): 101~105
[7] 李小芳, 陈青云. 墙体材料及其组合对日光温室墙体保温性能的影响. 中国生态农业学报, 2006, 14 (4): 185~189
[8] 孟力力, 杨其长, Gerard. P. A. Bot 等. 日光温室热环境模拟模型的构建. 农业工程学报, 2009, 25 (1): 164~170
[9] 佟国红, 李保明, David M Christopher 等. 用 CFD 方法模拟日光温室温度环境初探. 农业工程学报, 2007, 23 (7): 178~185
[10] 马承伟等. 日光温室墙体传热的一维差分模型与数值模拟. 农业工程学报, 2010, 6 (26): 231~235
[11] 王谦等. 冬季日光温室北墙内表面热流分析. 中国农业气象, 2010, 31 (2): 225~229
[12] 李建设等. 日光温室墙体与地面吸放热量测定分析. 农业工程学报, 2010, 4 (26): 231~235
[13] 白青. 土质墙体日光温室保温性及室内黄瓜群体结构参数分析: [硕士学位论文]. 宁夏: 宁夏大学: 2009

日光温室墙体材料对墙体温度分布及室内温度的影响

佟国红*

(沈阳农业大学水利学院,沈阳 110161)

摘要:日光温室的墙体材料影响温室的蓄热保温性能。采用实地测试试验和数值模拟的方法对同厚度不同材料墙体温度进行了研究。试验及模拟结果表明夜间复合墙中隔热层以内的砖墙及部分隔热层成为放热体,砖墙中靠近室内近1/3的墙体温度高于室内空气温度,而全苯板墙只有内表面附近略高于室内空气温度。白天复合墙、全砖墙及全苯板墙温度均低于室内空气温度,均为吸热体。研究还表明单一材料墙体的温室中,加气混凝土墙室内温度高于砖墙和钢筋混凝土墙温室,但钢筋混凝土墙温室全天的温度波动最小,热稳定性好;只改变复合材料墙体的蓄热材料,对室内的温度影响不明显。

关键词:日光温室;墙体;试验、CFD;温度

Temperature Variations for Various Wall Materials in Chinese Solar Greenhouses

Tong Guohong*

(College of Water Conservancy, Shenyang Agricultural University, Shenyang 110866, P. R. China)

Abstract: Wall materials play an important role in heat conservation for Chinese solar greenhouses. Wall temperature distributions and variations were measured and simulated. The results show that the thermal sources in the three types of greenhouses consisted of the inside layer of brick and part of the insulating layer for the layered wall, one-third of the inside wall for the brick wall and only the area just inside the wall surface for the Styrofoam wall. The wall temperatures inside the layered wall, inside the brick wall and inside the Styrofoam wall were below the air temperatures inside the greenhouses during the day with the thermal blanket unrolled. Predicted air temperatures for various wall designs show that for single walls, the daily average interior temperatures in the aerated concrete wall greenhouse were higher than in the brick wall and reinforced concrete wall greenhouses. However, the air temperature fluctuations were lower in the reinforced concrete wall greenhouse due to greater thermal storage capacity. The results also show that the temperatures in the layered wall greenhouses are quite similar.

Key words: Solar greenhouse; Wall; Measurement; CFD; Temperature

日光温室墙体在温室的蓄热保温方面起到非常重要的作用,因此近20年来对于日光温室墙体的研究一直是热点。中国专家、学者对于日光温室墙体的组成、墙体的厚度和高度等

---

* 作者简介:佟国红(1966—),女,副教授,博士,从事设施农业建筑与环境的研究。沈阳农业大学水利学院,110866。E-mail: guohongtong@yahoo.com.cn。

进行了多方面的探讨，陈端生等[1]通过测试土墙及复合墙体内的温度，发现土墙全天从室内吸热，而复合墙白天蓄热，夜间向室内放热，并指出日光温室较理想的墙体结构是内侧为蓄热层，中间为隔热层，外部为保温层。亢树华等[2]分别对采用4种不同隔热材料复合墙体、不同厚度土墙的温室内温度环境进行了测试研究，结果表明490mm厚砖墙内部隔热材料以珍珠岩最好，土墙以1m厚适宜。张立芸等[3]测试了复合墙分别为砖加苯板以及加气混凝土砌块加苯板温室的室内温度。李小芳和陈青云[4]比较了墙体材料的不同组合对室内温度环境的影响。马承伟等[5]以墙体夜间放热量作为评价日光温室墙体保温蓄热性能评价的指标对不同材料及做法的墙体进行了比较。杨建军等[6]在试验测试墙体表面及室内空气温度基础上，结合温室建造成本及土地利用率提出西北地区几个典型城市最佳土墙厚度。温祥珍等[7]通过模型研究表明墙体高度对日光温室内夜间气温的影响明显，随墙体高度增加，温室保温性改善。

1 材料与方法

为了探求墙体材料对墙体内部温度分布的影响，采用的研究方案包括实地测试试验方案和数值模拟方案。

1.1 试验方案

1.1.1 试验墙体方案

辽沈I型实验示范日光温室位于沈阳市郊，跨度为7.5m，长为89.25m，脊高为3.5m，后墙高为2.2m、厚为450mm，后坡仰角为32°。墙体所选两个方案分别为复合异质墙体（从内到外的顺序：240mm厚红砖墙+100mm厚聚苯板+240mm厚红砖墙），位于同一温室与之对比的墙体为450mm厚红砖墙墙体。

1.1.2 测试方案[8~9]

采用热敏电阻测试两个方案墙体的内部温度分布，测点距室内地坪1.2m，将测温探头按设定的距离砌入北墙中，两个方案墙体内测点1至测点6布置位置一致（图1），分别距室内墙体表面10mm，120mm，230mm，285mm，340mm和440mm，每小时采集一次数据。

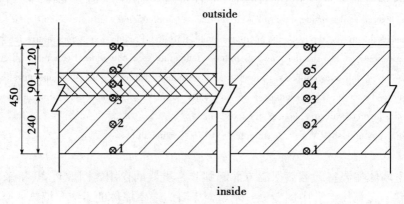

图1 墙体内测点分布（单位：mm）

Figure 1 Measurement positions across the sections of the wall with dimensions in mm

1.2 模拟方案
1.2.1 模拟墙体方案

日光温室跨度为12.0m，脊高5.5m，后墙高3.0m、厚0.6m，后坡水平投影2.5m，位于沈阳农业大学校园内（图2）。600mm厚复合异质墙体，由内向外材料依次为360mm红砖墙，50mm缀铝箔聚苯乙烯泡沫塑料板，20mm空气夹层，50mm缀铝箔聚苯乙烯泡沫塑料板，120mm红砖墙。

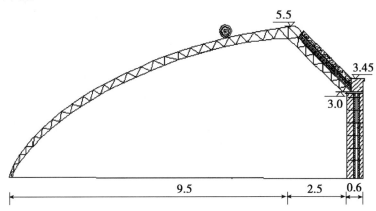

图2 模拟温室截面（单位：m）

Figure 2 Cross-section view of the simulated solar greenhouse with dimensions in m

模拟的墙体材料方案如下[10~11]：

（1）不同材料墙体 复合墙：由内向外材料依次为360mm红砖墙，50mm缀铝箔聚苯乙烯泡沫塑料板，20mm空气夹层，50mm缀铝箔聚苯乙烯泡沫塑料板，120mm红砖墙。该复合墙为12m跨日光温室原墙体。

厚重材料墙：600mm厚红砖墙。

轻质隔热材料墙：600mm厚聚苯乙烯泡沫塑料板（简称聚苯板）墙。

（2）单一材料墙体 将图2所示的0.6m厚墙体材料改变为单一墙体材料砖、钢筋混凝土及加气混凝土。

（3）复合材料墙体 砖复合墙体：由内向外材料依次为360mm红砖墙，50mm缀铝箔聚苯乙烯泡沫塑料板，20mm空气夹层，50mm缀铝箔聚苯乙烯泡沫塑料板，120mm红砖墙。该复合墙为12m跨日光温室原墙体。

加气混凝土复合墙体：由内向外材料依次为360mm加气混凝土，50mm缀铝箔聚苯乙烯泡沫塑料板，20mm空气夹层，50mm缀铝箔聚苯乙烯泡沫塑料板，120mm加气混凝土。

1.2.2 模拟方法[12~13]

（1）控制方程 在外界气候条件作用下，温室内的微气候环境发生着动态变化。本研究采用CFD软件（Fluent v.6.1）对温室内的温度环境进行二维非稳态模拟。通用的守恒方程为[14]：

$$\frac{\partial \rho\varphi}{\partial t} + div(\rho\varphi\vec{v}) = div(\Gamma_\varphi grad\varphi) + S_\varphi$$

式中：φ——为质量守恒、动量守恒、能量守恒方程的通用变量，φ为1时，该方程为连续方程；φ 为 u,v,w 时，该方程为动量方程；φ 为 T 时，该方程为能量方程。u,v,w——x,y,z 三个方向的速度，m/s；T——温度，K；ρ——流体密度，kg/m³；$\vec{v}$——速度向量，m/s；Γ_φ——广义扩散系数，m²/s；S_φ——源项，W/m³。

（2）边界条件 模拟时采用的边界条件为室外的气候条件（温度、湿度、风速、太阳辐射）以及温室下1.0m深土壤的温度。边界条件中考虑了温室结构外表面与天空的辐射热交换、温室结构外表面与室外气流的对流换热；考虑了室内各表面之间的辐射热交换、室内空气与室内各表面的对流换热；考虑了室内外空气的显热及潜热交换；考虑了冷凝及作物蒸腾产生的潜热影响。

2 结果与分析

2.1 同一温室不同墙体材料墙体内温度

图3为同一温室不同墙体材料墙体内温度分布及逐时变化情况。

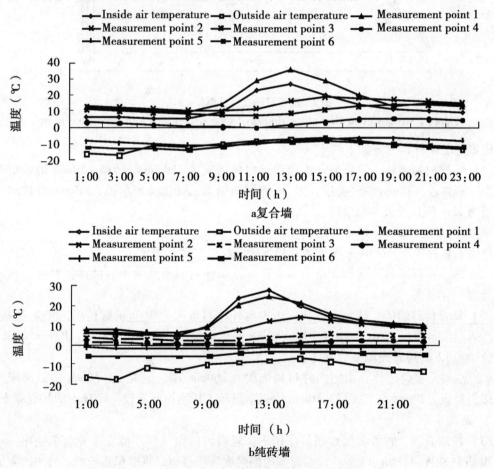

图3 同一温室不同墙体材料墙体内温度变化[9]

Figure 3 Temperature variations inside the two kind of walls[9]

复合墙和纯砖墙位于同一温室内，纯砖墙内表面与复合墙内表面（测点1）温度夜间均高于室内气温，且在 0：00～7：00 及 20：00～23：00 时间段二者温差为 3.4～4.2℃，平均达到 3.7℃，可见复合墙白天蓄热多，夜间与纯砖墙相比，向室内放出更多的热量；纯砖墙夜间只有内表面温度（测点 1）高于室内气温，而复合墙的测点 1、测点 2、测点 3 夜间温度均高于室内气温，说明聚苯板的隔热效果好，减少了热量的损失，使聚苯板以内的墙体夜间成为热源。

另外，由于聚苯板的隔热效果好，使复合墙外砖墙的温度均低于纯砖墙相应位置的温度。

2.2 不同材料墙体温度分布[10]

为分析墙体内温度分布及对室内的热贡献，选择一天中室内温度最低的 7：00 温度分布图（图 4）。为便于观察墙体内的温度变化，图 4 中有放大的后墙底脚处的详图。

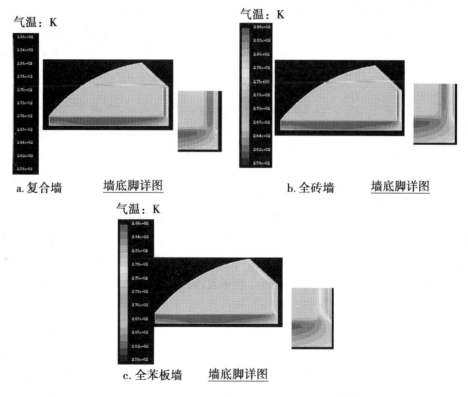

图 4 在 7：00 时不同墙体温室内的温度分布[10]

Figure 4　Temperatures for greenhouses with different walls at 7：00[10]

图 4 所示 7：00 复合墙、全砖墙及全苯板墙 1.0m 高度内表面温度分别为 283.4K，283.4K，280.6K，均高于其对应的室内空气温度，均向室内放热。因为复合墙内有隔热层，使得隔热层以内的砖墙及部分隔热层的温度均高于室内空气温度，成为放热体，并且隔热层以外的砖墙平均温度与全砖墙、全苯板墙相应部分平均温度相比低 5.9℃、1.7℃，说明隔热层阻隔墙体向外传热的作用非常明显。全砖墙及全苯板墙内的温度梯度不及复合墙明显，但由于砖墙的蓄热作用，使得靠近室内近 1/3 的墙体温度高于室内空气温度，而全苯板墙只有内表面附近略高于室内空气温度。复合墙、全砖墙及全苯板墙体均温为 278.5K，279.2K

及274.6K，均低于相应的室内空气温度，但各墙体表面温度均高于室内空气温度，是向室内放热，由此可见，不能用墙体的均温代替墙体温度来分析墙体的蓄、放热。

2.3 单一材料墙体及复合材料墙体对室内温度的影响[11]

单一材料墙体及复合材料墙体对室内温度的影响见图5。

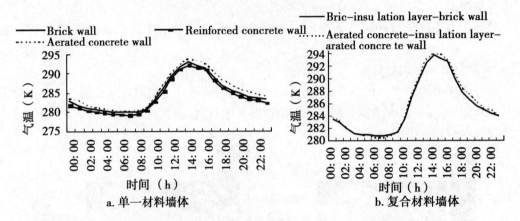

图5 单一材料墙体及复合材料墙体变化对室内温度的影响[11]

Figure 5 Predicted air temperatures inside solar greenhouses with single and layered walls[11]

图5a表明在三种单一材料墙体的温室中，加气混凝土墙室内温度高于砖墙和钢筋混凝土墙温室。加气混凝土墙温室温度与砖墙温室相比，全天平均高0.7℃；而加气混凝土墙温室温度与钢筋混凝土墙温室相比，全天平均高1.4℃。但图5a中还可看出钢筋混凝土墙温室全天的温度在三种单一材料墙体的温室中波动最小，热稳定性最好。

具有砖复合墙体与加气混凝土复合墙体的温室内温度相比较（图5b），除了在5:00~9:00时，砖复合墙体温室温度比加气混凝土复合墙体温室高出0.3℃以外，其他时间段，加气混凝土复合墙体温室温度均高于砖复合墙体温室，平均高约0.5℃，因此两者室内的温度变化不明显。

3 结论

综合不同墙体材料的试验及模拟结果，得出结论如下：

（1）夜间复合墙、全砖墙及全苯板墙内表面温度均高于室内空气温度，向室内放热。

（2）复合墙苯板以内的墙体夜间温度高于室内空气温度，成为夜间的热源。

（3）改变单一材料墙体的材料，可明显改变温室内的温度；但改变复合材料墙体的蓄热材料，对室内的温度影响不明显。

参考文献

[1] 陈端生，郑海山，刘步洲. 日光温室气象环境综合研究——墙体、覆盖物热效应研究初报. 农业工程学报，1990.6（2）：77~81

[2] 亢树华，戴雅东，房思墙等. 节能型日光温室墙体材料及结构研究. 中国蔬菜，1992，16：1~5

[3] 张立芸，徐刚毅，马成伟等. 日光温室新型墙体结构性能分析. 沈阳农业大学学报，2006，37（3）：459~462

[4] 李小芳，陈青云. 墙体材料及其组合对日光温室墙体保温性能的影响. 中国生态农业学报，2006，14（4）：

185~189

[5] 马承伟，卜云龙，籍秀红等. 日光温室墙体夜间放热量计算与保温蓄热性评价方法的研究. 上海交通大学学报（农业科学版），2008，26（5）：411~415

[6] 杨建军，邹志荣，张智等. 西北地区日光温室土墙厚度及保温性的优化. 农业工程学报，2009，25（8）：180~185

[7] 温祥珍，梁海燕，李亚灵等. 墙体高度对日光温室内夜间气温的影响. 中国生态农业学报，2009，17（5）：980~983

[8] 佟国红，王铁良，白义奎等. 日光温室墙体传热量计算及节能分析. 农业系统科学与综合研究，2003，19（2）：101~102，105

[9] 佟国红，王铁良，白义奎等. 日光温室墙体传热特性的研究. 农业工程学报，2003，19（3）：186~189

[10] 佟国红，Christopher David M. 墙体材料对日光温室温度环境影响的CFD模拟. 农业工程学报，2009，25（3）：153~157

[11] Tong Guohong, David M Christopher, Li Tianlai etc. Temperature variations inside Chinese solar greenhouses with external climatic conditions and enclosure materials. Int J Agric &LBiol Eng, 2008, 1（2）：21~26

[12] 佟国红，李保明，David M. Christopher等. 用CFD方法模拟日光温室温度环境初探. 农业工程学报，2007，23（7）：178~185

[13] Tong Guohong, Christopher David M., Li Baoming. Numerical modelling of temperature variations in a Chinese solar greenhouse. Computers and Electronics in Agriculture, 2009, 68（1）：129~139

[14] Versteeg H K, Malalasekera W. An Introduction to Computational Fluid Dynamics. England：Longman Group Ltd, London, 1995

山东寿光日光温室冬季热环境测试*

曹晏飞[1]**,张建宇[1],赵淑梅[1],马承伟[1],蒋程瑶[1],魏家鹏[2],桑毅振[2]

(1. 中国农业大学,北京 100083;2. 新世纪种苗有限公司,寿光 262700)

摘要:寿光是我国日光温室整体发展水平较高的地区,温室形式比较齐全,尤其是大跨度地面下沉型日光温室近年来有着广泛的应用。通过对传统非下沉型日光温室和新型大跨度下沉型日光温室的冬季温度、湿度环境连续测试,分析了典型晴天、阴天及连续阴雨雪天气条件下温室内部空气温度、湿度的变化规律,结果显示,在同样天气条件下,普通日光温室环境较下沉型日光温室更易受外界天气影响,昼夜温度波动较大,白天升温较快,最高温度也较高;而在连续阴雨雪天之后,下沉型日光温室的最低温度明显高于普通温室,显示下沉型日光温室具有较好的抗极端天气的能力。

关键词:下沉型日光温室;温度;湿度;测试

Thermal Environment Testing on Solar Greenhouse of Shandong Shouguang in Winter

Cao Yanfei[1], Zhang Jianyu[1], Zhao Shumei[1], Ma Chengwei[1], Jiang Chengyao[1], Wei Jiapeng[2], Sang Yizhen[2]

(1. *China Agricultural University*, *Beijing* 100083;
2. *New Century Seeding Co.*, *Ltd*, *Shouguang* 262700, *P. R. China*)

Abstract: Shouguang is developed in solar greenhouse in China and has many kinds of greenhouses; especially large-span sunken solar greenhouse is widely used in recent years. By continuously testing the temperature and humidity condition of traditional non-sunken solar greenhouse and new large-span sunken solar greenhouse in winter, analyzed the variation of the air temperature and humidity inside the greenhouse under the typical sunny, cloudy and continuously rainy or snowy condition. The results show that under the same weather condition, common solar greenhouse is more easily affected by outside factors whose temperature fluctuation between day and night is large and the maximum temperature is higher than other kinds of solar greenhouses. However sunken solar greenhouse has a good ability of anti-extreme weather that the minimum temperature of sunken solar greenhouse is higher than that of common solar greenhouse after continuously rainy and snowy weather.

Key words: Sunken solar greenhouse; Temperature; Humidity; Test

山东省寿光市是中国设施园艺整体发展水平较高的地区,设施蔬菜面积达到了 5.33 万 hm^2,其中 80% 为日光温室[1],因此也是中国日光温室发展较快、普及面积最广的地区之一。特别是近年来由当地农民根据经验摸索建造的大跨度地面下沉型日光温室,发展极为迅

---

* 基金项目:现代农业(大宗蔬菜)产业技术体系建设专项资金资助项目(Nycytx - 35 - gw24)。

** 第一作者简介:曹晏飞(1986—),男,湖南娄底,硕士,现从事设施农业工程研究工作。E-mail:bmxzbx@ 126. com。

速，有逐渐取代传统日光温室的趋势，因此受到广泛关注，一些学者也从开始不同角度对其进行了研究[2~7]，如王晓冬等[4]分析了下沉型日光温室对地温的影响，张峰等[6]对下沉型日光温室墙体的保温蓄热性能进行了探讨，佟国红等[7]利用模型计算了下沉型日光温室室内太阳能得热及温室向外部的散热，而系统测试下沉型日光温室实际温湿度环境状况的研究还比较少。为了从实际生产角度深入考察寿光地区典型日光温室内温湿度变化规律，本研究对不同形式日光温室进行了连续的现场测试，并在此基础上对晴天、阴天及连续阴雨雪天等典型气候条件下温室内温湿度变化规律进行了分析，为温室的建造设计及生产管理提供了参考。

1 材料与方法

1.1 试验温室

试验温室选择在山东省寿光市，测试时间为2010年至2011年两个冬季，其中，在2010年1月至3月期间，主要测试了一种普通日光温室和一种下沉1.5m的下沉型日光温室（分别编号为温室Ⅰ和温室Ⅱ），2010年12月至2011年3月期间则测试了一种普通日光温室以及下沉1.0m和1.6m两种下沉型日光温室（分别编号为温室1、温室2和温室3）。温室墙体均为土墙，采用EVA半无滴膜覆盖，外覆盖保温材料为草帘。详细构造参数见表1。

表1 测试温室的构造参数

温室构造参数	温室Ⅰ	温室Ⅱ	温室1	温室2	温室3
跨度（m）	7.6	12.0	8.5	11.0	12.5
长度（m）	47	52	62	88	69
脊高（m）	2.7	3.2	2.7	3.5	3.9
后墙高度（m）	1.95	2.5	1.75	2.8	3.0
墙体厚度	上下等厚，1.0m	上厚2.2m，下厚6.0m	上下等厚，1.0m	上厚2m，下厚5.0m	上厚2.2m，下厚6.0m
种植面下沉（m）	0	1.5	0	1.0	1.6
种植作物	辣椒	辣椒	辣椒	辣椒	黄瓜

1.2 试验方法

本试验主要考察各试验温室冬季室内温度、湿度环境状况，因此测试内容为室内外空气温度和相对湿度。为减少边界条件影响，室内测点布于温室长度及跨度方向的中部；同时为了使测试数据更具代表性，在长度方向距温室中线东西10m处各布一点，测点高度分别距地面0.8m和1.5m。室外设置一个测点。各个测点均采用了防辐射处理。测点布置如图1所示。测试仪器采用日本Esupekkumikku有限公司的RS-12温湿度记录仪，自动采集记录温度和湿度数据，数据采集记录间隔10min，测量精度为±0.3℃，±5%。

2 结果与分析

将室内两个测点同一时刻的测试数据求平均值，作为温室内温湿度的数值。由于试验测

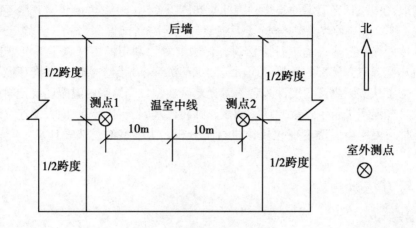

图 1　温室测点平面布置图

试周期较长，为进行针对性分析，特选取典型晴天、阴天以及连续阴雨雪天气条件下的测试结果进行讨论。根据试验期间的气象记录，2010 年 1 月 22 日为晴天，2010 年 3 月 22 日为阴天，2010 年 1 月 2 日至 1 月 5 日和 2011 年 2 月 26 日至 2 月 28 日为连续的阴雨雪天气。

2.1　典型晴天条件下的温湿度状况

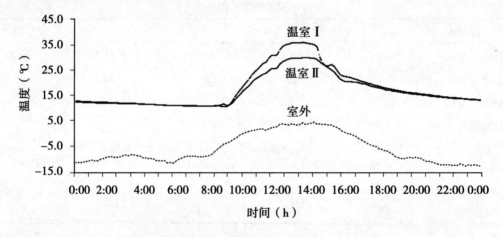

图 2　晴天室内外温度变化图（2010 年 1 月 22 日）

2.1.1　温度

由图 2 中可知，温室内外气温的日变化趋势一致，均昼高夜低。室外气温在夜间一直处于零度以下，5：30 达到 -11.3℃，然后气温开始回升，14：10 达到最高气温 4.7℃，室内外温差较大。温室Ⅰ室内温度在 9：00 时达到最低值 10.9℃，这是由于温室揭帘后，室内气温与外界气温之间的总热阻值减少，室内向室外散热量增加，而短时间内太阳辐射还较弱，从而导致室内气温下降。温室Ⅱ在 7：50 时达到最低温度 11.1℃，尽管在揭帘后，室内气温也有短时间的下降，但是下降幅度没有温室Ⅰ明显。同时室内气温较室外先达到最大值，温室Ⅰ在 13：30 时达到最高气温 36.0℃，温室Ⅱ在 13：50 时达到最高气温 30.1℃。然而辣椒在生长期适宜生长的最高温度为 28℃[8]，温室Ⅰ内室温超过 28℃ 的时间达到

3.3h,超过30℃的时间也达到了3h,这样的高温环境显然不利于作物的生长。

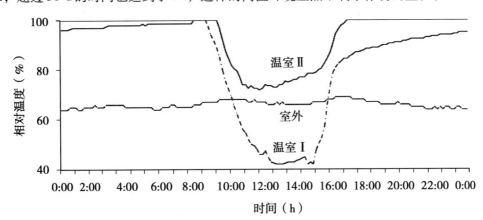

图3 晴天室内外相对湿度变化（2010年1月22日）

2.1.2 湿度

晴天室内外湿度状况如图3所示。室外空气相对湿度相对稳定,在60%~70%,平均为66%,与之相比,室内相对湿度波动较大,呈昼低夜高趋势。其中,盖帘期间,两个温室的相对湿度最高,均处于饱和状态;揭帘后,室内相对湿度开始降低,温室Ⅰ下降幅度较大,在12:40时达到最低为42%,而温室Ⅱ下降幅度较小,即相对湿度较高,11:40时达到最低时仍为72%。

2.2 典型阴天条件下的温湿度状况

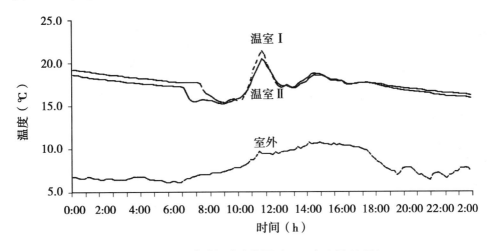

图4 阴天室内外温度变化图（2010年3月22日）

2.2.1 温度

阴天室内外温度日变化趋势与晴天基本相同（图4）,同样是昼高夜低。室外气温均处于0℃以上,在6:00达到最低值6.2℃,最高温度为10.8℃,出现在14:10,室内外温差较小。温室Ⅰ和温室Ⅱ的揭帘时间分别为8:00和7:00,尽管温室Ⅱ的揭帘时间早,但是两个温室的温度相差不大,且均在9:00时达到最低温度,分别为15.4℃和15.3℃。两个温室达到最高温度的时间也比室外要早,均出现在11:20时,分别为21.5℃和20.5℃。与晴天相比,阴天室内温度变化幅度小,这与陈端生[9]的研究相符,即光环境是影响温室气候环境的第一因素。

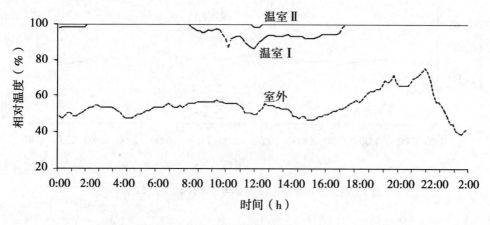

图5 阴天室内外相对湿度变化图(2010年3月22日)

2.2.2 湿度

与晴天相同,温室内相对湿度的变化趋势也是昼低夜高,盖帘期间的相对湿度均处于饱和状态(图5)。揭帘后,温室Ⅰ的相对湿度下降明显,最低值为87%,而温室Ⅱ的相对湿度几乎处于饱和状态,这样的高湿环境容易造成植物病害,不利于作物生长。

2.3 典型连续阴雨雪天气条件下温湿度状况

在温室的使用过程中,低温冻害经常在连续的阴雨雪天气时发生,连续阴雨雪天的天气状况最能考验温室是否能够维持适宜作物生长的温湿度环境,因此,对温室内温湿度在连续阴雨雪天时的变化进行分析显得极其重要。

2.3.1 温度

室内外空气温度在2010年1月2日至1月5日和2011年2月26日至2月28日的变化如图6和图7所示。

2010年1月2日、3日为小雪,4日阴天,5日为多云。从图6可知,室外空气温度一直处于0℃以下,在1月5日6:40达到最低值-15.4℃。受雨雪影响,温室在1月2日和3日均处于盖帘状态,室内温度不断降低,其中温室Ⅰ在1月4日7:00时的温度为8.6℃,温室Ⅱ同时刻温度为10.5℃,室内外温差分别为16.6℃和18.4℃。1月4日,由于阴天,温室升温幅度小,温室Ⅰ和温室Ⅱ在当天的最高温度分别为23.4℃和20.5℃。同时又受外界降温的影响,温室Ⅰ和温室Ⅱ在7:30达到最小值,滞后于室外最低温度出现的时间,其中温室Ⅱ的最低温度为8.8℃,比温室Ⅰ高1.9℃,说明温室Ⅱ抵抗连续阴雨雪天气的能力要强于温室Ⅰ。辣椒在生长期适宜生长的最低温度为10℃[8],在这段时间内,温室Ⅱ低于10℃的时间达到了9h,全部出现在1月5日,而温室Ⅰ低于10℃的时间长达25.5h,这样长时间的低温环境对于植物生长极为不利。

2011年2月26日为阴天,27日为小雪,28日多云。从图7可以看出,室外从26日下午开始降温,在28日4:40时达到最低温度-6.3℃。26日受阴天太阳辐射弱的影响,温

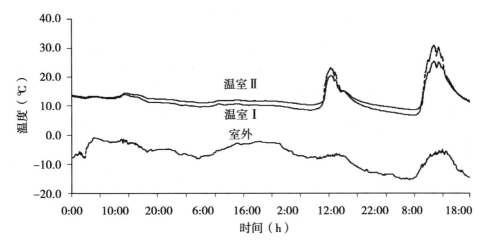

图6 空气温度在连续阴雨雪天的变化（2010年1月2日至1月5日）

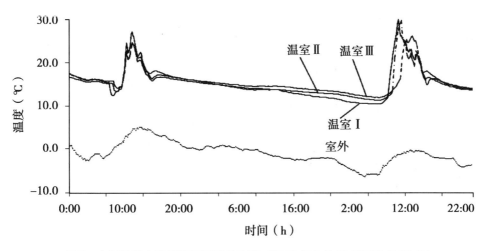

图7 空气温度在连续阴雨雪天的变化（2011年2月26日至2月28日）

室升温幅度较小，温室Ⅰ、温室Ⅱ、温室Ⅲ在当天的最高温度分别为27.3℃、24.7℃、24.4℃。27日降雪，全天没有揭帘，温室的温度不断降低，其中温室3达到的最低气温为12.0℃，比温室2高0.6℃，比温室1高1.5℃。

结果表明，在连续阴雨雪天气条件下，下沉深度较大的温室，最低温度值较高，保温蓄热性能也较好。温室室内温度受盖帘天数影响明显，在一天未揭帘之后，普通日光温室和下沉型日光温室的温度能够维持在作物生长的适宜温度范围内，而在两天未揭帘之后，普通日光温室的气温就会低于作物生长适宜温度的最低值，下沉型日光温室的气温仍然能够达到作物生长的适宜温度，但是如果继续出现阴天或降温，其气温也会降至适宜温度最低值以下。因此，在出现数天未揭帘时，尤其在1月份，应注意对温室进行补温。

2.3.2 湿度

室内外空气湿度在2010年1月2日至1月5日的变化如图8所示。盖帘期间，温室内的湿度同样处于饱和状态。揭帘后，温室内的相对湿度均不同程度降低，温室Ⅰ下降幅度比温室Ⅱ要大，在 4 日的最低相对湿度为 54%，比温室Ⅱ的最低值小 20%。

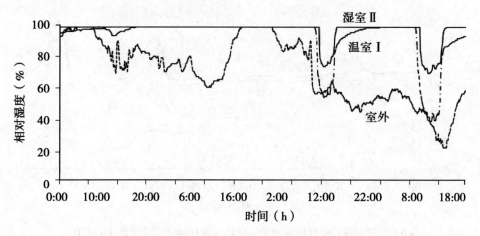

图 8　空气湿度在连续阴雨雪天的变化（2010 年 1 月 2 日至 1 月 5 日）

3　结论

（1）下沉型日光温室在冬季的保温蓄热性能要明显好于传统非下沉日光温室，尤其是在经历连续阴雨雪天后，下沉型日光温室的保温性能优势更加明显。

（2）与下沉型日光温室相比，普通日光温室的温度变化幅度大，受外界环境影响明显，晴天时，温室气温可能会超过作物适宜生长温度的最高值；连续阴雨雪天时，温室气温可能会低于作物适宜生长温度的最低值，因此应注意适时地降温和补温。

（3）在揭帘期间，不论什么样的天气状况，下沉型日光温室的相对湿度均要高于普通日光温室，而在盖帘期间，两种类型的温室相对湿度均处于饱和状态，应注意除湿。

参考文献

[1] 韩太利，魏家鹏．寿光新型日光温室的结构特点与推广应用．中国蔬菜．2010（13）：7~9
[2] 刘桂芝，李杰林，陈建国等．坑式日光温室增温保温试验研究．北方园艺．2004（2）：16~18
[3] 张春梅，陈修斌，王勤礼．河西走廊北部半地下式日光温室研究．西北园艺．2005（11）：5~6
[4] 王晓冬，张丽，王国强等．半地下式日光温室对地温的影响．新疆农机化．2007（4）：36~37
[5] 张峰，张林华，刘珊．日光节能温室下沉深度对其采光性能的影响．山东建筑大学学报．2008（06）：474~477
[6] 张峰，张林华．下沉式日光温室土质墙体的保温蓄热性能．可再生能源．2009，27（3）：18~20
[7] 佟国红，罗新兰，刘文合等．半地下式日光温室太阳能利用分析．沈阳农业大学学报．2010（03）：321~326
[8] 马承伟，苗香雯．农业生物环境工程．北京：中国农业出版社，2005
[9] 陈端生．日光温室的温度环境．农村实用工程技术．2003（5）：26~28

基于 ZigBee、3G 网络的温室无线远程植物生理生态监测系统

杨 玮[1]*，杨仁全[2]，周增产[1]，商守海[1]，李东星[1]，董明明[1]

（1. 北京京鹏环球科技股份有限公司，北京 100094；
2. 北京市科委政策法规处，北京 100035）

摘要：基于 ZigBee、3G 网络的温室无线远程植物生理生态监测系统，主要包括远程控制终端、通信链路和数据采集终端；其中远程控制终端包括连接了 3G 网络的 PC 机以及 PC 机内部运行存储数据的历史数据库、决策支持系统；通信链路，包括 ZigBee 近距离无线通讯网络以及 3G 远程无线网络；数据采集终端包括采集植物生理生态参数的各种传感器，如土壤温度传感器，土壤水分传感器，叶片温度传感器，光合作用传感器，植物茎秆生长速率传感器。ZigBee 子节点能够将采集到的土壤温度、土壤水分、叶片温度，光照强度以及植物茎秆生长数据通过 ZigBee 网络发送到 ZigBee 终端节点，终端节点的数据通过 3G 通信模块发送到指定服务器端。在服务器端能对以上数据分别提取、保存，并实时获取现场的各类数据，通过对各类数据的分析、处理判断做出预测、报警等服务，从而达到增产增收的效果。

关键词：温室；ZigBee；3G；植物生理生态

Wireless Remote Plant Eco Physiological Characteristics Monitoring System Based on ZigBee and 3G

（Yang wei[1], Yang Renquan[2], Zhou Zengchan[1], Shang Shouhai[1], Li Dongxing[1], Dong Mingming[1]）

（1. Beijing Kingpeng international Li – Tech Lorportion, Beijing 100094；2. policies and Regulations Department of Beijing Scientific Association, Beijing 100035, P. R. China）

Abstract: The system contains remote controller, communication links and data collection. The remote controller is made up of the PCs on 3G network and the software that manages the devices across the network. Communication links contains ZigBee short distance wireless network and 3G remote wireless network. Data collection contains all kinds of plant eco physiological characteristics sensors like soil temperature sensor, soil moisture sensor, leave temperature sensor, light content sensor and culm strength sensor. ZigBee endpoints send all data of soil temperature, soil moisture, leave temperature, light content and culm strength to ZigBee coordinator, and then all data are transferred from ZigBee coordinator to server though 3G network. Data are extracted, saved by server. Forecasting and warning are sent according to analyzing of data, so the system promotes the production.

Key words: Greenhouse; ZigBee; 3G; Eco physiological characteristics

目前，温室环境调控能力普遍不强，智能化水平低，植物生理生态信息获取与传输手段落后。传统的农业决策支持系统和管理软件以历史数据库、专家经验、知识模型为基础，缺

---

* 作者简介：杨玮（1981— ），女（蒙族），内蒙古人，北京京鹏环球科技股份有限公司，博士，从事电子技术在农业中的应用。E-mail：weiwei810311@163.com。

乏以实时信息为基础的农情监测功能，系统的决策缺乏实时性，滞后的信息很难适应农业生产的时变性特点。随着移动终端设备的普及，以及3G技术的不断完善，无线访问带宽不断提高，3G技术不断应用于各个行业领域，各行业利用先进的无线技术，将自己的领域不断创新，提出新的无线解决方案。3G技术作为一种先进的无线通信技术，比起以往的2G通信技术具有带宽明显增大、数据传输速度明显提升、数据传输更加安全可靠等优点。为各种远程实时数据的采集、设备的远程控制奠定了基础。

农田信息的分布式数据传输具有如下要求：数据采集或监控的网点多、数据更换数量不大、设备成本低、数据传输可靠性高、安全性高、设备体积小、不便放置较大的充电电池或者电源模块、用电池供电、地形复杂、需要较大的网络覆盖等特点。因此，要选择一种合适的传输标准，实现并得到最好性价比的无线传输系统。ZigBee这种近距离、低复杂度、低功耗、低数据速率、低成本的双向无线通信网络技术是满足农田信息数据传输的最佳选择。本课题进行的无线植物生理生态监测系统研发正是基于解决设施农业的这些关键技术问题而提出的。

1 系统硬件组成

基于ZigBee、3G网络的温室无线远程植物生理生态监测系统，主要包括远程控制终端，通信链路和数据采集终端；其中远程控制终端包括连接了3G网络的PC机以及PC机内部运行存储数据的历史数据库、决策支持系统；通信链路，包括ZigBee近距离无线通讯网络以及3G远程无线网络；数据采集终端包括采集植物生理生态参数的传感器，包括土壤温度传感器，土壤水分传感器，叶片温度传感器，光合作用传感器，植物茎秆生长速率传感器。整个装置的原理框图如图1所示。

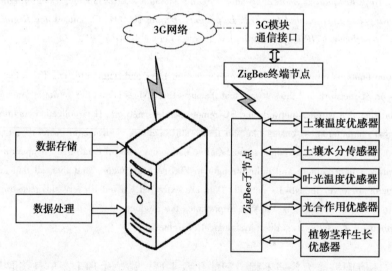

图1 系统原理框图

ZigBee子节点将采集到的土壤温度传感器、土壤水分传感器、叶片温度传感器，光合作用传感器以及植物茎秆生长传感器的数据通过ZigBee网络发送到ZigBee终端节点，终端节点的数据通过3G通信模块发送到指定服务器端。在服务器端能对以上数据分别提取、保存，并实时获取现场的各类数据，通过对各类数据的分析、处理判断做出预测、报警等服务，从而达到增产增收的效果。

1.1 3G模块

3G模块采用了CDMA2000 EVDO DTU。CDMA2000是国内3G移动通信的一种，有技术成熟、宽大等优点。目前电信的CDMA2000都采用HSDPA技术，下行速率为7.2Mbit/s今后还会升级到14.4Mbit/s。利用它进行网络传输时，首先需要通过PPP协议拨号接入互联网，然后才能进行数据的无线传输。PPP协议拨号是通过控制器发送AT指令操作3G模块来实现的。AT指令建立网络协议流程如图2所示。

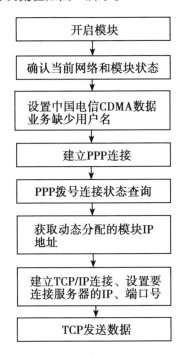

图2 AT指令建立网络协议流程

1.2 ZigBee无线模块

ZigBee无线模块采用JN5139。JN5139是一款兼容IEEE 802.15.4的低功耗、低成本的无线微型控制器。该模块内置一款32位RISC处理器，并集成有2.4GHz频段的IEEE802.15.4标准的无线收发器。微控制器内置64K的ROM存储集成点对点通信与网状通信的完整协议栈；利用此芯片开发的无线传感器网络节点成本低、功耗小，适用于电池长期供电。Zigbee节点分为协调器和子节点两种，协调器是整个ZigBee无线网络的管理员，负责管理子节点以及与3G远程无线网络进行通信。Zigbee子节点负责数据的实时采集发送，之后ZigBee主节点可以将采集到的所有子节点数据通过3G模块上传至上位机，用于对植物生长周期内综合生长环境的跟踪记录、查询和分析。

1.3 传感器模块

为采集温室内的环境因子，系统采用了土壤温度传感器、叶片温度传感器、土壤水分传感器、光照传感器以及茎秆生长传感器，传感器具体参数如表1所示。传感器采集环境因子信息之后经过A/D转换成标准的电压信号送入ZigBee控制器进行数据分析处理。

表1 五种传感器参数 Table1 Indexes of the sensors

传感器种类	技术参数
茎秆生长速率传感器	测量范围：0~5mm　适合茎秆直径：4~25mm 测量精度：0.01mm
叶片温度传感器	测量范围：0~50℃　精度：±0.2℃ 探头与叶片接触区域≤1mm
土壤温度传感器	测量范围：0~50℃　精度：±0.2℃
光照传感器	测量波长范围：400~700nm 测量范围：0~2 000μmol/m²·s 测量精度：±2% 温度漂移：10~35℃时<2%
土壤水分传感器	测量范围：0~100% 测量精度：±4%（重复性）

2　系统软件组成

2.1　协调器运行流程

协调器是整个ZigBee无线网络的管理员。节点上电后，首先进行网络的初始化操作，包括ZigBee协议栈的初始化及硬件外设的初始化；接着进行信道查询，选择合适的信道建立一个无信标网络，并设置网络的PAN ID，等待终端节点加入网络；在有终端节点加入网络之后，接收终端节点发送的传感器数据，之后通过串口控制3G模块转发数据到远程服务器端。

2.2　终端节点应用程序设计

网络系统中终端节点负责采集植物叶面温度、土壤温度、光合有效辐射、茎/果实/树木生长变化，以及土壤水分等，并将其传送到协调节点。节点上电时，首先进行初始化操作，包括ZigBee堆栈的初始化及硬件外设的初始化。接着进行信道查询，选择合适的网络等待加入；然后向该网络的协调节点发送请求加入，最后，在收到允许加入的确认之后加入网络，读取传感器数据，发送到协调器。数据完成并获得协调器发送的确认帧后马上进入休眠状态，直至下一次数据采集时刻的到来[14~15]。

2.3　数据服务器端

运行于远程数据服务器端的上位机程序是整个数据采集系统的主控部分，开发平台：windows XP操作系统；开发工具：Microsoft VisualC++6.0；数据库：Access 2003。主要功能包括远程控制、数据采集、数据存储、数据显示等。数据服务器端主要处理前端数据采集提供的各类数据，并针对目前3G网络不能直接访问的特点。系统提出一种新的解决办法，在服务器端通过添加DDNS服务，能够实现外网的访问，从而将3G网络跟现有的有线网络进行互联。DDNS—动态域名服务。DDNS是将用户的动态IP地址映射到一个固定的域名解析服务上，用户每次连接网络的时候客户端程序就会通过信息传递把该主机的动态IP地址传送给位于服务商主机上的服务器程序，服务器程序负责提供DDNS服务并实现动态域名解析。前端的数据采集终端,本身不能直接访问,然而由于服务器端提供 DDNS 服务,这样可以通过外网 IP 访问前端采集设备服务器端对数据进行分析,处理并作出判断。数据服务器端实现了很好的屏蔽作用,终端访问用户直接访问数据服务器即可,从而实现了随时随地实时访问,监测当前当地状况。

3 系统工作原理

基于 ZigBee、3G 网络的温室无线远程植物生理生态监测系统,系统主要包括三个部分:远程控制终端,通信链路和采集终端。控制终端是一台连接了 3G 网络的 PC 机,通过绑定的网络 IP 地址与 3G 模块通信;另外 PC 机内部运行着数据采集的历史数据库、决策支持系统,对采集模块发送的数据进行存储、处理和决策支持。通信链路即为 3G 网络和 ZigBee 无线网络,连接在公网的远程控制终端通过一个固定 IP 地址与 3G 网络相连。ZigBee 监测子节点集成了土壤温度传感器、土壤水分传感器、叶片温度传感器、植物茎秆生长速率测量传感器以及光合作用传感器。ZigBee 子节点采集的所有数据汇聚到 ZigBee 中心主节点后通过 3G 网络发送到远程控制端 PC 机内部进行数据的分析处理。通过本发明,可以实现以 ZigBee 无线模块为核心的温室无线网络远程植物生理生态监测系统,整个无线传感器网络节点分布于温室的各个测量点执行各数据的采集、预处理和发送等工作,最后由中心主节点把传送来的数据上传到上位机。该监测系统操作简单、功能丰富、便于温室环境的方便使用。

4 结语

通过将 Zigbee 技术的短距离无线通信和 3G 业务相结合,可以实现对数据采集的远程控制和传输。因而解决了信息传输的两个问题:
(1) 温室和管理者之间的远程,高速通信;
(2) 分布式的温室局域网通信。

参考文献

[1] 马铁军. 基于 GSM、GPS、GIS 技术的城市紧急救援指挥系统. 河北科学院学报,2008,(9):51~52
[2] 韩晓萍. GPRS 技术在电力远程抄表系统中的应用. 电子测量与仪器仪表学报,2005,(4):81~84
[3] 辛颖,谢光忠,蒋亚东. 基于 ZigBee 协议的温度湿度无线传感器网络. 传感器与微系统,2006,25(7):82~88
[4] 宫赤坤,陈翠英,毛罕平. 温室环境多变量模糊控制及其仿真. 农业机械学报,2000,31(6):52~54
[5] 乔晓军,张馨,王成等. 无线传感器网络在农业中的应用. 农业工程学报,2005,21(2):232~234
[6] 程雪,王彬,徐艳等. 网络技术在温室智能控制系统中的应用. 微计算机信息,2009,(01):46~47,67
[7] 刘涛,赵计生. 基于 ZigBee 技术的农田自动节水灌溉系统. 测控技术,2008,27(2):61~62
[8] 徐志如,崔继仁. 基于单片机的温室智能测控系统的设. 传感器与微系统,2006,(05):46~48

蔬菜变量喷药研究与试验

马 伟*，王 秀，郭建华

（国家农业信息化工程技术研究中心，北京 100097）

摘要：为了满足目前设施标准园建设对精准变量施药技术的迫切需求，设计开发了一种设施果蔬适用的精准变量喷药控制器。采用滑动电位器调节输出脉宽实现喷药泵电机的转速的控制，使用三种不同的喷头进行试验，控制器可控制喷药压力从 0.03M ~ 0.35MPa 范围内调节，喷药量在 600 ~ 2 100ml/min 范围内稳定变化，能够满足施药过程中的精准变量调节的需要。通过大量试验表明，分布均匀系数 DU 为 92.09% ~ 99.82%，均匀度 UC 为 93.95% ~ 99.22%。该喷药机能够有效满足实际生产中变量调节的需要。

关键词：单片机；施药；均匀性

Experimental Study on Variable Spraying for Vegetable

Ma Wei*, Wang Xiu, Guo Jianhua

(*Nationcl Engineering Research Center for information Technology in Agriculture, Beijing 100097, P. R. China*)

Abstract: In order to met the precision spraying urgent needs for the current developing facility standards greenhouse base, designed a precision variable spray controller for fruits and vegetables in facilities. Adjusted the pulse width output to achieve the speed of the pump motor to control spraying by sliding potentiometer. Tested with three different nozzles, it achieved regulation of spraying pressure range from 0.03M ~ 0.35MPa, and praying volume changes in the stability range 600 ~ 2 100ml/min. So it can met the variable adjustment process needs of the precise praying. By experiments show that the distribution coefficient (DU) is 92.09% ~ 99.82%, he uniformity (UC) is 93.95% ~ 99.22%.

Key words: SCM; Uniform; Electric spraying

随着一些重大食品安全事件的发生，"放心食品"问题越来越严峻，食品安全成为社会广泛关注的焦点。设施果蔬近年来发展迅速，产生巨大的经济价值，为农业经济的发展注入了新鲜的活力。用时设施农业由于其密闭的人工环境以及高附加值的特点，对农作物的品质提出了更高的要求。农药的高效喷洒是保证品质的重要环节，设施果蔬生产作业时的农药精准少量投入就成为广大种植户首要关心的问题。

1 原理

果蔬变量喷药机利用蓄电池作为动力，利用芯片作为主控制单元，利用压力传感器采集

---

* 作者简介：马伟，男，助理研究员，研究方向为精准农业智能装备技术，E-mail：maw516@sohu.com。

管路的压力,通过单片机实时进行控制调节。主要的工作原理如图 1 所示:

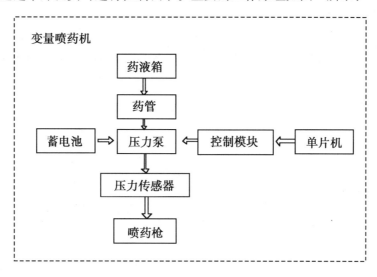

图 1 工作原理图

2 结构设计

京郊设施温室环境下道路宽度为 0.5m 左右,对施药设备的设计宽度提出了要求。同时出入温室大多留有台阶、起伏物,道路上也多有作物藤蔓,药液箱搬运不方便,必须有脚轮,考虑越障,脚轮的直径约在 20cm 左右。为避免狭小空间的磕碰撞击,控制系统应该嵌入式安装。设施施药由于空间密闭,湿度大,对施药的均匀性提出较高要求,杜绝喷头堵塞和滴漏现象,因此结构应有多重过滤。综合以上结构,经过计算后进行计算机辅助设计的三维零件组装图如图 2 所示。

图 2 计算机三维零件组装图

3 变量控制系统

555timer（定时器）是美国 Signetics 研制的一种集成电路定时器。在输入端设计有三个 5kΩ 的电阻。其功能主要有两个比较器决定，两个比较器的输出电压控制着 RS 触发端和放电管的状态。药量变量控制模块由 555 组成的触发器、光电耦合器和一个 MOS 管构成。

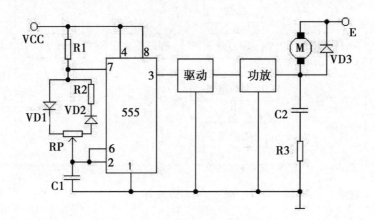

图 3　药量控制模块框图

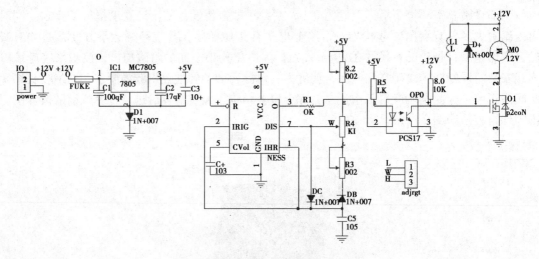

图 4　药量变量控制模块设计原理图

这是一个占空比可调的脉冲振荡器（图 3 和图 4）。输出脉冲驱动施药压力泵，脉冲占空比越大，施药压力泵电驱电流就越小，转速减慢；脉冲占空比越小施药压力泵电驱电流就越大，转速加快。因此调节电位器 RP 的数值可以调整电机的速度。如电极电驱电流不大于 200mA 时，可用 CB555 直接驱动；如电流大于 200mA，应增加驱动级和功放级。图中 VD3 是续流二极管。在功放管截止期间为电驱电流提供通路，既保证电驱电流的连续性，又防止电驱线圈的自感反电动势损坏功放管。电容 C2 和电阻 R3 是补偿网络，它可使负载呈电阻性。整个电路的脉冲频率选在 3～5kHz 之间。频率太低电机会抖动，太高时因占空比范围小使施药压力泵调速范围减小。该电路具有性能特点主要有：较硬的机械特性，较宽的调速范围，较快的动态响应过程，较好的启动性能，可靠性高，结构紧凑，输出电压：（0~12V），低速运行力矩大，限制速度 RO 调节，电流设置、限流保护，跟随性好，响应速度快。

4 方法和材料

本试验的目的是通过测定单位时间不同压力不同喷头的喷雾量，来确定该施药系统的流量稳定性，以及针对不同作物对象最佳的工作压力。通过对照目前广泛使用的喷药喷头，选择了三种喷头进行测试，通过电路的控制，施药测试的管路压力可以使用电位器进行精准调节。

试验的三种喷头分别是单喷头、两喷头和三喷头。在不同的压力模式下，不同重复试验施药时的流量稳定性和均匀性存在一定的差异，某个型号的喷头会有最佳的施药的压力，能够达到比较好的雾化效果和流量稳定性，本试验对于评价电动施药机的保持施药流量的稳定性，具有重要意义。

单喷头在 1、2 压力档位时，施药均匀，雾化很好。流量曲线如图 5 所示：

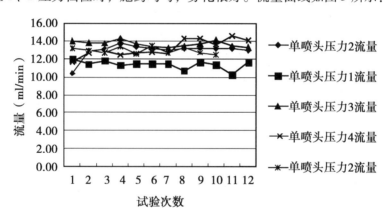

压力 1：0.03MPa 压力 2：0.06MPa 压力 3：0.14MPa 压力 4：0.15MPa
压力 5：0.18MPa

图 5 单喷头不同压力下的流量变化

将喷枪的换成二喷头（就是一个喷头有两个喷嘴），这个时候喷头的流量就会变大。相应需要的最佳压力就会变化。图 6 是二喷头的施药流量曲线图。图 7 是三喷头的施药流量曲线图。

结果分析：

从单喷头流量曲线图我们可以看出：

一档和二档压力是比较适合进行施药，而且调节药量非常明显，流量波动不是很大。二喷头流量曲线图表明一、二、三档相对均匀，施药流量非常稳定，施药流量和稳定压力之间具有很好的一致性关系，非常适合进行施药作业。

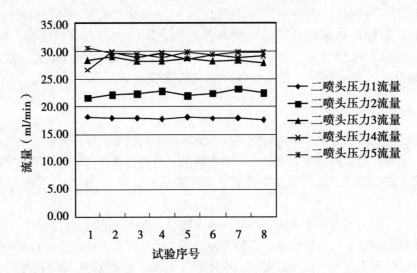

压力1：0.05MPa 压力2：0.1MPa 压力3：0.2MPa 压力4：0.3MPa 压力5：0.35MPa

图6 二喷头不同压力下的流量变化

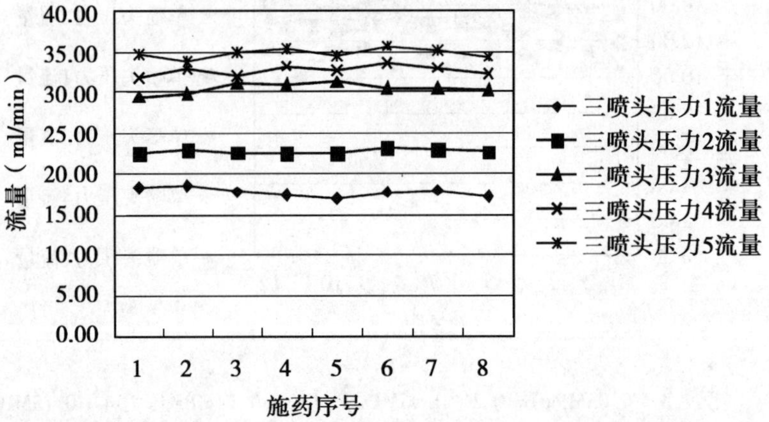

压力1：0.01MPa 压力2：0.03MPa 压力3：0.1MPa 压力4：0.15MPa 压力5：0.25MPa

图7 三喷头不同压力下的流量变化

5 结果分析

本研究参照相关文献，采用以下四个指标作为评价该施药机不同压力条件下各组重复试验的的流量稳定性和均匀性。

参照文献的均匀度 UC 计算为（表1、表2、表3和表4）：

$$UC = \left[1 - \frac{\frac{1}{n}\sum_{i=1}^{n}|q_i - q_{mean}|}{q_{mean}}\right] \times 100\% \quad (1)$$

式中：n 是样本数，q_{mean} 是样本均值，q_i 是第 i 个样本值。

分布均匀系数 DU 参照文献

$$DU = \frac{q_{lq}}{q_{mean}} \times 100\% \quad (2)$$

式中：q_{lq} 是所有样本中较小的 $\frac{1}{4}$ 样本的样本均值。

参照文献的变异系数 CV 计算为：

$$CV = \frac{s}{q_{mean}} \quad (3)$$

式中：s 是样本方差，q_{mean} 是样本均值。

参照文献的变化率 q_{var} 计算为：

$$q_{var} = \frac{q_{max} - q_{min}}{q_{max}} \quad (4)$$

式中：q_{min} 和 q_{max} 分别表示的是所有样本的最小样本值和最大样本值。q_{var} 可用来反映样本的偏差范围。体现施药量的波动情况。

表1 实验数据的评价指标表

评价指标		测试喷头		
		三喷头	单喷头	二喷头
Q_{var} 1 档压力	q_{mean}	12.96	17.89	17.74
		0.061	0.033	0.088
	CV	0.049	0.002	0.017
	UC%	96.68	99.22	97.65
	DU%	99.82	98.54	96.26
Q_{var} 2 档压力	q_{mean}	11.38	22.35	22.73
		0.133	0.071	0.021
	CV	0.018	0.0111	0.003
	UC%	97.22	98.41	99.04
	DU%	92.09	97.36	99.01
Q_{var} 3 档压力	q_{mean}	13.69	28.36	30.49
		0.061	0.039	0.048
	CV	0.007	0.0044	0.013
	UC%	98.08	99.11	98.45
	DU%	97.07	98.73	97.31

（续表）

评价指标		测试喷头		
		三喷头	单喷头	二喷头
Q_{var} 4档压力	q_{mean}	13.25	28.86	32.73
		0.146	0.036	0.054
	CV	0.059	0.034	0.012
	UC%	93.95	99.73	99.76
	DU%	94.61	95.66	97.55
Q_{var} 5档压力	q_{mean}		29.68	34.83
			0.035	0.051
	CV		0.009	0.009
	UC%		98.78	98.64
	DU%		98.78	99.37

表2 单喷头不同压力流量表

单喷头1档压力流量	1	2	3	4	5	6	7	8
时间（s）	11.89	11.55	11.23	12.11	11.2	10.75	11.34	11.26
流量（ml）	156	159	150	160	145	142	150	148
（ml/min）	13.12	13.77	13.36	13.21	12.95	13.21	13.23	13.14
单喷头2档压力流量	1	2	3	4	5	6	7	8
时间（s）	11.28	11.79	10.57	13.26	11.86	11.23	11.61	11.76
流量（ml）	135	135	125	150	136	129	133	126
（ml/min）	11.97	11.45	11.83	11.31	11.47	11.49	11.46	10.71
单喷头3档压力流量	1	2	3	4	5	6	7	8
时间（s）	11.43	11.21	11.26	12.06	10.94	11.12	11.31	10.4
流量（ml）	160	155	155	173	150	149	150	140
（ml/min）	14	13.83	13.77	14.34	13.71	13.4	13.26	13.46
单喷头4档压力流量	1	2	3	4	5	6	7	8
时间（s）	10.36	10.23	10.85	11.96	11.38	11.31	10.63	10.88
流量（ml）	121	130	140	149	145	145	134	156
（ml/min）	11.68	12.71	12.9	12.46	12.74	12.82	12.61	14.34

表3 二喷头不同压力流量表

二喷头压力1 流量	1	2	3	4	5	6	7	8
时间（s）	11.02	10.75	11.45	10.9	10.93	11.54	10.25	11.8
流量（ml）	200	193	205	193	197	206	184	207
（ml/min）	18.15	17.95	17.9	17.71	18.02	17.85	17.95	17.54
二喷头压力2 流量	1	2	3	4	5	6	7	8
时间（s）	10.58	10.61	10.54	11.29	10.88	11.27	11.51	11.11
流量（ml）	228	235	235	257	239	252	267	250
（ml/min）	21.55	22.15	22.3	22.76	21.97	22.36	23.2	22.5
二喷头压力3 流量	1	2	3	4	5	6	7	8
时间（s）	11.38	11.15	11.79	11.15	10.61	11.24	11.26	11.28
流量（ml）	323	323	332	315	305	317	319	314
（ml/min）	28.38	28.97	28.16	28.25	28.75	28.2	28.33	27.84
二喷头压力4 流量	1	2	3	4	5	6	7	8
时间（s）	11.15	11	11.21	10.8	10.71	10.53	10.9	11.01
流量（ml）	297	328	324	320	306	308	315	321
（ml/min）	26.64	29.82	28.9	29.63	28.57	29.25	28.9	29.16
二喷头压力5 流量	1	2	3	4	5	6	7	8
时间（s）	10.93	10.53	11.04	10	9.94	11.6	10.71	10.69
流量（ml）	334	312	325	288	297	341	320	319
（ml/min）	30.56	29.63	29.44	28.8	29.88	29.4	29.88	29.84

表4 三喷头不同压力流量表

三喷头压力1 流量	1	2	3	4	5	6	7	8
时间（s）	10.37	10.76	9.86	9.3	11.92	10.61	9.97	10.64
流量（ml）	190	200	175	162	202	188	179	183
（ml/min）	18.32	18.59	17.75	17.42	16.95	17.72	17.95	17.2
三喷头压力2 流量	1	2	3	4	5	6	7	8
时间（s）	10.61	9.92	10.4	9.92	10.44	10.17	9.84	9.93
流量（ml）	239	227	235	223	235	236	226	225
（ml/min）	22.53	22.88	22.6	22.48	22.51	23.21	22.97	22.66
三喷头压力3 流量	1	2	3	4	5	6	7	8
时间（s）	9.77	9.98	9.91	9.85	10.22	9.79	9.97	10.07

(续表)

三喷头压力1 流量	1	2	3	4	5	6	7	8
流量（ml）	288	298	308	305	320	298	304	305
（ml/min）	29.48	29.86	31.08	30.96	31.31	30.44	30.49	30.29
三喷头压力4 流量	1	2	3	4	5	6	7	8
时间（s）	10.69	10.06	10.3	11	10.46	9.99	9.75	9.99
流量（ml）	340	331	330	366	343	336	322	323
（ml/min）	31.81	32.9	32.04	33.27	32.79	33.63	33.03	32.33
三喷头压力5 流量	1	2	3	4	5	6	7	8
时间（s）	10.1	10.42	10.23	9.79	10.4	9.75	10.32	9.92
流量（ml）	350	353	358	346	359	348	363	341
（ml/min）	34.65	33.88	35	35.34	34.52	35.69	35.17	34.38

6 结论

试验证明该机性能稳定，在果蔬产业技术体系北京市创新团队设施设备功能研究室的推广示范中，得到较好评价。

相变蓄热技术应用于温室节能中的技术分析[*]

梁 浩[1][**],杨其长[2][***],方 慧[2]

(1. 中国农业大学水利与土木工程学院,北京,100083;
2. 中国农业科学院农业环境与可持续发展研究所,北京,100081)

摘要:相变蓄热是最有效的蓄热方法之一,它通过物态转变过程吸收和释放热量,解决太阳辐射能量昼夜变化所引起的时空分布不均的问题。本文综述了国内外相变材料用于温室蓄热的研究,分析了相变蓄热技术在设施农业中的技术优势和应用前景,并提出了需要进一步解决的问题。

关键词:相变材料;相变储能材料;设施园艺;温室

A review on Latent Heat Storage in Greenhouse Heating Application

Liang Hao[1]**, Yang Qichang[2]***, Fang Hui[2]

(1. College of Water Conservancy and Civil Engineering of CAU, Beijing, 100083, P. R. China;
2. Institute of Environment and Sustainable Development in Agriculture of CAAS, Beijing, 100081, P. R. China)

Abstract: Latent heat storage is of some superiority over sensible heat storage due to its high heat of fusion. Phase change materials (PCMs) can increase the thermal energy storage inside the greenhouse during the day and release it at night to satisfy the heating need of the greenhouse. Continuous research in this field has result in good development and satisfactory results. This paper reviewed the advances. Future application for the potential users is discussed with the development of PCMs.

Key words: PCMs; Latent heatstorage; Protected horticulture; Greenhouse

相变材料作为一种蓄热功能材料,利用物态转变过程中伴随的能量吸收和释放而进行能量迁移[1],实现能量在不同时空位置之间的转换。材料在相变过程中与环境进行能量交换,达到控制环境温度和能量利用的目的。与显热蓄热相比,相变蓄热具有储能密度高、体积小、温度控制恒定、节能效果显著、相变温度选择范围宽、易于控制等优点[2],因此在航天、建筑、纺织、农业及太阳能诸多领域都有非常好的应用前景[3]。

温室是光热利用型的农业生产建筑,光照昼夜变化规律决定了温室昼夜温度的大幅波

---

[*] 基金项目:北京市科委重点项目(D111100000811001);国家自然科学基金(31071833)。

[**] 作者简介:梁浩,男,(1983—),中国农业大学水利与土木工程学院,博士研究生。从事农业生物环境工程研究。

[***] 通讯作者:杨其长(1963—),男,博士,研究员,博士生导师,主要从事设施园艺环境工程研究。中国农业科学院农业环境与可持续发展研究所,北京 100081。E-mail:yangq@ieda.org.cn。

动，通过增加温室的蓄热能力，可以显著减弱这种波动，提高温室的热性能[4]。就日光温室而言，重质墙体结构就是通过增加墙体厚度提高温室的显热蓄热能力[5,6]。潜热蓄热的特点是轻质量，重热质，表1所反映的相变材料相比显热材料的优势，因此在设施农业领域有很广泛的研究和应用价值。

表1 潜热材料与显热材料蓄热优势比较

	岩石	水	正十八烷	$CaCl_2 \cdot 6H_2O$
密度（kg/m^3）	2 240	1 000	814	1 562
显热[$kJ/(kg \cdot K)$]	1.0	4.2	2.0	1.45
潜热（kJ/kg）	—	—	244（28℃）	190.8（29℃）
每10^6J所需质量（kg）	100	23.8	3.8	4.9
每10^6J所需体积（m^3）	44.6	23.8	4.7	3.1

1 相变蓄热材料的种类及其在温室蓄热中的研究进展

相变材料的选材范围非常广泛，考虑到温室蓄热的节能特点，相变温度在18~35℃的固液相变材料值得关注，包括无机水合盐、有机石蜡和低共熔混合物。20世纪70年代以来，这类物质在温室冬季蓄热中的效果得到了深入的研究。

表2 几种18~35℃适宜温室环境的相变材料

材料	熔点（℃）	导热系数 W/(m·K)	相变潜热（kJ/kg）	密度 kg/m^3
$CaCl_2 \cdot 6H_2O$	29	0.540（38.7℃）	190.8	1 562（32℃）
$Na_2SO_4 \cdot 10H_2O$	32	0.544	251.1	1 485（solid）
$Na_2CO_3 \cdot 10H_2O$	33	—	247	1 442
$Na_2HPO_4 \cdot 12H_2O$	35	—	280	1 522
66.6% $CaCl_2 \cdot 6H_2O$ + 33.3% $MgCl_2 \cdot 6H_2O$	25	—	127	1 590
十八烷	28	0.15	244	814
羊脂酸65mol% + 月桂酸35mol%	18.0	—	148	—
34% 硬脂酸 + 66% 羊脂酸	24	0.164（39.1℃）	147.7	888（25℃）

$CaCl_2 \cdot 6H_2O$（相变温度为29℃，相变潜热为290kJ/kg）是单一物质中最适于温室蓄热的无机盐类相变材料。它具有相变温度适中，相变潜热值高，传热良好以及成本低廉等优点，因此在早期研究中，大部分的温室加热试验采用了这种材料。V. P. Sethi[7]对使用$CaCl_2 \cdot 6H_2O$加热温室的研究进行了统计，每平方米使用10~13.7kg $CaCl_2 \cdot 6H_2O$蓄热，能够减少30%~75%的冬季加热能耗，节能效果明显。但它存在过冷与相分离现象，降低了材料的蓄热效果和耐久性，限制了其大范围的应用。然而近年来随着材料科学的发展，可对这一不足通过化学方法调节和改善，$CaCl_2 \cdot 6H_2O$依然是一种非常具有竞争力的相变储

能材料[8]。

有机石蜡（Paraffins）性质稳定，无过冷现象，同时具有很强的蓄热能力。传热系数低是石蜡相变材料的不足，因此，蓄热效果更多地依赖于科学的传热设计。传热强化后的石蜡蓄热系统，与岩石床温室加热系统相比，有较高的蓄热能力和节能效果[9]。

低共熔混合物（Eutectics）是通过两种或两种以上的材料按不同比例混合制备而成，通过混合形成共熔合金，能够获得不同相变温度和相变潜热的材料。低共熔混合物的相变温度低于所加入的任一材料的相变温度，潜热值也会有所变化，但可以丰富选材范围，非常具有研究价值。王宏丽[10]用 $Na_2SO_4 \cdot 10H_2O$ 和 $Na_2CO_4 \cdot 10H_2O$ 制备了适用于温室使用的水合盐共熔材料，相变潜热值高，性能稳定；Sari 等[11]用豆蔻酸和棕榈酸制备了有机共熔合金，用于温室加热系统，实现了高效温室蓄热。

2 相变材料在温室中蓄热应用的特征

光热作用是相变材料蓄积热量的主要方式。太阳直射光通过温室覆盖材料透射到温室内部，分别照射到植物、土壤和围护结构表面，被植物吸收的光照转化为营养物质，而照射到其他表面的光照，一部分通过一次或多次反射蓄积起来，其余则反射到温室之外，无法得到利用。增加温室蓄热密度，提高温室内部的光能利用效率，能有效改善温室温度环境，节约能源。

将集热性更好，蓄热密度更大的系统配置到温室中，相当于增加了一个内部热源，明显改善温室热环境。以日光温室而例，相变墙体与普通墙体相比，不但能够明显降低墙体厚度（图1），而且能够提高其蓄热能力（图2）。

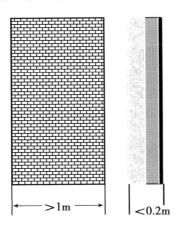

图1 相变蓄热单元与普通墙体厚度比较

3 相变蓄热材料的选择与封装方法

用于温室蓄热的相变材料应该具有以下特点[12]：①相变温度适宜；②相变潜热值高；③导热良好；④原料价格便宜，易于购买；⑤性能稳定，无过冷和相分离现象；⑥体积膨胀率低；⑦安全无毒、不燃等。目前所能选择的相变材料都存在一些不足之处，相变材料科学的发展也主要致力于克服这些缺点，使其能够更安全稳定的应用于工程实际。

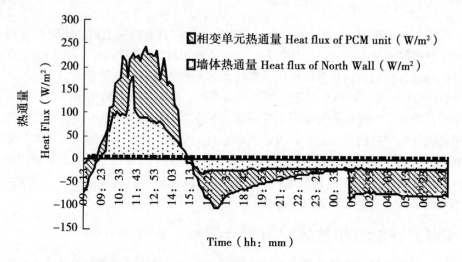

图2 相变蓄热单元与普通墙体24h蓄放热热流量比较

无论是用于温室潜热加热系统还是相变蓄热围护结构，除了材料要满足以上几个条件外，还必须解决好封装的问题。由于固－液相变材料液体状态下具有流动性，必须进行密封处理，以防止泄漏、腐蚀或污染环境。目前所采用的封装方法主要有宏封装和定形封装。

宏封装方法的原理简单，将材料放入容器中使其能够保持形状，不外泄；同时增加了材料与周围环境的接触面积，有助于热量的及时交换。因此早期研究使用较多，塑料袋封装、管道封装以及金属箱体封装都有所尝试[13]。张勇等将相变材料用聚乙烯塑料薄膜封装，加入到空心砖的孔隙中，用于温室墙体的蓄热并进行了实验[14]。

除了宏封装，定型封装方法在于围护结构结合上时有一定的优势。定形封装的相变材料由形不定的固液相变物质和形定的载体支撑材料按比例组合而成，发生相变前后保持形态不变。制备定形相变材料的一种方法是利用聚乙烯、聚酰胺、树脂等高分子材料的成膜特性，通过微胶囊工艺将相变材料包覆起来形成相变微胶囊，蓄热、传热和机械性能较好，能直接与水泥砂浆、石膏等材料混合，制成砌块或墙板，具有很好的应用前景[15]。另一种方法是膨胀石墨、珍珠岩、高岭土、二氧化硅等具有多孔结构的物质，通过对液态相变材料的吸附作用可制备定形相变材料，安全稳定，同时对传热起到了很好的强化作用[16]。陈超[17]以石蜡为核心制作成这种吸附型定型相变材料，按比例与砂浆混合涂抹于温室蓄热后墙，结果表明能有效增加墙体蓄热，也避免了相变材料的泄漏。

4 相变传热过程的描述与计算机模拟

4.1 相变材料传热的强化

相变材料蓄放热的快慢和多少受到很多因素的影响。主要有材料本身的特性及热媒体两个方面。由于多数相变材料的热传导性比较差，因此提高传热率至关重要。相应选择增加材料的比表面积、增大热媒的流动性的方法来强化传热。用风扇驱动空气，加速相变单元蓄积和释放热量的方法在研究中使用较多，Borlard[18]的试验系统在温室内蓄热，风机与相变材料强制换热，蓄积的能量与消耗的电能之比为4.0～4.7；太阳能利用率占温室总太阳辐射为 6% ~10%。能减少 40% 的能源消耗；Huseyin[19] 的研究中也有类似的节能效果，不同之处是他将蓄热系统安放在了温室之外。

4.2 相变过程的模拟

相变传热过程模拟的复杂性在于：第一，存在随时间移动的固/液界面，它的位置随时间变化，且是非线形的，求解难度较大；第二，液相中的自然对流使传热机理复杂化[20]；第三，相变引起的体积变化和固相中形成的空穴使传热过程复杂化。

目前解决相变问题的解法分为两种。一种是以温度为变量，分别在固、液两相和固/液界面上建立能量方程，用有限差分法或有限元法求解。另一种方法是以焓为变量，建立起满足整个求解域的能量方程。由于焓包含了潜热，相变的影响自动地包含在焓形式的能量方程中，从而不需要考虑固/液移动界面，相变材料对温室昼夜温度波动的计算过程得以简化[21]。

5 结语

相变材料以其不可替代的优势越来越受到人们广泛的重视，同时，随着材料科学的发展，相变材料的选择性更加广泛，新的封装技术和强化传热技术不断发展。因此在农业领域对相变材料的开发也应该不断深入：

（1）进一步筛选适于温室的相变材料和封装方法，不仅要解决好材料稳定性以及固－液相变材料的流动控制问题，还要进一步深入蓄热系统与温室围护结构一体化的研究。

（2）考虑新型相变材料的成本、价格，某些新工艺开发的相变材料或蓄热系统已逐步成熟并商业化，但是昂贵的价格与其在实现在农业中应用还有距离。

（3）在相变储能技术研究中，计算机数值模拟是不可缺少的手段。采用合适的数学模型，将在对温室环境研究及相变蓄热系统开发方面将起到关键性作用。

参考文献

[1] 张仁元. 相变材料与相变储能技术. 北京：科学出版社，2009
[2] 梁辰，固全英. 相变储能技术的研究和发展. 建筑节能，2007，35（12）：41~44
[3] 张寅平，胡汉平，孔祥东. 相变储能—理论与应用. 合肥：中国科学技术大学出版社，1996
[4] 周长吉. 现代温室工程. 北京：化学工业出版社，2003
[5] 马承伟，卜云龙，籍秀红，陆海，邹岚，王影，李睿. 日光温室墙体夜间放热量计算与保温蓄热性评价方法的研究. 上海交通大学学报（农业科学版）. 2008, 10：411~415
[6] 杨仁全，马承伟，刘水丽，周增产，刘文玺，吴松. 日光温室墙体保温蓄热性能模拟分析. 上海交通大学学报（农业科学版）. 2008, 10：449~453
[7] V. P. Sethi, S. K. Sharma. Survey and evaluation of heating technologies for worldwide agricultural greenhouse applications. Solar Energy, 82（2008），832~859
[8] V. V. Tyagi, D. Buddhi, Thermal cycle testing of calcium chloride hexahydrate as a possible PCM forlatent heat storage. Solar Energy Materials & Solar Cells. 92（2008），891~899
[9] H. H. Ozturk; A. Bascetincelik. Energy and Exergy Ef?ciency of a Packed-bed Heat Storage Unit for Greenhouse Heating. Biosystems Engineering（2003）86（2），231~245
[10] 王宏丽，李凯，王剑，张立明. 适于温室生产的无机盐复合相变材料热性能的测试. 西北农林科技大学学报（自然科学版）. 2008（3）：141~144

[11] Sari, A., 2003. Thermal characteristics of a eutectic mixture of myristic and palmitic acids as phase change material for heating applications. Applied Thermal Engineering 23 (8), 1 005 ~ 1 017

[12] 王宏丽，邹志荣，陈红武，张勇．温室中应用相变储热技术的研究进展．农业工程学报，2008（6）：304 ~ 307

[13] Ahmet Kurklo. Energy storage applications in greenhouses by means of phase change materials (PCMs): a review. Renewable Energy, 1998 (1), 89 ~ 103

[14] 张勇，邹志荣，李建明，胡晓辉．日光温室相变空心砌块的制备及功效．农业工程学报 2010（2）：263 ~ 267

[15] 李祎，于航，刘淑娟相变材料微胶囊的国内外研究现状．能源技术．2007（2）：4 ~ 10

[16] 吕学文，考宏涛，李敏．基于复合相变储能材料的研究进展．材料科学与工程学报，2010（10）：797 ~ 800

[17] 陈超，果海凤，周玮相变墙体材料在温室大棚中的实验研究．太阳能学报，2009，03：287 ~ 293

[18] T. Boulard, E. Razafinjohany, A. Baille, A. Jaffrin and B. Fabre. Performance of a greenhouse heating system with a phase change material. Agricultural and Forest Meteorology, 1990 (52): 303 ~ 318

[19] Hüseyin Benli, Aydin Durmus. Performance analysis of a latent heat storage system with phase change material for new designed solar collectors in greenhouse heating. Solar Energy, Volume 83, Issue 12, 2009 (12), 2 109 ~ 2 119

[20] 王哲斌，许淑惠，严颖．石蜡相变蓄热过程数值模拟．北京建筑工程学院学报，2008（2）：10 ~ 13

[21] Atyah Najjar, Afif Hasan. Modeling of greenhouse with PCM energy storage. Energy Conversion and Management, Volume 49, Issue 11, 2008 (11), 3 338 ~ 3 342

MATLAB 和 VB 在温室环境模型构建中的混合编程*

孟力力[1]**,张 义[2],杨其长[2]***

(1. 江苏省农业科学院蔬菜研究所,南京 210014;
2. 中国农业科学院农业环境与可持续发展研究所,北京 100081)

摘要:在日光温室热环境模拟模型的软件开发中,VB 软件因其良好的人机界面被选为主控界面,在前台运行,为了弥补其计算能力差的缺点,选用具有强大工程计算和图像图形处理能力的 MATLAB 软件作为计算机处理工具,在后台运行,以实现优势互补。VB 和 MATLAB 软件的混合编程成为关键。本文就是针对此问题详细介绍一项基于组件对象模型(COM)的 MATLAB 与 VB 混合编程方法,以 COM Builder 为转换工具,将 MATLAB 函数文件转换为 COM 组件,在 VB 程序中调用这个组件进行参数赋值。建立一个日光温室可视化模拟模型软件。

关键词:VB;MATLAB;COM Builder;可视化模型

Hybrid Programming with MATLAB and VB in Building Visual Simulation Model for Thermal Environment in Chinese Solar Greenhouse

Meng Lili[1], Zhang Yi[2], Yang Qichang[2]

(1. *Institute of Vegetable Crops*, *Jiangsu Academy of Agricultural Sciences*, *Nanjing* 210014, *P. R. China*;
2. *Institute of Environment and Sustainable Development in Agriculture*, *Chinese Academy of Agricultural Sciences*, *Beijing* 100081, *P. R. China*)

Abstract: In the development of visual simulation model of thermal environment in Chinese solar greenhouse, VB was selected as host interface because of its good human-machine interface, Matlab was selected as computer-processing tools because of its strong engineering calculations and image processing capabilities. The function of software exploited by user can be greatly strengthened through the union. The key was hybrid programming with VB and Matlab. A hybrid programming method based on COM Builder was introduced in the paper. In order to develop the VB program without Matlab environment, the method based on COM is introduced in detail. By mean of Matlab COM builder, Matlab's function files can be converted into a COM component, in the VB program to call this component parameter assignment. which is applied in VB. The realization of visual simulation model of greenhouse thermal environment indicates the method.

Key words: VB; MATLAB; COM Builder; Visual model

---

* 基金项目:国家"863"基金资助项目(2006AA10Z260)。
** 作者简介:孟力力(1982—),女,硕士,研究方向为温室环境建模。江苏省农业科学院蔬菜研究所,南京 210014。E-mail:menglili90@163.com。
*** 通讯作者:杨其长(1963—),男,研究员,博士生导师,从事设施农业方向的研究。中国农业科学院农业环境与可持续发展研究所,北京 100081。E-mail:yangq@cjac.org.cn。

在日光温室热环境理论模型实现中,前人已采取了多种方法:李元哲等[1]根据微气候模型编制成的 FORTRAN 语言程序;陈青云等[2]用 QUICK-BASIC 编制了求解软件。输入温室结构参数、物性参数和逐时室外气象参数,即可求出各内表面温度、室内空气温度和湿度;于威等[3]利用 FORTUNE 编写程序实现了辽沈 I 型日光温室内热平衡的建立及数值模拟;邓玲黎等[4]通过自编的回归软件进行回归计算后,得出夏季室内温度的预测模型;综上所述,MATLAB 和 VB 混合编程,实现优势互补,是模型开发研究中的一个新方法。使用 VB 开发模型主界面,完成模型参数初始赋值、系统参数更改、数据汇总等工作。而 MATLAB 作为后台应用程序,包含了十几个工具箱,集数值分析、矩阵运算、信号处理和图形显示于一体,具有强大的高性能可视化的数值计算功能,将计算后的数据及图像传输给 VB 软件,进行人机界面开发。

因此,如何通过 VB 和 MATLAB 之间的混合编程实现日光温室热环境模拟模型,在前人工作的基础上,已经实现利用软件进行复杂计算得到结果的目的,只是在源程序中更改参数值比较复杂,容易出错。作者开发出具有可视界面的软件,更改参数赋值简洁明了,这是本次研究重点。

1 MATLAB 和 VB 软件的混合编程方法

目前,MATLAB 和 VB 软件的混合编程方法有以下几种方法[5]:①动态数据交换(DDE):DDE 是 Dynamic Data Exchange 的缩写,是 Windows 过程之间的通信机制,使用 Windows 消息和共享的内存,使相互合作的应用程序能够交换数据。其中,数据的提供者为 DDE 的服务端,数据的接受者为 DDE 的客户端。②ActiveX 自动化服务技术。将 MATLAB 作为服务器,接收通过引擎传来的数据和指令信息并进行相应的处理,然后将结果经过引擎返回给发送请求的客户机;③中间文件传递法:用一种自己熟悉的高级语言如 VB 编写前端用户交互界面,收集必要的参数信息,并保存在一个中间文件(如 temp.txt)中,然后利用异步程序调用方式执行 MATLAB 程序。④动态链接库(DLL 文件):Matcom 是一个从 MATLAB 到 C++ 的编译器,MathTools 公司利用 Matcom 技术编写了 Mideva 工具软件,可以借用 C++ 编译器将 MATLAB 下的 M 文件转换为可被 VB 调用的 DLL 文件。⑤COM 组件技术:COM 是 component object module 的简称,它是一种通用的对象接口,支持 COM 的程序称为 COM 组件。从 MATLAB6.5 版本开始,MATLAB 提供了 COM 生成器。COM 生成器是实现 MATLAB 独立应用的一种新途径。它能把 MATLAB 开发的程序做成组件,这些组件作为独立的 COM 对象,可以直接被 VB、VC++ 等支持 COM 的语言调用。

在上述五种方法中,前三种调用过程复杂,且无法脱离 MATLAB 环境,这会直接影响所开发软件的独立运行和发布。Mideva 虽然能脱离 MATLAB 环境,但已经停止发行。COM 组件不但可以脱离 MATLAB 环境,而且在 VB 访问该组件时,无需进行环境切换,因此可以获得最快的运行速度。

2 日光温室热环境理论模型和求解方法

VB 软件在设计开发的 Windows 应用程序方面界面友好,方便快捷。操作人员可以直接在界面上直接进行模型初始参数赋值、系统参数更改,另一方面可送到 MATLAB 中进行复杂的运算处理,处理后的结果再送回VB,进行计算结果及图像显示的汇总,形成一个具有可视化界面的模型软件。

现就温室热环境模拟模型开发中的混合编程方法展开叙述。

2.1 日光温室热环境理论模型

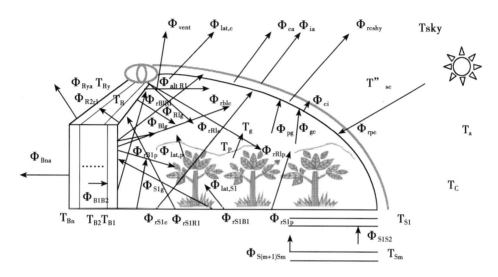

图1 日光温室中的热流量
Figure 1 Heat fluxes in the CSG

温室热环境是一个相当复杂的过程,如图1所示。本试验建立的日光温室热环境模拟模型[6],定量描述了日光温室内的太阳辐射、对流换热、辐射换热、热传导、自然通风和水分相变带来的潜热对日光温室热环境的影响,根据质能平衡和传热学理论,得到一组关于覆盖物C、室内空气g、温室分层后墙B1、B2……Bn、分层地面土壤S1、S2……Sm、分层后坡R1、R2……Ry和作物p热平衡的微分方程组。方程组包含诸多可变化参数如:温室的结构参数、室外的温度值参数、墙体、土壤的热性能参数等,未知数为覆盖物C、室内空气g、温室分层后墙B1、B2……Bn、分层地面土壤S1、S2……Sm、分层后坡R1、R2……Ry和作物p的温度值。

在本模型中,特做如下简化条件:

(1) 温室空气、覆盖物、墙体各层、后坡各层、土壤各层视作均匀的;

(2) 根据传热学理论可知,如平面板壁的高(长)度和宽度是厚度的8~10倍,按一维导热处理时,其计算误差不大于1%[7]。因此,在本模型中忽略墙体沿长度和高度方向的导热,即只考虑沿厚度方向的一维导热;

(3) 由于土壤沿跨度和长度方向的温度梯度远小于垂直方向的温度梯度,在计算土壤传热时,只考虑纵向导热;

(4) 作物的热物理性质和水一样。

冬季日光温室的热交换包括太阳辐射、热辐射、热传导、对流热交换、水蒸气蒸发冷凝等基本形式。日光温室热量的主要来源是太阳辐射,室内空气热量的得失主要有以下几种途径:一是与温室围护结构(即覆盖材料、地面、后坡和后墙)的对流换热;二是与室内作物叶片之间的对流换热；三是通过温室通风和缝隙漏气与室外空气进行热量交换；四是植物的蒸腾及土壤蒸发引起的潜热变化。

2.1.1 覆盖材料的热平衡方程

包括覆盖材料吸收的太阳辐射，覆盖材料与土壤、墙体、后坡、植物进行长波辐射换热，与室内和室外空气进行对流放热，还以辐射形式与大气层 sky 进行热交换：

$$C_{apth,c}\frac{dT_c}{dt} = \Phi_{gc} - \Phi_{ca} + a_c A_{gc} I'' - \Phi_{rcsky} + \Phi_{rpc} + rB1c + \Phi_{rs1c} + \Phi_{lat,c}$$

$$= F_c(T_a, T_{b1}, T_c, T_g, T_p, T_{R1}, T_{s1}, T_{sky}, I''_{sc}) \tag{1}$$

Φ_{12} 代表温室各部分之间的传导热流量，与各部分温度有关 $\Phi_{12} = f(T_1, T_2)$，Φ_{r12} 代表温室各部分之间的辐射热流量，与各部分温度有关 $\Phi_{r12} = f_r(T_1, T_2)$，$\Phi_{lat,c} = \Delta H k_{gc} A_c (C_g - C_c)$ 为薄膜内表面水分相变引起的潜热变化，C 代表各部分的水蒸汽浓度，与各部分的温度有关：$C = f_c(T)$。因此整个微分方程可以写为：

$$C_{apth,c}\frac{dTc}{d\tau} = F_c(Ta, TB1, Tc, Tg, Tp, TR1, Ts1, Tsky, I''sc)$$

2.1.2 温室空气的热平衡方程

温室空气主要与土壤表面、后坡表面、墙体表面、覆盖材料和植物进行对流交换，并通过通风换气和外界空气进行热交换：

$$C_{apth,a}\frac{dT_g}{dt} = -\Phi_{gc} - \Phi_{ci} + \Phi_{pg} + \Phi_{S1g} + \Phi_{B1g} + \Phi_{R1g} + \Phi_{vent}$$

$$= F_g(T_a, T_{B1}, T_c, T_g, T_p, T_{R1}, T_{s1}, T_{sky}, V) \tag{2}$$

其中 $\Phi_{vent} = f_{vent}(V)$ 是一个与外界风速有关的函数。

2.1.3 温室后墙的热平衡方程

将其沿厚度方向划共分为 N 层，从内到外分别是 1、2、…、n 层，并认为每个分层温度与热物理性质参数均匀一致。墙体的热过程包括：各层墙体之间的热传导，吸收的太阳辐射，与室内空气的对流热交换，与室内各个面的长波辐射热交换，与室外空气长波辐射热交换。各层的能量平衡方程为：

第一层墙体：

$$C_{apth,B1}\frac{dT_{B1}}{dt} = -\Phi_{rB1p} - \Phi_{rB1c} - \Phi_{rB1S1} - \Phi_{rB1R1} + a_{B1}A_{B1}I''_{sc} - \Phi_{B1g} + \Phi_{B1B2}$$

$$= F_{B1}(T_a, T_{B1}, T_{B2}, T_c, T_g, T_p, T_{R1}, T_{s1}, T_{sky}, I''_{sc}) \tag{3}$$

第二层到第 n-1 层墙体：

$$C_{apth,B(n-1)}\frac{dT_{B(n-1)}}{dt} = \Phi_{BnB(n-1)} - \Phi_{B(n-1)B(n-2)}$$

$$= F_{B(n-1)}(T_{B(n-2)}, T_{B(n-1)}, T_{Bn}) \tag{4}$$

靠近外界的最后一层墙体（第 n 层）的热交换过程包括与天空的辐射传热，与外界空气的对流换热、对太阳辐射的吸收，各层之间的热传导：

$$C_{apth,Bn}\frac{dT_{Bn}}{dt} = -\Phi_{rBnsky} - \Phi_{Bna} + a_{Bn}A_{Bn}I''_{sc} - \Phi_{BnB(n-1)}$$

$$= F_{Bn}(T_a, T_{Bn}, T_{B(n-1)}, I''_{sc}) \qquad (5)$$

2.1.4 温室土壤的热平衡

根据上述简化，室内土壤传热为一维导热，将其按深度方向共分为 M 层，分别为 1、2、…、m 层。第一层要考虑土壤与温室各部分的辐射传热、太阳辐射、土壤与空气的对流热传导，相邻土壤层间的热传导，土壤表面冷凝水的潜热；其余各层只考虑各层间的热传导情况。

$$C_{apth,S1}\frac{dT_{S1}}{dt} = -\Phi_{rS1p} - \Phi_{rS1c} - \Phi_{rS1B1} - \Phi_{rS1R1} + a_{S1}A_{S1}I''_{sc} - \Phi_{S1g} + \Phi_{S2S1} - \Phi_{lat,S1}$$
$$= F_{S1}(T_a, T_{B1}, T_c, T_g, T_p, T_{R1}, T_{S1}, T_{S2}, I''_{sc}) \qquad (6)$$

$$C_{apth,Sm}\frac{dT_{Sm}}{dt} = \Phi_{SmS(m-1)} - \Phi_{S(m-1)S(m-2)} = F_{Sm}(T_{S(m-2)}, T_{S(m-1)}, T_{Sm}) \qquad (7)$$

2.1.5 温室后坡的热平衡

后坡多为异质轻型复合结构，假设各不同材质层是均匀的，按不同材质划共分为 Y 层，分别为 1、2、…、y 层，考虑表层与温室其他部分的换热过程和其余各层间的热传导过程。

第一层：

$$C_{apth,R1}\frac{dT_{R1}}{dt} = -\Phi_{rR1p} - \Phi_{rR1c} - \Phi_{rR1S1} - \Phi_{rR1B1} - \Phi_{R1g} + \Phi_{R2R1} + a_{R1}A_{R1}I''_{sc} + \Phi_{lat,R1}$$
$$= F_{R1}(T_{B1}, T_c, T_g, T_p, T_{R1}, T_{R2}, T_{S1}, I''_{sc}) \qquad (8)$$

对于第二层到第 y-1 层的热平衡方程，只存在各层之间的热传导：

$$C_{apth,R(y-1)}\frac{dT_{R(y-1)}}{dt} = \Phi_{RyR(y-1)} - \Phi_{R(y-1)R(y-2)} = F_{R(y-1)}(T_{R(y-2)}, T_{R(y-1)}, T_{Ry}) \qquad (9)$$

后坡最外面一层（第 y 层）的热平衡方程，包括与室外空气的对流热交换、与天空温度的辐射热交换、对太阳辐射的吸收和相邻各层的热传导过程：

$$C_{apth,Ry}\frac{dT_{Ry}}{dt} = -\Phi_{rRysky} - \Phi_{Rya} + a_{Ry}A_{Ry}I''_{sc} - \Phi_{RyR(y-1)}$$
$$= F_{Ry}(T_a, T_{R(y-1)}, T_{Ry}, T_{sky}, I''_{sc}) \qquad (10)$$

2.1.6 温室作物的热平衡

温室作物与其他部分的热交换和吸收的太阳辐射，以及作物叶片上水蒸气引起的潜热变化：

$$C_{apth,p}\frac{dT_p}{dt} = -\Phi_{pg} - \Phi_{rpc} + \Phi_{rS1p} + \Phi_{rB1p} + \Phi_{rR1p} + \alpha_p A_p I''_{SC} + \Phi_{lat,p}$$
$$= F_p(T_a, T_{B1}, T_g, T_{R1}, T_{S1}, I''_{sc}) \qquad (11)$$

2.2 求解方法

将上述分析所得热平衡方程式联立得到一个微分方程组。方程组中方程式比较多，而且是非线性的，因此要借助 MATLAB 的计算函数 ode 进行求解。如果已知温室的结构参数、温室各部分热物性参数、外界气象因子参数，给定计算步长，可以逐时算出温室分层后墙温度 TB1，TB2，…，TBn，覆盖材料温度 Tc，温室空气温度 Tg，温室作物温度 Tp，温室分层后坡温度 TR1，TR2，…，TRy，温室分层土壤温度 TS1，TS2，…，TSm。

MATLAB 使用龙格－库塔－芬尔格（Runge-Kutta-Fehlberg）方法来解 ODE 问题。此方

法最大的优点是：这些点的间距由解的本身来决定。当解比较平滑时，区间内使用的点数少一些，在解变化很快时，区间内使用较多的点。在求解微分方程组时 Ode45 被推荐为首选方法。在用 odesolver（ode45，ode15s，…）来解微分方程的时候，最基本的用法是：［t, y］= odesolver（odefun，tspan，y0）；这里的 odefun 是待求的微分方程。

本模型中含有多个模型参数，通常要通过改变模型参数来观察模型动态的变化。那么如何在调用 odesolver 的时候传递参数呢？如果通过更改源程序来实现参数值变化，则程序可靠性及通用性降低。本文研究的 VB 与 MATLAB 混合编程中，参数赋值在 VB 建立的主界面实现，这些参数值在调用 MATLAB 程序时由 COM 组件传递，大大增加了程序的稳定性，最大程度保证计算结果的精确。具体方法如下所述。

3 模型软件的实现

3.1 编写 MATLAB 程序（即，M 文件）

每个程序的规范要求是以函数形式出现，不管是有参或是无参。

模型中的函数为有参函数 CSG_model（W1，W2，…，Cs6）.m（因函数参数较多，本文程序中不一一列出）：运行时，输入参数值，对微分方程组进行运算，最后绘出模拟值变化曲线。

源程序部分代码如下：

function CSG_model（W1，W2，…，Cs6）

x0 =［9.6, 6.2, 10, 12, 7.5, 10, 0.005, 12, 12.6, 13.8, 8, 4, 3, 2, -1, -2］;% 赋给微分方程组初始值

Ta = load（'Book41.txt'）;% 读取外界气象参数

n = length（Ta）;

tspan =［0：300：(n-1)*300］';% 确定时间步长

［t, x］= ode45（CSGghs, tspan, x0,［］, W1, W2, …, Cs6）;% 求解微分方程组

figure;

disp（［'Tg Tc Tp Ts1 TB1 TR Cg Ts2 Ts3 Ts4 TB2 TB3 TB4 TB5 TB6 TB7'］）;

disp（［x（:, 1）, x（:, 2）, x（:, 3）, x（:, 4）, x（:, 5）, x（:, 6）, x（:, 7）, x（:, 8）, x（:, 9）,

x（:, 10）, x（:, 11）, x（:, 12）, x（:, 13）, x（:, 14）, x（:, 15）, x（:, 16）］）;% 显示模拟值

plot（t, x［:, 1］, '——r'）% 绘出各预测值随时间变化的曲线

hold on

plot（t, x［:, 2］, '-b'）

hold on

…

plot（t, x［:, 16］, '-c'）

hold on

box on, legend（'Tg', 'Tc', 'Tp', 'Ts1', 'TB1', 'TR', 'Cg', 'Ta',

'Ts2','Ts3','Ts4','TB2','TB3','TB4','TB5','TB6','TB7')%标注图例
function xdot = CSGghs(t, x, W1, W2, …, Cs6);%描述微分方程组的函数

3.2 创建工程

在 MATLAB 命令行中输入 comtool 命令，调用 COM 生成器，出现主窗口图 2。在"File"菜单中选择"New Project"选项，打开"New Project Setting"对话框。添加组件名（Component Name）CSG、类名（Class Name）MATLABmodle、版本号，如果调用 MATLAB 绘图命令，则需要 MATLAB 提供的 C/C++ 图形库，这时在编译器选项（Compiler Options）中必须选中"使用句柄图形库（Use Handle Graphic Library）"、生成调试版本号、显示详细编译信息。设置完成后，按"OK"按钮，将设置保存在一个工程中。

3.3 管理 M 文件和 MEX 文件

单击"Add File"按钮，载入已经在 MATLAB 环境下编译通过的 M 函数文件 CSG_model.m，M 文件中的函数名和类中的方法名相一致（图3）。

图 2　MATLAB COM Builder 主窗口
Figure 2　Main window of MATLAB COM Builder

图 3　新工程设置对话框
Figure 3　The new project settings dialog

3.4 生成组件

单击"Build"按钮。待编译完成，COM 组件在计算机中自动注册。如果要将此组件在另一计算机使用，选择菜单"Component/Package Component"，得到 distrib 文件夹，它包含表1所示的几个文件。在目标计算机上运行此文件便可完成对 COM 组件的注册。

表 1　应用 MATLAB COM Builder 生成的文件
Table 1　The generated files of MATLAB COM Builder

文件（File）	功能（Purpose）
< componentname >.exe	自解压可执行程序
_install.bat	自解压可执行程序所需调用的批处理文件
< componentname_projectversion >.dll	已编译成功的 DLL
mglinstaller.exe	安装 MATLAB 数学和图形库
mwcomutil.dll	COM Builder 应用库
mwregsvr.exe	在目标计算机上注册 DLL

3.5 在 VB 中调用 COM 组件

调用 VB 系统，选择新建工程，打开"Project"——"Reference"对话框，选中 CSG1.0 Type Library，将在 MATLAB 中生成的 COM 组件引入 VB 中。在代码中创建和定义新类。本例只讨论窗体级引用，窗体中设置两个命令按钮，command1（调用 MATLAB 程序）、command2（退出）；界面还有若干 Text 按钮，用于给函数参数赋值。主界面如图 4 所示。

VB 源程序部分代码如下：

```
Private CSG1 As CSG..MATLABmodel   '窗体级变量，定义类实例
Private Sub Form_ Load（）          '窗体装载时，创建新的类变量 CSGl
On Error GoTo handle_ error
Set CSG1 = New CSG.MATLABmodel
handle_ error：
MsgBox（Err.description）
End Sub
Dim Ta   As Double            '定义输入参数的变量类型
Dim W1 As Single
Dim W2 As Single
……
Dim Cs6 As Single
W1 = Val（Text1.Text）         '在 VB 中给参数赋值
W2 = Val（Text2.Text）
……
Cs6 = Val（Text60.Text）
CommonDialog1.ShowOpen
Open CommonDialog1.FileName For Input As #1    '打开文件
n = 0
Do While Not EOF（1）
n = n + 1
Input #1，Ta       '读取外界温度变化值
Loop
Close #1          '关闭文件
Call CSG1.CSG_ MATLAB（W1，W2，…，Cs6）    '带参调用.M 文件
End Sub
```

单击调用按钮后将出现如图 5 所示的结果。

运行结果将可以在脱离 MATLAB 环境的情况下，在 VB 界面上轻松调用 MATLAB 文件，同时具有较快的运行速度。

4 结论与讨论

本试验通过一种基于组件对象模型的 COM 组件技术，利用具有良好人机界面的 VB 和

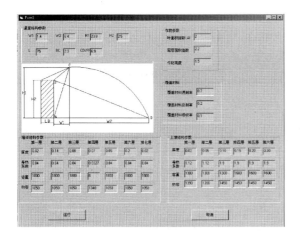
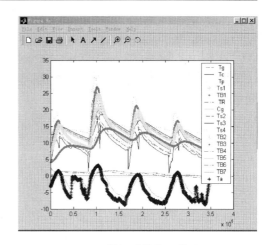

图 4　VB 用户主界面
Figure 4　The VB User interface

图 5　模拟计算结果曲线
Figure 5　The simulating value curve

具有强大计算能力的 MATLAB 混合编程,开发出具有可视化界面的日光温室热环境模拟模型软件。具体方法为：以 COM Builder 为转换工具,将 MATLAB 函数文件转换为 COM 组件,在 VB 程序中调用这个组件进行参数赋值。通过试验验证,各部分温度预测值准确率可达 0.66 以上,模拟结果与实测结果基本相符。该模型软件的主界面清晰明了,赋值操作简便,可以模拟不同条件下的温室热环境,为日光温室结构优化和模型软件的商业化开发提供了理论计算依据。在下一步的研究工作中,可以对温室参数进行分类细化,形成不同的子界面,丰富软件功能与界面表达,使模型更加完善。

参考文献

[1] 李元哲,吴德让,于竹. 日光温室微气候的模拟与实验研究. 农业工程学报,1994,10(1)：130~136
　　Li Yuanzhe, Wu Derang, Yu Zhu. Simulation and test research of micrometeorology environment in a sun-light greenhouse. Transactions of The Chinese Society of Agricultural Engineering, 1994, 10 (1): 130~136. (in Chinese with English abstract)

[2] 陈青云,汪政富. 节能型日光温室热环境的动态模拟. 中国农业大学学报,1996,1(1)：67~72
　　Chen Qingyun, Wang Zhengfu. Dynamic simulation of sun-light greenhouse thermal environment. Journal of China Agricultural University, 1996, 1 (1): 67~72. (in Chinese with English abstract)

[3] 于威,王铁良,任冰洁等. 辽沈 I 型日光温室内热平衡的建立及数值模拟. 节能,2005,8：22~24
　　Yu Wei, Wang Tieliang, Ren Bingjie, et al. Research on numberical simulation and heat balance of solar greenhouse-type liaoshen. Energy Conservation, 2005, 8: 22~24

[4] 邓玲黎,李百军,毛罕平. 长江中下游地区温室内温湿度预测模型的研究. 农业工程学报,2004,20(1)：263~266
　　Deng Lingli, Li Baijun, Mao Hanping. Forecasting model of inter temperature and humidity for intelligent greenhouse in the Middle and Lower Reaches of the Yangtze River. Transactions of the CSAE, 2004, 20 (1): 263~266. (in Chinese with English abstract)

[5] MATLAB COM Builder user's guide. The Mathworks Inc, 2002

[6] 孟力力,杨其长,Gerard. P. A. Bot 等. 日光温室热环境模拟模型的构建. 农业工程学报,2009,25(1)：164~170
　　Meng Lili, Yang Qichang, Gerard. P. A. Bot, et al. Visual simulation model for thermal environment in Chinese solar greenhouse. Transactions of the CSAE, 2009, 25 (1): 164~170. (in Chinese with English abstract)

[7] 彦启森,赵庆珠. 建筑热过程. 北京:中国建筑工业出版社,1986
　　Yan Qisen, Zhao Qingzhu. Building Thermal Process. Beijing: China Building Industry Press, 1986. (in Chinese)
[8] 岳玉芳,尤忠生,张玉双. 基于COM的VB与MATLAB混合编程. 计算机工程与设计,2005,26(1):61~65
　　Yue Yufang, You Zhongsheng, Zhang Yushuang. Hybrid programming with VB and MATLAB based on COM. Computer Engineering and Design. 2005, 26 (1): 61~65

下挖式节能日光温室采光优化设计*

李清明[1]**,艾希珍[1],于贤昌[2]***

(1. 山东农业大学园艺作物生物学农业部重点开放实验室,作物生物学国家重点实验室,
山东 泰安 271018; 2. 中国农业科学院蔬菜花卉研究所,北京 100081)

摘要:采光优化设计是节能日光温室结构优化设计的关键和改进温室环境性能最经济有效的方式。本试验以寿光市脊高4.0m、跨度10m的下挖式节能日光温室为例,对中国北方地区下挖式节能日光温室合理采光时段屋面角以及温室方位角进行了理论计算与初步分析,并利用计算线阴影的原理,通过计算不同时刻下挖壁面在温室内形成的阴影区域和整个温室形成的阴影区域,提出了中国北方地区下挖式节能日光温室适宜的下挖深度和温室间距建议,为目前下挖式节能日光温室的设计建造提供理论依据和技术参数。

关键字:下挖式;节能日光温室;采光设计;阴影区域

Lighting Optimization Design of Sunken-type Energy-saving Sunlight Greenhouse

Li Qingming[1]**, Ai Xizhen[1], Yu Xianchang[2]***

(1. Agriculture Ministry Key Laboratory of Horticultural Crop Biology, State Key Laboratory of
Crop Biology, Shandong Agricultural University, Tai'an, Shandong 271018, P. R. China;
2. Institute of Vegetable and Flower, Chinese Academy of Agricultural Sciences, Beijing, 100081, P. R. China)

Abstract: Lighting optimization design is very important for structure optimization design of energy-saving sunlight greenhouse, and it is the most cost-effective means to improve greenhouse environmental performance. In this paper, sunken-type energy-saving sunlight greenhouse with 4.0m ridge height and 10m span in Shouguang city was taken for example. The front roof angle within reasonable lighting time and azimuth of sunken-type energy-saving sunlight greenhouse in northern China were calculated and analyzed in theory. Using the principle of calculating the line shadow, through calculated the indoor shaded regions of the greenhouse formed by the front sunken wall and the north shaded regions formed by front greenhouse at different times of the day, the feasible sink depth and spacing of sunken energy-saving sunlight greenhouse in north China were put forward. It can provide a theoretical basis and technical parameters for the current design and construction of sunken energy-saving sunlight greenhouse.

Key words: Sunken-type; Energy-saving sunlight greenhouse; Lighting design; Shaded region

从2009年12月7日至18日在哥本哈根召开的气候变化峰会可以明显的看到,如何实现节能减排是目前全世界无论发达国家还是发展中国家普遍关注的问题[1]。太阳能是一种蕴藏量巨大、永不枯竭的洁净能源,是各种可再生能源中最重要的基本能源[2]。20世纪80

---

* 基金项目:国家"十一·五"科技支撑计划项目(2008BADA6B05);山东省农业重大应用技术创新专项;山东农业大学青年科技创新基金项目;山东农业大学博士后流动站资助项目(71999)。
** 作者简介:李清明,博士,副教授,主要从事设施环境调控与栽培生态生理方面的教学与科研工作。
*** 通讯作者 Author for correspondence:E-mail: xcyu1962@163.com。

年代在中国北方地区迅速发展起来的节能日光温室是有效利用太阳能的最主要的设施类型，由于节能日光温室建造和运行成本低，适合中国国情，现已成为中国北方地区设施农业生产的主体设施，对中国"菜篮子"工程发挥着极为重要的作用[3]。

太阳辐射不仅是节能日光温室内气温和土壤温度的主要热源，也是植物进行光合作用等基本生理活动的能量源和形态建成及控制生长等过程的信息源[4]，温室内的光量及其分布，制约着温室的生产潜力[5]，因此，采光优化设计是节能日光温室结构优化的关键和改进温室环境性能最经济有效的方式。目前中国节能日光温室结构优化与环境调控的研究在理论和应用上都取得了很大的进展[6~11]，研究主要集中在最优采光屋面角和采光屋面形状的确定，提高温室保温性能的结构设计以及保证温室强度条件下低建造成本的最优结构设计等方面[3]。

近年来在山东寿光出现了厚墙大跨下挖式节能日光温室，由于其地温比较稳定，保温性好，热状况明显优于普通节能日光温室[12]，因而发展较快，现已成为该地区节能日光温室的主流结构类型，面积迅速扩大。但是，目前对这种下挖式节能日光温室结构缺乏系统的理论研究，各地主要依靠模仿和有限的实践经验而建造，大部分下挖深度过深而前后排温室间距又过小，造成室内阴影面积过大、前排温室阴影部分遮挡后排温室的采光面，影响温室的温光性能。因此，本文在对下挖式节能日光温室的采光进行理论计算与分析的基础上，利用计算线阴影的原理，计算不同时刻下挖壁面在室内形成的阴影区域以及整个温室形成的阴影区域，以确定适宜的下挖深度和温室间距，为目前下挖式节能日光温室的设计建造提供理论依据和技术参数。

1 下挖式节能日光温室合理采光时段屋面角优化设计

采光屋面角（β，指温室采光屋面与地平面的夹角）在很大程度上决定着太阳直射光透过采光屋面透明覆盖材料的透过率，因而是节能日光温室设计和建造中的关键[3]。太阳直射光的透过率与太阳入射角（i，太阳射线与采光面法线之间的夹角）有关，入射角 i 越小，透光率越大。所以确定节能日光温室合理采光时段屋面角（β_m）以保证日光能有合理的入射角 i 是采光设计的关键之一。

从理论上讲，入射角 i 为 0°（日光垂直照射于采光屋面）时，日光反射率为 0 而透过率为最大值 1。但是在冬季太阳高度角（h）很小，要使入射角 i 为 0°，则采光屋面角 β 必须很大，这样的温室采光虽然较好，但是脊高和后墙则相应增高，既浪费建材，又不利于保温。而且由于太阳位置随季节和时刻的变化而变化，日光也不可能始终垂直照射采光屋面，因此在生产上并不现实。入射角 i 在 0°~40°范围内，随入射角 i 增大，日光透过率略有下降，但不显著。入射角 i 为 40°时的反射损失仅为 3.4%，对采光影响不大[13]。另外，根据合理采光时段理论，节能日光温室的合理采光时段应保持午间 4h 以上[14]。因此，为了最大程度地为温室作物生长提供最佳温光条件，一般以太阳高度角最小、光照最弱的冬至日为计算日，以日光在上午 10：00 时的入射角 i 为 40°来计算节能日光温室的合理采光时段屋面角 β_m，按式（1）计算[15~16]：

$$\cos i = \cos\beta_m \times \sin h + \sin\beta_m \times \cos h \times \cos\varepsilon \quad (1)$$

式中：i 为太阳入射角，ε 为采光屋面太阳方位角（图1），即采光屋面上某点和太阳之间的连线在水平面上的投影与采光屋面法线在水平面上的投影之间的夹角（$\varepsilon = \alpha - \gamma$，$\gamma$ 为节能日光温室方位角）。采光屋面朝向正南时 $\gamma = 0$，则 $\varepsilon = \alpha$，即采光屋面太阳方位角等于太阳方位角。

笔者以采光屋面朝向正南（$\gamma = 0$）的日光温室为例，计算我国北方各地冬至日（$\delta = -23.45°$）10：00 时（$\omega = -30°$）的太阳高度角 h 和太阳方位角 α，再将 $i = 40°$ 代入式（1），通过计算机试算，可分别得到我国北方各地合理采光时段屋面角 β_m，如表 1 所示。

表 1　中国北方地区冬至日 10：00 时太阳位置及合理采光时段屋面角
Table 1　Sun location at 10：00 and front roof angle within reasonable lighting time in winter solstice in north China

城市	纬度 φ (°)	太阳高度角 h (°)	太阳方位角 α (°)	合理采光时段屋面角 β_m (°)
西安	34.25	25.6	-30.6	30.4
郑州	34.73	25.2	-30.5	30.9
兰州	36.05	24.1	-30.2	32.2
西宁	36.62	23.6	-30.0	32.8
济南	36.67	23.6	-30.0	32.8
寿光	36.88	23.4	-30.0	33.0
太原	37.85	22.5	-29.8	34.0
银川	38.47	22.0	-29.7	34.6
北京	39.90	20.8	-29.4	36.1
呼和浩特	40.80	20.0	-29.2	37.0
沈阳	41.80	19.1	-29.0	38.0
乌鲁木齐	43.77	17.4	-28.7	39.9
长春	43.92	17.2	-28.7	40.1
哈尔滨	45.75	15.6	-28.4	41.9

由表 1 可以看出，不同地区由于地理位置不同，合理采光时段屋面角 β_m 也不同。随着纬度 φ 的增加，10：00 时太阳方位角 α 逐渐减小（绝对值，负号表示午前），也就是说该时刻太阳位置越偏向正南方向，说明一天中日照时数越少。太阳高度角 h 也随着纬度 φ 的增加而逐渐减小，必须增加采光屋面角来减小入射角进而增加透光率，以获得更多的太阳辐射热量，因此合理采光时段屋面角 β_m 必然随着纬度 φ 的增加而逐渐增大。

2　下挖式节能日光温室下沉深度优化设计

照射到地球表面的太阳光线可视为一束平行光，不同方向的光线，将产生不同形状的阴影[17]，因此同样结构参数的节能日光温室在不同的地理位置、季节和时间产生的阴影区域也就不同。对下挖式节能日光温室采光而言，在屋面结构尺寸、方位、骨架材料和覆盖材料等确定的情况下，在某一时刻其自身骨架、立柱以及后屋面等在室内产生的阴影就基本确定了，那么下壁面深度就成为室内地面阴影面积的决定因素。因此，确定适宜的下深度对下挖式节能日光温室的采光和蓄热保温就显得尤为重要。

我们假设温室下挖壁面是由无数有规律的直线组成的平面，那么其在室内形成的阴影则由各种平行四边形组成，这样就可以得出任意时刻室内阴影区的大小和位置。反过来，如果确定了某一时刻室内规定的最大阴影面积，则可求出该时刻下沉壁面的下沉深度。

如图 1 所示，AB 为下挖壁面上任意一条垂直线段，在计算阴影时把下挖壁面看作是由无数条线段 AB 的平行线组成的矩形面，只要知道某个时刻 AB 的投影位置及投影长度，用连线方法就能得到该时刻整个矩形区域的阴影。AB 的高度即下沉深度为 H_o，其在某个时刻的阴影为 BC，则矩形阴影区域的高为 BD，即阴影长度 D_o，那么整个阴影区域的面积即可由 D_o 与温室长度的乘积得到。阴影长度 D_o 按式（2）计算[18]：

$$D_o = \frac{H_o}{\tan\lambda} \tag{2}$$

式中：λ 为太阳高度角 h 的侧影角（太阳高度角在地平面的一个垂直表面法线方向上的投影角），λ 可按式（3）计算[18]：

$$\tan\lambda = \frac{\tan h}{\cos\varepsilon} = \frac{\tan h}{\cos(\alpha - \gamma)} \tag{3}$$

将式（3）代入式（2），可得：

$$D_o = H_o \times \frac{\cos(\alpha - \gamma)}{\tan h} \tag{4}$$

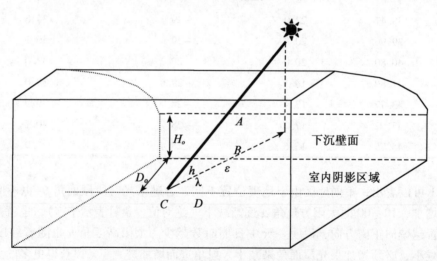

图 1　节能日光温室下沉壁面在室内产生阴影示意图
Figure 1　The indoor shadow region formed by sunken wall of energy-saving sunlight greenhouse

由式（4）可以看出，阴影长度 D_o 与太阳高度角 h 的正切函数成反比，因此太阳高度角 h 最小的冬至日阴影长度 D_o 最大，阴影区域面积也就最大。如果仅按冬至日计算阴影区域面积的话，那么适宜的下沉深度则很小，发挥不了下挖式节能日光温室的高保温蓄热效应的优势。因此，我们提出以冬季温度较低的 11 月至翌年 2 月的平均值来计算阴影区域面积。以寿光市采光屋面朝向正南（$\gamma = 0$）的下挖式节能日光温室为例，计算下沉深度分别为 0.5 ~

1.5m、11月至翌年2月合理采光时段10：00～14：00时的平均阴影长度，如表2所示。

表2 寿光市各下沉深度的下沉壁面在室内产生的平均阴影长度日变化
Table. 2 Diurnal variation of indoor average shadow length formed by sunken wall
of energy-saving sunlight greenhouse in Shouguang city （单位：m）

下沉深度 (m)	时间								
	10：00	10：30	11：00	11：30	12：00	12：30	13：00	13：30	14：00
0.5	0.84	0.80	0.77	0.76	0.75	0.76	0.77	0.80	0.84
0.6	1.00	0.95	0.92	0.91	0.90	0.91	0.92	0.95	1.00
0.7	1.17	1.11	1.08	1.06	1.05	1.06	1.08	1.11	1.17
0.8	1.34	1.27	1.23	1.21	1.20	1.21	1.23	1.27	1.34
0.9	1.51	1.43	1.39	1.36	1.35	1.36	1.39	1.43	1.51
1.0	1.67	1.59	1.54	1.51	1.50	1.51	1.54	1.59	1.67
1.1	1.84	1.75	1.69	1.66	1.65	1.66	1.69	1.75	1.84
1.2	2.01	1.91	1.85	1.81	1.80	1.81	1.85	1.91	2.01
1.3	2.18	2.07	2.00	1.96	1.95	1.96	2.00	2.07	2.18
1.4	2.34	2.23	2.16	2.12	2.10	2.12	2.16	2.23	2.34
1.5	2.51	2.39	2.31	2.27	2.25	2.27	2.31	2.39	2.51

由表2可以看出，随着温室下深挖度的增加，各下挖壁面在室内形成的阴影长度明显增大。阴影长度以午时12：00最小，且午前、午后各时刻阴影长度关于午时对称。越接近中午，阴影长度的变化幅度越小，而越远离中午，变化幅度越大，这与太阳高度角和太阳方位角的日变化幅度相一致。对于下挖式节能日光温室而言，其自身骨架材料和覆盖材料等在室内产生的阴影率约15%左右，在室内总阴影率不大于30%的情况下，可兼顾温室的采光与蓄热保温性能[19]。据此，下挖式节能日光温室下壁面在室内产生的阴影率不宜超过15%。以寿光市跨度为10m的下挖式节能日光温室为例，要使合理采光时段10：00～14：00时内各时刻下沉壁面产生的阴影率均不超过15%，则其阴影长度不能大于1.5m（1.5/10 = 15%）。由表2可以看出，下沉深度不超过0.8m时，在合理采光时段内阴影率均小于15%，因此，寿光市跨度为10m的下挖式节能日光温室适宜的下沉深度为0.8m。以下沉壁面在室内产生的阴影率不宜超过15%为约束条件，适宜的下沉深度则随温室的跨度的变化相应地发生变化。

采用同样的方法，可以计算得到我国北方地区10m跨度的下挖式节能日光温室适宜的下沉深度，如表3所示。

表3 我国北方地区10m跨度下挖式节能日光温室适宜的下沉深度和温室间距
Table. 3 The feasible sink depth and spacing of sunken energy-saving sunlight greenhouse with 10m span in north China

城市	纬度 φ (°)	太阳高度角 h (°)	太阳方位角 α (°)	适宜下沉深度 (m)	温室间距 B (m)
西安	34.25	25.6	-30.6	0.9	3.0
郑州	34.73	25.2	-30.5	0.9	3.1
兰州	36.05	24.1	-30.2	0.9	3.5
西宁	36.62	23.6	-30.0	0.9	3.7
济南	36.67	23.6	-30.0	0.9	3.7
寿光	36.88	23.4	-30.0	0.8	3.8
太原	37.85	22.5	-29.8	0.8	4.2
银川	38.47	22.0	-29.7	0.8	4.4
北京	39.90	20.8	-29.4	0.7	5.0
呼和浩特	40.80	20.0	-29.2	0.7	5.4
沈阳	41.80	19.1	-29.0	0.7	5.9
乌鲁木齐	43.77	17.4	-28.7	0.6	7.0
长春	43.92	17.2	-28.7	0.6	7.1
哈尔滨	45.75	15.6	-28.4	0.6	8.4

由表3可以看出，随纬度的增加，不同地区下挖式节能日光温室的适宜下沉深度有逐渐降低的趋势。因此在设计建造下挖式节能日光温室时，必须要根据当地的地理位置、气候、温室结构与建筑材料等条件，以室内产生的阴影率为依据来确定适宜的下沉深度，切不可盲目下挖过深，影响温室的采光和土壤蓄热性能，对室内作物的生长发育带来不利影响。

3 下挖式节能日光温室间距的计算

在生产实践中，节能日光温室大多集中连片建造，因此前后排温室的间距对温室的采光以及土地利用率影响较大。根据合理采光时段理论，一般以太阳高度角最小、阴影最长的冬至日10：00时前后栋温室不相互遮荫为依据来计算温室间距 B（图2），按式（5）计算[2]：

$$B = L_1 - L_2 - D = H_1 \times \frac{\cos(\alpha - \gamma)}{\tan h} - L_2 - D \tag{5}$$

式中：H_1 为温室保温覆盖材料卷起后至地平面的高度，L_1 为前栋温室屋脊在水平面上的投影到后栋温室前地角距离，L_2 为温室后屋面投影（温室屋脊在水平面上投影到室内后墙的距离），D 为温室后墙底部厚度。L_1 按式（6）计算[2]：

$$L_1 = H_1 \times \frac{\cos(\alpha - \gamma)}{\tan h} \tag{6}$$

以跨度为10m、脊高 H_1 为4.0m、后墙底厚 D 为3.0m、后屋面投影 L_2 为1.2m、朝向正南（$\gamma = 0$）的下挖式节能日光温室为例（目前寿光普遍使用的温室结构参数），计算中国北方地区适宜的节能日光温室间距（冬至日 $\sigma = -23.45°$，10：00时 $\omega = -30°$），结果

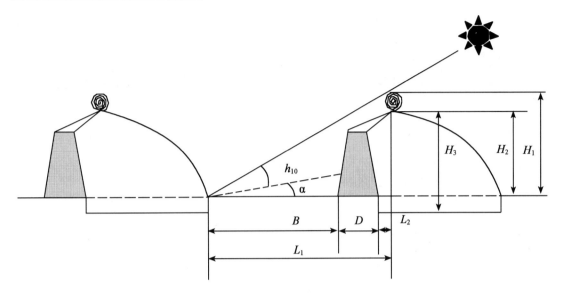

图2 下挖式节能日光温室间距示意图
Figure 2 The sketch of sunken energy-saving sunlight greenhouse spacing

如表3所示。

由表3可以看出，各地区适宜的温室间距也与纬度和太阳位置密切相关。与下沉壁面产生阴影的原理相似，纬度越高的地区，太阳高度角越小，前排温室产生的阴影就越长，因而前后排温室的间距需相应加大。

4 讨论与结论

与普通节能日光温室相比，下挖式节能日光温室可在很大程度上阻断后墙、山墙的传热散热，保温比加大，缝隙放热和土壤横向传热的热量损耗大大减少，还可避风储热，热容量较大[12]。因此，在一定范围内，节能日光温室下沉越深，温室保温效果越好，地温越稳定，夜间气温也越高（数据未列出）。而且下挖式节能日光温室在高跨比一定的情况下，其地平面上高度 H_1 降低，则前后栋温室的间距 B 也随之降低，因而可相应提高土地利用率。部分下挖式节能日光温室在室外前地角处采用二次下挖技术，这样在下沉深度相同的情况下，可增加室内前地角处的进光量，减少阴影区域面积。从以上计算结果可以看出，室内阴影面积随着下沉深度的增大而增大，若阴影面积过大，则会对室内土壤蓄热及作物的生长发育带来不利影响。适宜的下沉深度取决于温室所处的地理位置、太阳高度角、太阳方位角、温室采光屋面角和方位角、温室跨度以及温室建筑材料自身阴影率等，还与室内地下水位高低、室外排水等诸多因素有关。因此必须因地制宜，选择适宜的下挖深度，使室内获得较多的太阳辐射的同时，提高温室的蓄热保温性能。

目前，山东寿光下挖式节能日光温室的下挖深度达1.2~1.5m，按照上述理论计算，下沉深度偏大。温室间距也大小不一，大多不符合要求。另外，其侧墙和后墙基本都是土墙堆砌而成，由于机械施工等原因，造成墙体太厚，部分墙底厚达6~7m，导致建造成本增加，土地利用率下降。随土地结构的调整，新建温室的跨度和长度也有逐年增加的趋势，部分地区出现跨度达 16m、长度达 200m 以上的"巨型"节能日光温室。因此，下挖式节能日光温室的优化设计相当复杂，本文只对其采光设计进行了初步探讨，其适宜的墙体厚度、跨度以及高跨比、长度等结构参数都有待进一步进行理论分析和试验验证，对其保温设计、室内光热环境、建筑材料和结构受力等方面也有待进一步研究，以促进我国设施农业的持续健康发展。

参考文献

[1] http://news.xinhuanet.com/world/2009-12/19/content_12668033.htm
[2] 罗运俊，何梓年，王长贵. 太阳能利用技术. 北京：化学工业出版社，2005
[3] 刘志杰，郑文刚，胡清华等. 中国日光温室结构优化研究现状及发展趋势. 中国农学通报，2007，23（2）：449～453
[4] 闫毅，王瑞梅，陈丽梅等. 北方地区日光温室的设计与环境控制. 安徽农业科学，2008，35：15741～15743
[5] 吴毅明，曹永华，孙忠富. 温室采光设计的理论分析方法. 农业工程学报，1992，9：73～80
[6] 陈青云，汪政富. 节能型日光温室热环境的动态模拟. 中国农业大学学报，1996，1：67～72
[7] 高志奎，武占会，任士福等. 经济型节能日光温室的设计与温光性能. 河北农业大学学报，2004（5）：27～30
[8] 李小芳，陈青云. 日光温室山墙对室内太阳直接辐射得热量的影响. 农业工程学报，2004（5）：241～245
[9] 刘建，周长吉. 日光温室结构优化的研究进展与发展方向. 内蒙古农业大学学报（自然科学版），2007，28，3：264～268
[10] 马承伟，卜云龙，籍秀红等. 日光温室墙体夜间放热量计算与保温蓄热性评价方法的研究. 上海交通大学学报（农业科学版），2008，(5)：411～415
[11] 马承伟，王莉，丁小明等. 温室通风设计规范中通风量计算理论及方法体系的构建. 上海交通大学学报（农业科学版），2008，(5)：416～419+423
[12] 刘桂芝，李杰林，陈建国等. 坑式日光温室增温保温试验研究. 北方园艺，2004，（2）：16～18
[13] 董福奎. 日光温室的采光设计. 现代农业，2006，4：18～19
[14] 张真和，李健伟. 优化日光温室结构性能的途径和措施. 农村实用工程技术，1996，（9）：12
[15] Cooper, P. I. Absorption of radiation in solar stills. Solar Energy, 1969, (12): 333～346
[16] Duffie, J. A. and W. A. Beckman. Solar Engineering of Thermal Processes. 3rd edition ed. 2006, New York, the United States: John Wiley and Sons Inc
[17] 谢泳. 建筑物实体模型阴影的生成与应用. 西安科技大学学报，2003，23（4）：471～474
[18] 卜毅. 建筑日照设计（第二版）. 北京：中国建筑工业出版社，1988
[19] 张峰，张林华，刘珊. 日光节能温室下沉深度对其采光性能的影响. 山东建筑大学学报，2008，23（6）：474～477

植物工厂炼苗系统及其配套装备[*]

刘文玺[1][**]，张晓慧[1]，周增产[1,2][***]，卓杰强[1]，商守海[1]，李东星[1]，李秀刚[1]

（1. 北京京鹏环球科技股份有限公司，北京　100094；
2. 北京农业机械研究所，北京　100096）

摘要：本文主要针对植物工厂炼苗系统及其配套装备的研究设计，此系统内部包括循环式栽培床、环境控制系统和营养液循环再利用系统组成。环境控制系统主要控制炼苗系统内的温度、湿度、CO_2 浓度及光照；循环式苗床可根据作物生长的不同阶段，实现栽培床循环式自动输送；营养液循环再利用系统主要是自动实现营养液的循环、供给及消毒杀菌作用，改善植物根系周围的环境。植物工厂炼苗系统及其配套装备减少了作物生产用工量，提高了劳动生产率，降低了劳动成本，同时由于栽培床紧密排列，可通过相互作用力移动，因此增加了植物工厂有效使用面积，提高了植物工厂空间的利用率。

关键词：植物工厂；炼苗；循环栽培床；环境控制；营养液循环

Plant factory hardening seedling system and its equipments

Liu Wenxi[1][**], Zhang Xiaohui[1], Zhou Zengchan[1,2][***], Zhuo Jieqiang[1],
Shang Shouhai[1], Li Dongxing[1], Li Xiugang[1]

(1. *Beijing Jingpeng International Hi-Tech Corporation*, *Beijing* 100094, *P. R. China*
2. *Beijing Agriculture Machinery Institute*, *Beijing* 100096, *P. R. China*)

Abstract: This article is mainly for the researching and designing the plant factory hardening seedling system and its equipments, this system includes environmental controlling system, circulating cultivation bed and nutrient solution recirculation system. The environmental controlling system mainly control the temperature, humidity, CO_2 and the illumination of the hardening seedling system; according to the different growing period of the plants, the circulating cultivation bed could realize the cultivating bed to circular automatic convey; The nutrient solution recirculation system could mainly automatic realize the nutrient solution circling and having the bactericidal action, which could improve the root environment. Plant factory hardening seedling system and its equipments could reduce the labor quantity, improve the labor productivity, debase the work cost, and because the cultivation bed is tight arrangement and conveying by the interactions, it could increase the using area of the plant factory, and improve the space utilization rate.

Key words: Plant factory; Hardening seedling; Circulating cultivation bed; Environmental control; Nutrient solution recirculation

---

[*] 基金项目：北京市科技计划项目（D08060500450000）。

[**] 第一作者简介：刘文玺（1974— ），男，工程师，主要从事农业工程方面的工作。E-mail：kuayue3008@126.com。

[***] 通讯作者：周增产（1966— ），博士，教授级高工，主要从事农业工程研究。E-mail：zengchan@sina.com。

目前，中国设施内幼苗繁育较多，而大部分优质品种繁育多采用组织繁殖。为了提高优秀品种的繁殖系数和速度，炼苗是其中较为关键的一环，它直接关系到组培苗能否尽快地应用到生产中去。试管苗移栽过程较为复杂，盲目移栽稍有不慎，就会造成大批组培苗死亡。因此，研究炼苗系统及其配套装备尤为重要，尤其是组培苗与炼苗室中小气候环境诸要素之间的相互关系。本文从炼苗系统及其配套装备的研究来探讨现代化炼苗系统的应用[1]。

1 植物工厂炼苗系统结构组成及功能

1.1 炼苗系统的结构组成

炼苗系统主要由循环式栽培床、环境控制系统和营养液循环再利用装备组成。环境控制系统主要控制炼苗系统内的温度、湿度、CO_2浓度及光照；循环式苗床可根据作物生长的不同阶段，实现栽培床循环式自动输送；营养液循环再利用装备主要是自动实现营养液的循环、供给及消毒杀菌作用，改善植物根系周围的环境[3]。

1.2 炼苗系统功能

植物工厂炼苗系统应具备以下几大功能：

（1）栽培环境的精确控制。作物从播种到长出幼苗后，需要移栽进行大面积生产。然而幼苗移栽到大的生长环境中，往往由于大环境生长管理条件较粗放，使幼苗很难适应生产。因此炼苗系统可提供一个适宜的生长环境，使幼苗的生长得到锻炼和适应。

（2）适宜的栽培床。栽培床主要用于穴盘育苗和盆栽作物，有利于通风透气和排水，减少病虫害的发生[2]。植物工厂炼苗系统采用循环式栽培床，该床体可根据作物不同的生长阶段，自动移动，同时具有很好的供、排水功能。

（3）利于植物生长的营养环境。组培苗和人工育苗都是在无菌条件、营养充足、适宜温度与光照的环境中逐渐长大，这个过程是完全人为创造的条件，一旦移栽种植，所接触的湿度相对干燥，昼夜温差加大，营养的供应靠根系吸收，幼苗在短期内很难适应[4]。因此，植物工厂炼苗系统采用营养液循环再利用装置，可自动检测和供给营养液，为幼苗提供良好的营养环境。

2 循环式栽培床组成及运行原理

2.1 循环式栽培床组成

循环式栽培床输送装置包括循环式栽培床架、分段式栽培槽、开间支撑导轨、横向位移驱动装置、纵向位移驱动装置、气动驱动装置等，通过PLC控制器和传感器控制纵向位移驱动装置实现栽培床纵向自动输送，横向位移驱动装置实现栽培床横向自动输送，气动驱动装置实现栽培床横向和纵向输送的自动转换（图1和图2）。

循环式栽培床内配置栽培槽，采用潮汐式营养液循环灌溉。栽培槽分为三段，分别为供水段、循环中间段、回水段。栽培槽回水段尾部设有两个下凹的池形结构作为营养液回流池，池形结构的底面还装有营养液回流阀，池形结构的上方覆盖有栅状过滤网壳。循环中间段用于连接供水段与回水段。供水段与供水管道相连接，用于供给营养液和水。如图3和图4所示，由左向右分别是回水段、中间段和供水段。

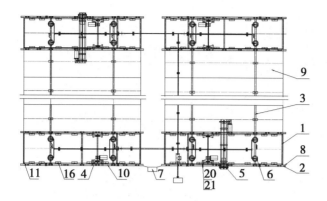

1. 循环式栽培床输送装置；2. 循环式栽培床架；3. 开间支撑导轨；4. 横向位移驱动装置；5. 纵向位移驱动装置；6. 气动驱动装置；7. PLC 控制器；8. 传感器；9. 栽培床；10. 横向轨道梁；11. 轨道梁连接杆；12. 气动驱动纵向连接杆；13. 气动驱动升降导向板；14. 支撑矩形管；15. 脚支撑；16. 栽培床输送胶轮；17. 气动驱动支撑杆；18. 支撑板；19. 角件连接件；20. T 型螺栓；21. 法兰螺母

图 1　循环式栽培床示意图

图 2　循环式栽培床

2.2　运行原理

如图 5 所示：

（1）4 处没有栽培床。

（2）3 处和 4 处气动装置上升，3 处推动装置起动，开始推动一个栽培床由 3→4，3 处推动装置回位，触碰传感器使 3 和 4 处气动装置下降。

（3）2 处滚动装置起动，将栽培床由 2→3 处，栽培床到 3 处后触碰传感器，起动 1 和 2 的气动装置上升，1 处推动装置起动，栽培床从 1→2 处，到位后触碰传感器 1 和 2 气动装置下降，1 处推动装置回位。

（4）4 处大苗从栽培床移走后，轨道滚动装置起动，4 处空的栽培床从 4→1 处，开始装小苗，4 处空出一个床位。

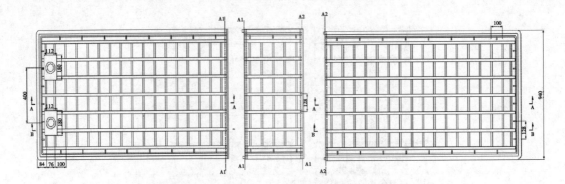

图 3 分段式栽培槽简图

图 4 循环式栽培床控制箱

3 炼苗系统的环境控制

炼苗是在环境精确可控的情况下，采取放风、降温、适当控水等措施对幼苗强行锻炼的过程，使其定植后能够迅速适应栽培环境条件，缩短缓苗时间，增强对新环境条件的抵抗能力。植物工厂炼苗系统的环境调控：

针对不同作物及品种，炼苗系统的环境调控也不一样。

温度控制：采用温度传感器进行温度检测，系统采用一套恒温恒湿可调节式空调机组进行温度调节。植物组培苗大致可以分为喜温、喜凉两大类。喜温植物整个生育期要求较高的温度条件，一般适温范围在 22～28℃，喜凉植物要求温度偏低，一般适宜温室范围在 18～25℃[5]。

光照控制：光照供给系统包括自然光和人工光两种。一般作物采用自然光源便可满足作物生长发育的需要，然而对于光照要求较高的作物可以采用人工光源进行补光。

湿度控制：采用温度传感器进行湿度检测，系统采用一套恒温恒湿可调节式空调机组进行湿度。一般作物生长发育适宜温度范围在 60%～80% 之间[6]。

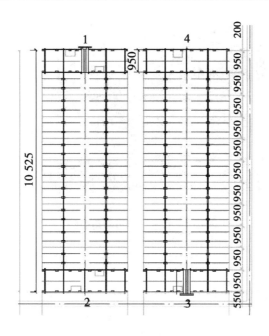

图 5　工作原理示意图

4　炼苗系统内营养液循环再利用装备

营养液循环再利用装备是由营养液循环系统、操作控制系统、紫外消毒系统和增氧装置组成。

4.1　营养液循环系统

营养液循环利用装置中的营养液池是贮存和供应栽培床营养液的容器，母液罐 A、B、C、D 中的溶液在电磁阀门的控制下流入营养液池。母液罐 A 盛放钙盐母液，母液罐 B 盛放磷酸盐母液，母液罐 C 盛放微肥，母液罐 D 盛放酸碱缓冲液。水处理装置将自来水转化成纯净水注入营养液中。水泵将营养液池的营养液经过滤器和紫外线消毒器输送到栽培床以供作物需求。回流管路则将栽培床内的营养液回流至营养液池，从而形成一个循环系统。如图 6 和图 7 所示。

4.2　操作控制系统

操作控制系统由软件程序来控制水流和营养配比。整套系统是由软件、硬件、传输设备、传感器、环境控制、灌溉控制及营养控制组成。

自动检测和控制装置由高低液位传感器、溶解氧传感器、pH 传感器、EC 传感器、温度传感器、加热棒和计算机组成，不同的传感器用隔板隔开以防止相互干扰。系统在检测软件和控制软件的支持下通过各种传感器检测到各种信号，经过阻抗变换和放大得到相应的电压信号，由数据采集卡送到计算机进行处理，根据控制算法产生控制信号并由 I：O 端口输出，经驱动电路驱动执行机构，完成营养液的加温以及各种离子浓度的在线检测和控制[7~8]。

为了能检测 pH 和 EC 值，系统设计的传感器接口可以采集传感器模拟量，并将数值显示在液晶上，通过人机交互界面，可以随时查看系统工作时的 pH 和 EC 值，并且数据可以

图6 营养液循环再利用装备之一

自动保存,可以查看历史数据文档。外接的光照传感器和水分传感器信号采集到控制器中,可以作为营养液供给程序的参考值[9]。

4.3 多段式消毒处理系统和增氧装置

炼苗系统中,营养液循环再利用系统核心装置首段为砂率段,可减轻后级处理负荷,中段利用波长为 225～275nm 的紫外线对微生物的强烈杀灭的作用,对原水中的微生物进行杀灭;末段导入羟基自由基 OH 进行最终杀菌[10]。其特点是:杀菌速度快,不改变水的物理、化学性质,不增加水的嗅味,不产生对人体有害的卤代甲烷化合物,无副作用;水处理筒体采用进口优质不锈钢,可满足 1.2MPa 的工作压力,具有防锈、强度高、无金属离子污染、设备表面易于清洁等优点;耗能低,连续使用寿命可达 3 000h 以上,并配有可靠的镇流装备;模块化电控装置,功能齐全,定时标准,与水处理筒体采用一体化设计,具有安装方便,操作简单,安全可靠。

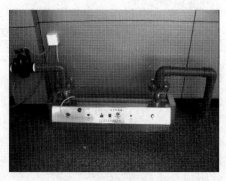

图7 营养液循环再利用装备之二

增氧装置主要用于为营养液内提供溶解氧,溶解氧可以改善水培植物根系周围环境,防止植物根系因缺氧而产生烂根和中毒现象。

5 结语

植物工厂炼苗系统及配套装备的研发有效地改变了传统农业的种植方式。其内部环境控制系统可对温度、相对湿度等进行适度的控制,营养液循环再利用装备可自动检测营养液

pH 和 EC 值，同时实现了营养液的自动配比和供给，循环式栽培床输送装置循环可根据作物生长的不同阶段，实现栽培床循环式自动输送。因此，植物工厂炼苗系统的研发，增强了从育苗过后到移栽成活中间环节对环境的适应能力，特别是组培苗的移栽，得到了更好的栽培环境，从而提高了幼苗的成活率。同时，该系统减少了农业生产用工量，提高了农业劳动生产率，降低了劳动成本。由于栽培床紧密排列，通过相互作用力移动，增加室内有效使用面积，提高了室内空间的利用率。

参考文献

[1] 申茂向等．荷兰设施农业的考察与中国工厂化农业建设的思考．农业工程学报，2000，16（5）：1~7
[2] 韦民，李文付，陈建丽等．GMJ-3 型滚动式育苗床研制报告．广西林业科技，1992，21（3）：95~97
[3] 杨其长，刘文科，管道平．植物无粮组培快繁装置及其环境控制系统的研制．中国农业科技导报，2007，9（4）：79~84
[4] 郝会军，陈美霞，巨荣峰等．半夏组培苗炼苗技术研究．安徽农业科学，2008，36（5）：1918~1919
[5] 蒋峰，韩先花，贺超英．温室远程监控系统的研究．农机化研究，2009，（8）：87~89
[6] 邓璐娟，张侃谕，龚幼民 智能控制技术在农业工程中的应用 现代化农业，2003（12）：1~3
[7] 杨万龙，张贤瑞，刘春来等．温室滴灌施肥智能化控制系统研制．节水灌溉，2004，5：27~30
[8] 高国涛，戈振扬，张云伟等．控制环境植物生产系统中营养液的检测与控制技术．农业装备技术，2006，32（1）：20~23
[9] 杜建福，马明建．营养液检测与控制技术概况．山东工程学院学报，2002，16（1）：69~72
[10] 刘伟，陈殿奎等．无土栽培营养液消毒技术研究与应用．农业工程学报，2005，21（12）：121~124

日光温室结构与环境因子相关性研究初探

王克安，杨 宁，吕晓惠，王 伟，张卫华，柴秀乾

（山东省农业科学院蔬菜研究所，山东济南 250100）

摘要：针对山东Ⅳ型、Ⅴ型日光温室，研究了气温、风速对日光温室内温度、湿度、CO_2浓度的影响。结果表明：晴天时，气温与室内温度呈极显著正相关、与室内湿度和CO_2浓度极显著负相关；风速与室内温度呈显著负相关、与室内湿度和CO_2浓度显著正相关。阴天时，室外温度与室内温度极显著正相关，与室内湿度极显著负相关；风速与室内温度、湿度相关性不显著。

关键词：日光温室；结构；环境因子

The Relationship of Greenhouses Structure and Environment Factors

Wang Ke'an, Yang Ning, Lu Xiaohui, Wang Wei, Zhang Weihua, Cai Xieqian

(*Vegetable Research Lnstituate, ShanDong Academy of Agricul Tural Sciences, Jinan 250100, P. R. China*)

Abstract: Shandong Ⅳ and Ⅴ greenhouses were chosen in this test, using temperature and humidity automatic recording instruments to record the change of environmental factors. Research the relationship of air temperature, wind speed as well as temperature, humidity, CO_2 concentration in greenhouses. The main results showed: In a sunny day, air temperature had significant positive relationship with greenhouse temperature, and had negative relationship with humidity and CO_2 concentration in greenhouse; wind speed had negative relationship with greenhouse temperature, and had positive relationship with humidity, CO_2 concentration in greenhouse. In a cloudy day, temperature had significant positive relationship with temperature in greenhouse, and had negative relationship with indoor humidity; wind speed had subtle relationship with air temperature and indoor humidity.

Key words: Greenhouses; Structure; Environment factors

设施农业是在充分利用自然环境条件的基础上，用人工控制环境因子如温度、湿度、光照、CO_2浓度等的方法来获得作物最佳生长条件，从而达到增加作物产量、改善品质、延长生长季节的目的[1]。日光温室是中国北方蔬菜冬季生产的主要设施之一，其环境因子一般是指温度、湿度、光照、CO_2，湿度是温室蔬菜生产的重要环境因素，直接影响到蔬菜病害的发病率和病情指数；温度、光照是影响蔬菜作物生长发育及产量、品质的主要因子之一；CO_2（气肥）是植物光合作用所必需的物质基础[2~4]。温室结构是影响温度、湿度及光照的重要因素之一，而可人工控制的通风措施是影响湿度、温度和CO_2浓度三大环境因子的重要手段之一[5~7]，本试验人工记录天气情况、通风时间、通风面积，并借助温室环境控制记录仪对两种山东新型日光温室环境因子变化实时记录，通过分析，初步探讨不同结构日光温室内环境因子与室外环境因子的相关性。

1 材料与方法

1.1 试验材料

试验在山东省农业科学院蔬菜研究所试验基地内进行,选择栽培越冬茬黄瓜的山东Ⅳ型(1号)、Ⅴ型(2号)日光温室各一栋,使用WR-4温室环境记录控制仪全天候记录温室内外环境因子变化。

山东Ⅳ型:脊高4m,前跨9m,采光屋面参考角23.9°,后墙高3m,后屋面仰角45°。

山东Ⅴ型:脊高4.3m,前跨10m,有立柱,采光屋面参考角23.3°,后墙高3.3m,后屋面仰角45°。

1.2 试验方法

分别在两栋日光温室内中间部位设置温、湿度、CO_2浓度检测探头,在温室外平行风口距离2m处设置温、湿度、风速检测探头。每隔10min记录一次。人工记录天气、通风时间、通风面积等情况。

2 结果与分析

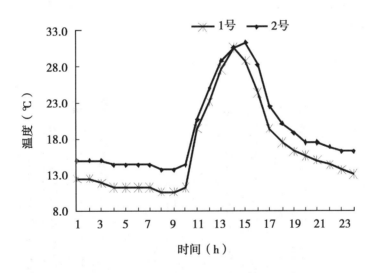

图1 晴天不同结构日光温室温度的影响

2.1 不同天气对日光温室温度、湿度的影响

由图1可知,晴天时,1号和2号日光温室全天的平均温度分别为19.8℃和22.6℃,在1:00~9:00时,两温室温度随时间的推移,呈降低趋势,但变化不大,日出前达到最低值,此后略有回升,揭苫后,受光照影响温度迅速回升,其中1号温室在13:00左右达到最高,为30℃,此时室外温度也达到最高;2号温室在15:00达到最高值,为30.7℃,两个温室的最高温度稍有差异,随后受通风影响两个温室的温度又迅速下降,盖苫后降温速率明显降低,23:00时后,温度变化缓慢,受外界温度变化影响,略有下降。全天整个变化过程,1号温室的温度均比2号温室的低。

由图2可知,阴天时,1号和2号日光温室的平均温度分别为9.8℃和13.1℃。阴天两

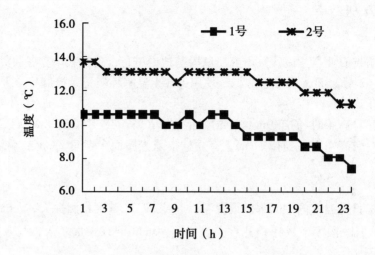

图2　阴天不同结构日光温室温度的影响

温室的温度呈波动下降趋势，两日光温室全天最低温度分别为7.8℃和10.5℃。1号温室的温度均比2号温室低，其变化范围较2号温室大。

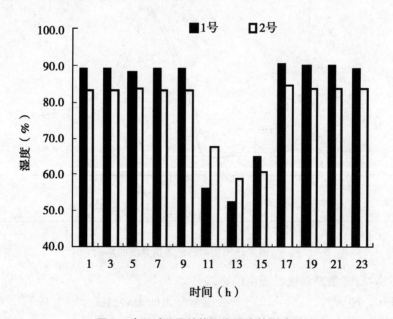

图3　晴天对不同结构温室湿度的影响

由图3可知，晴天15：00时，两日光温室b点湿度均达到最低值，1号和2号日光温室分别为57%和61%。在11：00~13：00时，1号日光温室由于通风口面积较大，湿度低于2号日光温室，其他时间均高于2号日光温室。1号和2号日光温室的平均湿度分别为79%和72%。由图4可知，阴天时，1号和2号日光温室的湿度基本保持稳定，平均湿度分别为89%和83%，在整个变化过程中1号日光温室的湿度均比2号高。

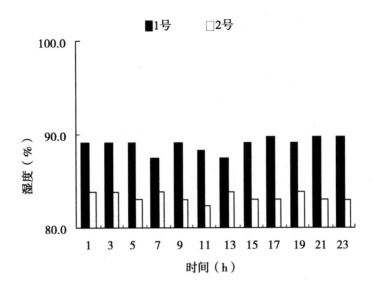

图 4 阴天对不同结构温室湿度的影响

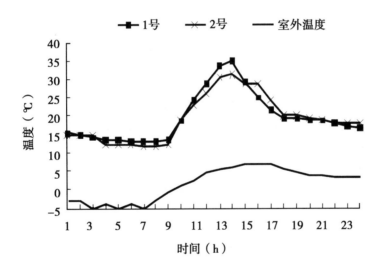

图 5 室外气温对不同结构日光温室温度的影响

2.2 室外气温对日光温室温度、湿度的影响

由图5可知,1号和2号日光温室全天的平均温度分别为24.9℃和21.3℃。两日光温室全天的最低温度出现在凌晨7:00左右,最低温度分别为13.9℃和12.8℃。两日光温室的温度均从上午9:00日光温室揭苫后迅速上升,至14:00 1号温室达到全天最高温度35.1℃,比2号温室最高温度高3.1℃。14:00~18:00时两日光温室的温度均迅速下降至19.8℃,夜间受室外气温影响室内温度亦呈缓慢下降趋势。

由图6可知,两日光温室的平均湿度分别为74.63%和75.10%,1号和2号温室相差不

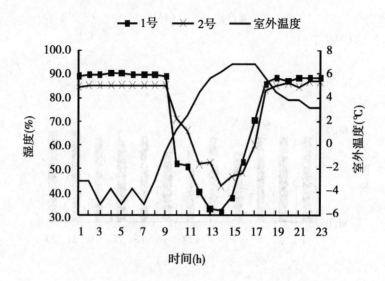

图6 室外气温对不同结构日光温室湿度的影响

大,夜间1号湿度略高于2号。揭苫后随室外气温的升高室内湿度迅速下降,14:00时后,随室外气温降低,室内湿度开始迅速升高。18:00时后基本趋于平稳。

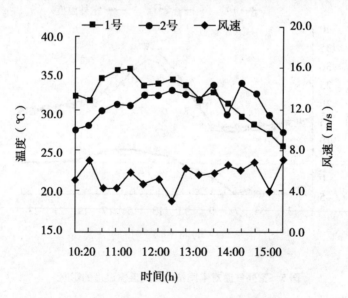

图7 风速对不同结构温室温度的影响

2.3 风速对日光温室温度、湿度的影响

由图7可知,在10:00～12:00时这一通风时段,1号和2号日光温室平均温度分别为31.35℃和30.66℃,时段内两日光温室温度变化幅度1号稍高于2号,风速平均值为5.36m/s,风速在该时段内变化比较平稳,温室初通风,室内温度受其他因素影响较大,在通风积累一定时间后,由于空气流动,使得室内的温度也开始缓慢下降。

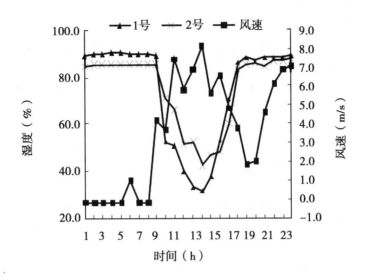

图 8 风速对不同结构温室湿度的影响

由图 8 可知,由于受日光温室外覆盖物的影响,夜间和日间温室不通风时段风力与室内湿度无明显相关性。通风后,风力对湿度影响较大,两者呈负相关关系。

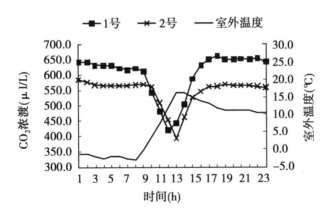

图 9 气温对不同结构日光温室内 CO_2 浓度变化

2.4 室外气温对日光温室 CO_2 浓度影响

由图 9 可知,全天 1 号和 2 号日光温室的 CO_2 浓度平均值分别为 602.67μl/L、554.63μl/L,差值为 57.04μl/L,全天室外气温平均值为 6.64℃。室外温度在 0:00~8:00 时基本处于水平变化范围,在 -5~0℃ 内变化,两日光温室内 CO_2 浓度维持 640μl/L、570μl/L 左右,日出后室外温度开始迅速升高,到 13:00 达到全天最高的 17℃,而室内 CO_2 浓度则迅速降低,1 号温室在 12:00 降至全天最低值 419μl/L,2 号温室在 13:00 达到最低值 393μl/L,在 14:00~18:00 时时段随外界气温的降低迅速上升,18:00 后基本趋于稳定。

2.5 风速对日光温室 CO_2 浓度的影响

由图 10 可知,盖苫不通风时,CO_2 浓度变化不明显,但 1 号温室的 CO_2 浓度要明显高于 2 号温室,揭苫后,风力对其影响明显加强,CO_2 浓度迅速降低,且 1 号温室要比 2 号温室的 CO_2 浓度降低更快,在 12:00 时就达到最低值 419μl/L,而 2 号温室稍慢,在 13:00 达到最低,418μl/L,随后室内的 CO_2 浓度开始回升,1 号温室的升高速度明显比 2 号温室快。

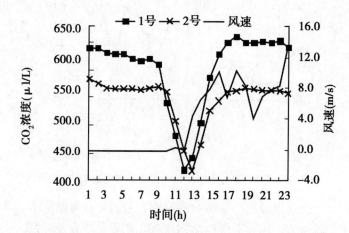

图 10 风力对不同结构日光温室 CO_2 浓度的影响

2.6 日光温室内环境因子与室外环境因子的相关关系

表 1 1 号温室内环境因子与室外环境因子的相关系数

		室内温度	室内湿度	室内 CO_2 浓度
晴天	室外温度	0.975	-0.993	-0.888
	风速	-0.627	0.504	0.713
阴天	室外温度	0.952	-0.890	
	风速	0.194	0.087	

表 2 2 号温室内环境因子与室外环境因子的相关系数

		室内温度	室内湿度	室内 CO_2 浓度
晴天	室外温度	0.993	-0.989	-0.887
	风速	-0.572	0.543	0.678
阴天	室外温度	0.961	-0.884	
	风速	0.165	0.091	

将室外温度和风速作为影响室内温度、湿度及 CO_2 浓度的变量,对数据进行标准差标准化转换处理并进行相关性分析。由表 1、表 2 可知,晴天时,两日光温室室外温度与室内

温度极显著正相关，与室内湿度和 CO_2 浓度极显著负相关；风速与室内温度显著负相关，与室内湿度和 CO_2 浓度显著正相关。阴天时，室外温度与室内温度极显著正相关；风速与室内温度、室内湿度相关性均不显著。

3 结论

经过对以上图表进行分析，汇总在不同天气条件下对不同结构日光温室内各环境因子的影响程度，可以初步得出以下规律：

晴天时，室外温度与室内温度呈极显著正相关，1号温室升温、降湿比2号温室快，平均温度2号>1号，平均湿度1号>2号；阴天时室内外温度、湿度均变化幅度不大。这可能是因为2号日光温室相对于1号日光温室跨度大，脊高高，采光角度小，整体空间容积大，蓄温能力强。

在温室通风时段，风速与室内温度呈显著负相关关系、与室内湿度和 CO_2 浓度显著正相关，对1号日光温室温度的影响大于2号温室，对1号温室湿度的影响小于2号日光温室；随通风时间的延长，1号温室 CO_2 浓度的下降速度较快，最低值低于2号温室，但1号和2号温室 CO_2 浓度的平均值差异较小。

参考文献

[1] 何启伟, 卢育华. 山东新型日光温室蔬菜系统技术研究与实践. 济南：山东人民出版社, 2002
[2] 程纯枢. 中国的气候与农业. 北京：气象出版社, 1991
[3] 王宇欣, 段红平. 设施园艺工程与栽培技术. 北京：化学工业出版社, 2008
[4] 韩世栋. 蔬菜冬暖型日光温室建造和高效栽培技术. 北京：中国农业出版社, 2008
[5] 周长吉. 现代化设施农业发展概况. 北京：机械工业出版社, 2002
[6] 张福墁. 设施园艺学. 北京：中国农业大学出版社, 2001
[7] 齐飞. 论温室产品的"低质—低价"趋向对温室行业的影响. 农业工程学报, 2002.18（增）
[8] 杨冬艳, 郭文忠, 张丽娟等. 不同结构日光温室冬季温光环境测试分析. 温室园艺, 2010 (2)：18~21
[9] 杨振超, 邹志荣. 不同结构类型节能日光温室内温、湿度比较研究. 陕西农业科学, 2002 (3)：25~27

LED 照明的发展潜力

Johann Buck 王 贺（译）*

LED 技术已经面世一段时间了，但尚未成为北美园艺市场的主流。以下为那些希望进行转变的用户提供一些注意事项。

"住在遍布光亮的屋内"。

——Cornelius Celsus（古罗马学者）

这是一句非常有趣的话，如果将其置于讲述植物生长的文章中时更是如此。我们可以为植物开出适合的矿质营养和温度的处方。也可以在必要时提供生长调节素。我们可以控制温室环境并投入各种元素以确保作物的生长来实现目标。

那么如何解决照明问题呢？植物所接受的光照主要来自太阳，但会受到若干因素的制约，包括日照时间和高度角、云量和温室结构的阴影等。种植者多年来一直采用补光技术，以弥补这些不足。通过补光，可延长白昼时间，可以实现夜间时间干预，必要时可完全阻止光照。然而，现有的补光灯的局限性是光谱质量。设想一下，除了提供优化营养和温度外，你还能够为植物提供适合的光量和光质，那会是什么效果呢？所谓适合的光线，我们指的是光合有效辐射（PAR），即光线波长介于 400~700nm 之间，并以光量子通量密度，即以每平方米每秒通过的微摩尔 [$\mu mol/(m^2 \cdot s)$] 来度量。

为了尽可能简单地描述，我们不讲光敏色素和光照比。一般来说，为实现光合作用，植物能利用的光线主要位于光合有效辐射的蓝色（450nm）和红色（660nm）的区域。想象一下一种灯具照射的光线中含有更多适合植物所需的光线，比较而言另一种是会产生多余光线的灯具。现在，再想象一下你的植物放置在充满了适合光线的房间内，而完全没有日光。

这就是在博览会和专业会议上已引发了大量讨论的"新型"辅助光源，其目标就是提供适合的光线。发光二极管（LED）也许是你已经听到过很多次的名词了。LED 可用于诸如汽车前灯和尾灯、交通灯、甚至是便携式手电筒等众多用途。LED 甚至还被试验作为无线局域网的替代品。简单地说，LED 正变得越来越常见。相比于灯泡，LED 更类似于一块电脑芯片，因为它是一个固态半导体设备。据估计，2008 年 LED 占据全球 7% 的市场份额，而到 2010 年，这一数字增加到了 20%。到 2020 年，预计 LED 将占据全球照明市场 75% 的份额。尽管 LED 也许看起来还是一个相对较新的光源，但事实上早在 20 世纪 20 年代就已发明了 LED。然而，直到 1960 年初期才开发出了可见光（380~780nm）版本，颜色为红色（~660nm）。20 世纪 70 年代则开发出了其他的波长（绿色、黄色、橙色，甚至蓝色）。首个高亮度蓝色 LED 在 1990 年初期才完成开发。

任何 LED 照射光线的颜色都取决于半导体材料和组成 LED 的杂质。在蓝色 LED 上添加磷光层之后就制造出了白色 LED。除了能提供特别的短波段外，相比传统的光源，LED 还

---

* 王贺，男，飞利浦，设施农业照明，业务发展高级经理. E-mail：Henry.Wang@philips.com。

带来了其他方面的积极影响。

1　优势与挑战

效率和寿命是两个最常谈及的 LED 优点。LED 比白炽灯和荧光灯更高效，基本上与高压钠灯的效率相同。白色 LED 的效率稍低，因为磷光层需要与基色反应以产生白色光。因而将其他现有的园艺灯具换成 LED 灯具将会节省大量成本。和传统灯具不同，LED 灯一般不会"烧坏"。而且，LED 所使用的度量标准是"使用寿命"，即光输出下降到最佳运行环境下的初始最大强度的一个百分比所需的时间（h）。种植者一般会在光输出下降到低于 90% 时更换灯具。因此，种植者期望的 LED 灯的使用寿命在 90% 效率以上运行 25 000h。LED 较长的使用期限降低了所需的采购、废弃及比如更换灯管相关的劳动力成本。

LED 的另一个优点是可以经常开关灯，而无需预热时间。它们照射时的辐射热很小或几乎没有。然而，在二极管的连接处中也会有热量的损失。当 LED 在高密度下使用时，热量输出会非常大。和荧光灯不同的是，LED 不含诸如水银等有害物质。如果外保护损坏，它们也不会产生有害的紫外线辐射。由于其体积较小，LED 使灯具设计变得更灵活。

在温室中大量推广 LED 设备的主要难点在于成本。虽然初期成本可能比其他辅助光源来得高，但 LED 的投资回报期更短，因其能效更高、使用寿命更长。其次，LED 的成本会不断降低，而包括飞利浦照明在内的公司则会为园艺的应用提供价格合适的 LED 产品。在园艺照明中采用 LED 的主要技术障碍在于种植所需波段输出的光效较低。尽管目前暂时还不可能用 LED 去代替高压钠灯，但有消息称这一情况将在三五年内发生。不管如何，LED 模组已经能提供植物需要的波段，并且已应用在园艺种植中。有的 LED 灯具拥有标准 E27 灯口配置，可在现有灯座上进行直接更换。无需额外调整。

2　园艺应用

控制光谱质量对种植业而言是非常有用的，这在采用广谱的光源时很难实现。LED 可制造窄谱波段，并已制出了能被植物高度吸收的颜色。还记得光合有效辐射（PAR）吗？由于种植者只需要能被植物高度吸收的颜色，因此 LED 灯需做到去除无效波段，从而减少生产无效光照和能量。应该做到可对光谱进行定制，甚至能调整到满足作物需求和能控制光周期或生长周期。因此，必须特别强调的是，LED 的应用需要根据种植的作物和种植者的目标而有步骤地进行。

为此，你要问自己几个问题：

你对于采用 LED 有什么样的期望？你是否对以下的情况感兴趣：更快地生长、更高植物质量、在没有日光的情况下生长、多层种植或减少作物在温室中生长的时间？你是否对于拉伸幼苗的下胚轴以实现更好的嫁接感兴趣？回想一下，LED 照射时的辐射热很少或几乎没有，从而可把它安装在离植物顶部很近的地方。这样就可减少被植物拦截的暗区，从而提高光照使用效率。减少热辐射还带来了一个考量：空气温度的降低。高压钠灯会向下方的作物排出部分辐射热，而 LED 不会。这会如何影响温室内的生长环境？也许当 LED 开始逐渐在市场上取代高压钠灯时我们会得到答案。

LED 可更贴近植物的特性可为种植者提供多样的选择，如包括组培在内的作物多层种植。多层种植时也可以使用其他的光源，然而，采用 LED 时，种植者在同样的单位空间可以安装更多的层架。例如，有一个种植者在换用 LED 之后，可将照明能耗减少 50% 以上，并将多层种植空间增加 33%，而不增加额外的种植用地。

3 注意事项

LED 的一些特性令其成为在园艺界具有吸引力的辅助照明光源。采用 LED 可以进行控制光谱合成，从而产生更多有效光照、减少无用光照。LED 可提供较高的光输出，同时热辐射低。它们的体积较小，从而在设计和安装上更具灵活性。最后也是同样重要的一点就是，它比其他辅助光源更特别持久耐用，且具有更高的能效。

免责声明：操作需小心谨慎。有些试图涉足园艺业的照明企业并不了解农作物的种植。这些企业仅把温室照明看作是可增加市场份额的一个领域。因此你在考虑使用 LED 时，应向了解该行业和作物种植的相关人员进行咨询。选择一个合适的合作伙伴为您的温室运行和目标进行专门的循序渐进的设计。

要不断发问。了解照明企业为其产品在作物种植上投入有多少。当然，固态照明技术的前景是非常光明的。现在就着手在您的经营中试用 LED。它们可帮助您增加种植面积、提高利用效率、扩大生产规模、增加产量和提高质量。

Johann Buck 博士 是 Hort Americas 有限责任公司的技术服务经理。您可通过邮件 jbuck@hortamericas.com 与他取得联系。

设施栽培理论技术

植物照明中绿光比例对不同莴苣生长之影响

张明毅[1],方 炜[2],邬家琪[3]

(1. 国立宜兰大学生物机电工程学系; 2. 国立台湾大学生物产业机电工程学系;
3. 国立宜兰大学园艺学系)

摘要:完全环控型植物工厂中,植物照明的人工光源必须是高发光效率、低耗能、低产热,以利应用于作物立体栽培层架,而高亮度发光二极体十分符合此需求。目前于室内照明的应用领域上,白光发光二极体在提升发光效率与降低成本上已取得显著成果,但针对植物栽培的应用则刚起步,相关研究专注于光合作用相关的红蓝光配比为主,多半不讨论绿光的功能。本研究在以红蓝配比为主体的光谱中加入不同比例的绿光,探讨对莴苣育苗期及养成期生长之影响。实验结果显示,在光量维持于 $150 \mu mol \cdot m^{-2} s^{-1}$ 的情形下,室内照明用的白光 LED 灯具或萤光灯管并非最佳的植物照明光源。红蓝光质配比为 85:15 有助于莴苣生长,在整体及蓝光光量不变的前提下,加入 25% 绿光能更进一步提升莴苣的产量。但当绿光比例大于 50% 时作物成长趋缓。

关键字:光质;植物工厂;发光二极体

Effects of the Green Light Proportion of Plant Illumination on the Growth of Different Lettuces

Chang Mingyih[1], Fang Wei[2], Wu Chiachyi[3]

(1. Dept. of Biomechatronic Engineering, National ILan University; 2. Dept. of Bio-Industrial Mechatronics Engineering, National Taiwan University; 3. Dept. of Horticulture, National ILan University)

Abstract: Artificial lights with high luminous efficacy, low power consumption and low heat discharge are required for the multi-layer cultivation beds used in the closed plant factory. Highpower light-emitting diodes are the suitable light source to meet this requirement. The indoor illuminous applications by using white light LED has proved remarkable achievements in improving the luminous efficacy and reducing the production costs. However, the application for the plant cultivation is just beginning. Most of the research in plant illumination focused on the red and blue sprectal ratio related to photosynthesis and mostly ignored the effects of green light. This study was to investigate the effects on the lettuce growth by adding different green light proportion to plant illumation. The results show that the white LED light or fluorescent tube for indoor illumination is not the optimal selection for plant cultivation. The red/blue spectral ratio of 85:15 helped the growth of lettuce at light intensity of $150 \mu mol \cdot m^{-2} \cdot s^{-1}$. Maintaining the blue light proportion and adding 25% of green light would further enhance the production of lettuce. However, the green light proportion greater than 50% will retard the plant growth.

Key words: Light quality; Plant factory; Light-emmiting diode

* 作者简介:张明毅,国立宜兰大学生物机电工程学系,E-mail:mgchang@ niu. edu. tw;方炜,台湾大学生物产业机电工程学系教授,E-mail:weifang@ ntu. edu. tw。

1 绪论

近年可稳定量产的植物工厂化栽培的概念日益受到重视，如何有效提升农业生产力与创造高品质、新鲜、洁净的农产品是重要课题。在作物栽培中，光、水、温度是最重要的因素，阳光利用型的植物工厂直接利用阳光，但日照量因时因地有极大变异，作物生产仍然受外界环境因素牵制。且受限于光线的方向，无法立体化栽培，需要广大的设施面积。完全控制型植物工厂是以人工光源取代阳光，不再受制于外界环境，且立体化栽培得以落实，可大幅缩小设施面积。然适用于立体化栽培的人工光源的选择比一般温室补光的光源限制更多，除需要低耗能外，低发热也是重要的考虑。近年高功率发光二极体（LED）技术突破，发光效率亦迎头赶上萤光灯管，而生产成本亦大幅下降，俨然已成为下一代照明光源的主流。然而市面上仍以满足室内照明需求的白光 LED 为主，此为针对人眼对光的敏感度开发的产品，主要光谱分布中有一大部分在绿光的波段范围中，针对植物生产开发的专用 LED 光源仍属罕见。

植物光合作用的有效波长在于可见光范围，叶绿素的吸收峰在红蓝光波段，蓝光也对气孔道度开闭及色素生合成有密切关系。叶绿素因在绿光波段几乎不吸收，被认为与植物的生长关连较小。是以目前在 LED 做为植物照明光源研究的主流仍以红蓝混色配比为主[1~6]。

红蓝混色光照在绿色植株上最大的问题在于呈色为暗紫色或趋近黑色，在此光环境下，人眼的敏感程度很差，叶面的异常色泽或病斑并无法用肉眼察觉。由于肉眼对绿光的敏感度极高，加入少部分的绿光便可察觉，植物栽培中的红蓝光光量配比加入 5% 的绿光便可改善此一问题，但对于莴苣生长则无显著影响[7]。近年研究显示绿光对于植物生理的影响远比想像中复杂，原则上是抑制红蓝光产生的效应[8]。但增加 10% 绿光比例对于部分莴苣品种可增加其地上部鲜重[9]，更进一步将绿光比例加至 24%，发现莴苣的叶面积、地上部鲜、干重显著优于只有红蓝混光的处理[10]，显示绿光对于作物的生长应有其正面功效。

本研究之目的在于利用 LED 红蓝混光之外加入不同比例的绿光，做为莴苣栽培光源，探讨对于莴苣幼苗及成株生长的影响，以做为未来开发植物灯具时光质调控的依据。

2 实验设备与方法

本研究分两个部分探讨绿光比例对莴苣（*Lactuca sativa*）生长的影响：①不同品种莴苣育苗栽培试验。②波士顿莴苣不同生长期栽培试验。

2.1 绿光比例对不同品种莴苣苗栽培试验

使用 5 种光质的处理进行莴苣栽培，各处理的光量均维持相同，使用光量计（LI-250A，LI-COR，USA）在灯下 20 cm 量测，均为 150 $\mu mol \cdot m^{-2} s^{-1}$。基本上以 RGB LED 三色混光调整绿光比例，以 0G、25G、50G 分别代表绿光占总光量的 0%、25%、50% 的三种处理，光源为 RGB 三晶 LED 灯板（亿光），以 PWM（Pulse-width modulation）控制器分别调控 RGB 光量输出。CW 处理为 T8 冷白 LED（5 500K，光茵）。对照组 FL 为 T5 白光萤光灯管（4 000K，力玛）。

使用光谱能量分析仪（USB4000，Ocean optics，USA）确认各处理的光质配比光谱分析。

图1为此5种处理的光谱图,其中R、G、B LED的峰值分别在645nm、525nm、462nm(图1a,b,c)。而冷白光LED(CW)的发光原理为蓝光LED晶片发光激发黄色萤光粉出现平滑光谱,约在550nm有最大峰值,另一个小峰值即出现在蓝光范围(图1d)。萤光灯管(FL)的光谱如图1e,为多波峰的光谱型式。

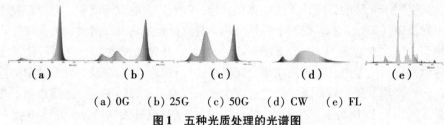

(a) 0G (b) 25G (c) 50G (d) CW (e) FL

图1 五种光质处理的光谱图

表1为此五种光质处理的RGB的光量配比,其中蓝光均调控在14%左右,0G的处理组虽在绿光波段有3%的比例,但此部分的波长其实非常接近600 nm。而50G及FL的绿光比例相同为50%,CW的绿光更高达64%。

表1 五种光质处理的RGB光量配比(总光量150 $\mu mol \cdot m^{-2}s^{-1}$)

光质处理	光量比例(%)		
	R(600~700nm)	G(500~600nm)	B(400~500nm)
0G	84	3	13
25G	61	25	14
50G	35	51	14
CW	22	64	14
FL	36	50	14

本试验使用3种莴苣品种:"翠花"、"翠妹"及红色的"红橡"。每个品种每种光质处理样本20株,于浸种后第18天育苗期结束后,进行生长调查,探讨绿光比例对不同品种莴苣育苗的影响,量测地上部鲜重、干重、叶片数、总叶面积、叶绿素SPAD值。

2.2 绿光比例对波士顿莴苣不同生长期栽培试验

本试验使用的设备同前述实验,各处理的光量均为150 $\mu mol \cdot m^{-2}s^{-1}$。而RGB的光量配比也和前述实验略同(如表2),但将原有的25G(绿光比例25%)的处理组提升至40G(绿光比例38%),原有的50G(绿光比例51%)的处理组提升至60G(绿光比例60%),探讨绿光比例再增加10%~15%的影响。

表2 波士顿莴苣光质处理配比（总光量150 $\mu mol \cdot m^{-2}s^{-1}$）

光质处理	光量比例（%）		
	R（600~700nm）	G（500~600nm）	B（400~500nm）
0G	87	3	10
40G	50	38	12
60G	25	60	15
CW	22	64	14
FL	36	50	14

本实验使用半结球莴苣品种-波士顿莴苣，进行两阶段栽培模式：第一阶段为育苗期，将子叶展开的莴苣苗移入栽培密度为180株/m²的育苗浮板。在浸种后第18天结束，随机取样20株进行生长调查。并另随机选取19株移入栽培密度为40株/m²之定植浮板，进入第二阶段养成期，进行光质处理至第39天收获，进行生长调查，量测地上部鲜重、干重、叶片数、总叶面积。

2.3 莴苣栽培条件

本试验之莴苣采水耕栽培，其操作流程如下：

在萌芽阶段先将莴苣种子置于清水中浸种5h后，去除浮起之种子，将种子植入充分浸泡清水的泡棉介质，移入植物生长箱内，以CW 5500K LED持续提供光照，光量为100 $\mu mol \cdot m^{-2}s^{-1}$，环境温度设定20℃。

莴苣约在第8天子叶完全展开后，移入栽培环控室内，日/夜温设定为25/18℃，光照时间16h，日间CO_2浓度维持在1 200μl/L。采用Hyponex#1养液，EC值调整至1.2 mS/cm。每座立体水耕栽培层架共有四层植床，下方设有养液回水槽，以沉水泵持续将养液送至最上层栽培植床，由上而下依序流经各层，再回到养液水槽内，沉水泵维持运作以确保养液浓度及温度均一。养液水槽内有热交换管，以冰水机控制养液温度在22℃。

2.4 调查项目与分析方法

（1）叶绿素含量：选取植株最大叶片，使用叶绿素计（SPAD-502，Minolta，Japan）夹取不同位置3点平均，测量叶绿素含量SPAD值。

（2）鲜重、干重：植株去除泡棉及根部后，量测其地上部鲜重。再置入烘箱，温度设定75℃，烘干48h后取出，测量其干重。

（3）叶片数、叶面积：植株地上部去除子叶及未完全展开之新叶后，计算其叶数量。再将叶片置入叶面积仪（LI-3100，LI-COR，USA）量测其叶面积。

（4）本试验分析使用SPSS软体进行邓肯氏多变域分析（$p < 0.05$）。

3 结果与讨论

3.1 绿光比例对不同品种莴苣苗生长之影响

本试验探讨不同绿光比例对于三种不同品种莴苣育苗期的影响，表3为不同绿光比例对于莴苣幼苗叶绿素值、叶片数及叶面积的实验结果。在叶绿素SPAD值方面，三种莴苣苗均在0G及25G的处理组表现显著优于50G、FL、CW，显示50%以上的绿光不利于叶绿素的合成。绿光比例对于三个品种的叶片数的影响较不一致，但 0G 的处理组均有最佳的表现。而在叶面积方面，50G 处理组在三个品种中均显著低于其他处理，代表绿光比例 50%，对于叶片发育有不良影响。但是 FL 及 CW 处理组同样绿光比例大于 50%，但叶面积反而比 50G 好。

表3　不同绿光比例对于叶莴苣幼苗叶绿素值、叶片数及叶面积之影响

光质处理	翠花莴苣			翠妹莴苣			红橡莴苣		
	SPAD	叶片数	叶面积（cm²）	SPAD	叶片数	叶面积（cm²）	SPAD	叶片数	叶面积（cm²）
0G	18.5 a	4.7 a	76.9 a	23.4 a	6.2 a	64.1 a	25.1 a	5.7 a	47.8 ab
25G	19.0 a	4.6 a	76.7 a	24.4 a	5.2 b	47.0 b	25.9 a	4.8 cd	40.3 bc
50G	18.0 a	4.3 b	54.7 b	21.7 b	4.8 b	35.0 c	23.8 b	4.6 d	34.9 c
FL	14.9 c	4.7 a	75.7 a	19.5 c	5.1 b	43.0 bc	19.3 d	5.2 bc	50.1 a
CW	16.5 b	4.9 a	70.7 a	21.2 b	5.1 b	36.0 bc	20.8 c	5.5 ab	51.1 a

Means followed by the different letters in each column are significantly different at 5% level by Duncan's Multiple Range Test.

表4 显示不同绿光比例对于莴苣幼苗地上部鲜、干重的影响，25G 处理在三种品种中都属最佳的处理，尤其在"翠花"、"翠妹"两种绿色莴苣中显著优于 50G、FL、CW 三种绿光大于 50% 的处理。而"红橡"为红色莴苣，叶片内色素组成和绿色莴苣不同，对于绿光的利用效率可能也有不同，因此在干重的差异上不如绿色莴苣显著。

表4　不同绿光比例对于莴苣幼苗地上部鲜、干重的影响

光质处理	翠花莴苣		翠妹莴苣		红橡莴苣	
	鲜重（g）	干重（g）	鲜重（g）	干重（g）	鲜重（g）	干重（g）
0G	2.82 a	0.147 2 b	2.36 a	0.106 8 a	1.68 a	0.085 8 ab
25G	3.15 a	0.181 3 a	1.83 b	0.095 7 a	1.46 ab	0.093 0 a
50G	2.07 b	0.116 3 c	1.16 c	0.063 8 b	1.22 b	0.070 9 b
FL	2.57 ab	0.117 3 c	1.36 b	0.055 3 b	1.44 ab	0.076 0 ab
CW	2.58 ab	0.119 8 c	1.15 c	0.050 4 b	1.66 a	0.078 9 ab

Means followed by the different letters in each column are significantly different at 5% level by Duncan's Multiple Range Test.

3.2　绿光比例对波士顿莴苣不同生长期成长之影响

表5 为不同绿光比例对波士顿莴苣的两阶段栽培的试验结果。在育苗期中，由红蓝混色的 0G 处理组有最大的叶面积及鲜重，而 40G 的处理组稍微差一点，可能是绿光比例由 25% 提升至 40% 所致。绿光比例大于 50% 的三个处理组（60G、FL、CW）鲜重及叶面积仍然最低，和前一试验的两种绿色莴苣有相同趋势，证明过高比例的绿光会延迟莴苣苗成长。

表 5　不同绿光比例对波士顿莴苣育苗期及收获产量之影响

光质处理	育苗期结束（第 18 天）			收获（第 39 天）	
	鲜重（g）	叶片数	叶面积（cm^2）	鲜重（g）	干重（g）
0G	3.4 a	6.8 a	93.9 a	122.7 b	4.32 b
40G	2.9 b	6.5 ab	86.6 a	143.5 a	4.93 a
60G	2.3 d	6.0 ab	72.9 b	119.9 b	4.38 b
FL	2.4 cd	5.8 b	67.1 b	93.2 c	3.68 c
CW	2.7 bc	6.0 ab	69.7 b	93.0 c	3.91 bc

Means followed by the different letters in each column are significantly different at 5% level by Duncan's Multiple Range Test.

在最后收获阶段时，40G 处理组反而在鲜重及干重都显著优于其他处理组，甚至超过红蓝混光的 0G 处理（表 5），此趋势与 Kim et al.（2004）的研究相符[10]。由于莴苣在育成阶段后期型态变大，相邻的两株作物可能接触甚至部分叶片相互遮荫，推测由于光线中的红蓝光部分可能在上位叶已被吸收利用，而绿光由于吸收率差，反而可能经由穿透或反射，进入下位叶区域，以利光合作用进行，反提升整体的光合作用效率。但是绿光比例超过 50% 对成株的生长有负面影响。

4　结论

目前白光 LED 灯管以做为室内照明灯具为主，市场需求量大，但其绿光比例高达 50% 以上，适合人眼对光线的敏感波长范围，但并不合适做为植物的栽培光源。本研究证实直接以高发光效率的白光 LED 做为植物照明光源并未能发挥最佳效益。

本研究亦证明在作物栽培中绿光确有其功效，仅需改变其比例即可。以光合作用所需的红蓝光配比为主体加入 25% 绿光（RGB 60∶25∶15）的组合能有效栽培莴苣。目前针对植物栽培所需光质开发的专用灯仍然罕见，仅少量生产，价格较高，且其光量远低于白光 LED。可用高发光效率的白光 LED 提供作物所需的光量，利用红光 LED 灯管搭配白光灯管使用，将绿光比例调整至 25% 左右，即可做为植物栽培用途。

参考文献

[1] Yanagi, T., Okamoto, K., Takita, S. Effects of blue, red, and blue/red lights of two different PPF levels on growth and morphogenesis of lettuce plants. Acta Hort. 1996, 440: 117~122

[2] Okamoto, K., Yanagi, T., and Kondo, S. Growth and morphogenesis of lettuce seedlings raised under different combinations of red and blue light. Acta Hort. 1997, 435: 149~158

[3] Goins, G. D. Performance of salad-type plants grown under narrowspectrum light-emitting diodes in a controlled environment. Proceedings of Bioastronautics Investigators´ Workshop. Galveston, TX, 2001

[4] Fang, W. and Jao, R. C. Development of a flexible lighting system for plant related research using super bright red and blue light-emitting diodes. Acta Hort. 2002, 578: 133~139

[5] Brazaitytė, A., Ulinskaitė, R., Duchovskis, P., Samuolienė, G., Siksnianien, J. B., Jankauskienė, J., Sabqievienė, G., Baranouskis, K., Stanienė, G., Tamulaitis, G., Bliznikas, Z., and Zukauskas, A. Optimization of lighting spectrum for photosynthetic system and productivity of lettuce by using light-emitting diodes. Acta Hort. 2006, 711: 183~188

[6] Pinho, P., Lukkala, R., Sarkka, L., Tetri, E., Tahvonen, R., and Halonen, L. Evaluation of lettuce growth under multi-

spectral-component supplemental solid state lighting in greenhouse Environment. International Review of Electrical Engineering. 2007, 2 (6): 854~860

[7] Kim, H. H., Wheeler, R. M., Sager, J. C., and Goins, G. D. A comparison of growth and photosynthetic characteristics of lettuce grown under red and blue light-emitting diodes (LEDs) with and without supplemental green LEDs. Acta Hort. 2004, 659: 467~475

[8] Folta, K. and Maruhnich, S. Green light: a signal to slow down or stop. J. Exp. Bot. 2007. 58: 3 099~3 111

[9] Fang, W., Wu, C. C., and Chang, M. Y. LED as light source for baby leaves production in an environmental controlled chamber. The 4th International Symposium on Machinery and Mechatronics for Agricultural and Biosystems Engineering. Taichung, Taiwan. 2008

[10] Kim, H. H., Goins, G. D., Wheeler, R. M., and Sager, J. C. Green-light supplementation for enhanced lettuce growth under red-and blue-light-emitting diodes. HortSci. 2004, 39 (7): 1 617~1 622

Production Pattern and Water Use Efficiency of Tomato Crops

Li YaLing

(*College of Horticulture, Shanxi Agricultural University, 030801, Taigu, Shanxi, P. R. China*)

Abstract: Lack of fresh water is one of the biggest problems for the living of human being in the world. Fresh water consumed in the agriculture is the most part, about 80.7% of the total water application. Among of this, 91.5% is used for irrigation of crop growing. Reducing water consumption for agricultural irrigation or increasing water use efficiency is the aim of sustainable agriculture. It is certificated from practice in China that the yield of tomato increased from 2.3kg/m^2 in open field to about 10kg/m^2 in protected cultivation, while (litres of irrigation water used for one kg product) fresh tomatoes produced per cubic meter of water is increased from 5~20kg, up to 50kg for drip irrigation under mulch. In the Netherlands, fresh tomato yield produced per cubic meter of water is 65~68kg by using closed growing system in the glasshouse. It is suggested that the development of sustainable agriculture must rely on the progress of scientific technology: increase production level, make the best use of fresh water or save water resource.

摘要：淡水资源缺乏是世界上人类生存的最大问题之一。农业用水是消耗淡水资源最大的部分，占总供水量的80.7%，农业用水的91.5%是用于种植业灌溉，减少灌溉耗水或者说提高作物的水分利用效率，是农业生产可持续发展的目标之一。实践证明，在中国采用设施栽培，番茄的产量从露地条件下的2 000kg/667m^{2}提高到7 000kg/667m^2，生产1kg番茄的用水量从200L降低为95L，在温室内采用膜下滴灌技术更能减少到40~45L。荷兰采用无土栽培和营养液循环利用技术，生产1kg番茄的用水量仅为15L。依靠技术的进步，提高生产技术，才能充分利用和节约水资源，走农业可持续发展的道路。

China is one of the countries for lacking fresh water, with the average water resource of 2 260 m^3, far less than the average of 10 796 m^3 of world level (Pi, 2000). Agricultural irrigation is the most part for consuming fresh water in China, accounting for about 80.7% of total fresh water use in the whole country. With the water consumption in agriculture, about 91.5% are used for irrigation of crop growing. Vegetable is one of the crops for consuming the most water per unit area. Therefore, choosing suitable irrigation methods and saving water is of stratagem significance for sustainable vegetable industry.

Greenhouse growing in the Netherlands keeps ahead in the world level. Analysis of its growing pattern, management ways, installations, construction and constitute of greenhouse area, we found that each hectare of greenhouse normally has 1 500 m^3 rain basin, which could supply water of 75% used for greenhouse, although soil land is very expensive. Vegetable production in the greenhouse is more than 90% with rockwool cultivation. Drain water, after supplying plant, is recycled. With this way, tomato yield in the Netherlands is higher up to 60kg/m^2, and water and fertilizer use efficiency are highly increased. The experience from the Netherland tells us: it is possible to increasing unit production level and increasing water use efficiency in the greenhouse by the improvement of techniques. This is of strategy instruction significance for tomato production in China.

Table 1 Tomato yield and water use efficiency under greenhouse condition

Site, year, season	Drip irrigation			Furrow irrigation			Related literature
	Tomato yield (kg/m^2)	Water use efficiency (kg/m^3)	Water used per kg tomato (L/kg)	Tomato yield (kg/m^2)	Water use efficiency (kg/m^3)	Water used per kg tomato (L/kg)	
Beijing, NKY, Aug.-Oct. 2004	11~12	23~24	42~44	9.18	5.0	200	Liu et al., 2005
Beijing, CAAS, summer	12.8~14.4	15~21	47.4~66.3	—	—	—	Gao et al., 2004
Beijing, Mar.-Jun. 2001	—	—	—	—	6.86	145.7	Liu et al., 2002
Beijing, Shunyi, Mar.-Sep. 1998	9.2	48	20.8	—	—	—	He et al., 2001
NW, YL, Mar.-Jul. 1997~1998	—	—	—	4.7~5.3	12~17	58~81	Li et al., 2000
NW, YL, Oct.-Jul. 2004	—	—	—	19~20	3.6~4.9	194~275	Zhou et al., 2006
SYAU, Apr.-Aug. 1999	7.6~8.1	23~26	38~43	6.76	11.67	85.7	Yang et al., 2004
SYAU, Apr.-Jul. 1998	—	—	—	6~7.5	24~34	30~41	Qi et al., 2000
SYAU, May.-Jul. 2005	8.6~12	58~78	17.24~12.82	—	—	—	Zhang et al., 2006
SYAU, Aug.-Oct. 2000	3.4~3.8	23~26.3	43~38	—	—	—	Zhuge et al., 2002
JL, CC, May.-Aug. 1995	12.5	17~23	43~59	11.4	10.45	96	Yu et al., 1996
GS, DX, Oct.-Jul. 2001	5.5~8.3	22~28	35~45	8.2	20	49.1	Tang, 2004
HB, Xinji, Jan.-Jun. 1997~1999	6.3	21	47.6	5.1	6.73	148.5	Han et al., 2003

1 Tomato production situation in China today

Tomato (*Lycopersicon esculentum* Mill) is one of the main vegetable crops in China. According to estimation of FAO (FAOSTAT), the area of tomato production in China in the year 2008 is 1 450 000ha. Besides 70 000ha for processing tomato (such as tomato juice in XinJiang province, Institute of Agro-economy and development in CAAS), most are growing for fresh market in domestic. In the 80~90's 20th century, tomato was grown mostly in open field. Because of short growing season, low yield disturbed by plant diseases and insect pests, by hailstone, rain and so on, tomato growing is gradually transferred to greenhouse since 1990.

At present, with the increase of requirement by the increase of human living level and with the progress of production level in protected cultivation, year-round growing and supply of tomato has been realized. According to Tomato Association in CAHS, tomato growing area for protected cultivation in China is about 600 000ha, 1/3 of vegetable growing area for protected cultivation. However, tomato yield is only 2.3kg/m^2 in open field, according to FAO (FAOSTAT) estimated calculation, 7~10kg/m^2 in solar greenhouse for the average (Table 1).

2 Production and irrigation patterns of protected tomato cultivation

Soil planting is still main growing pattern for protected tomato cultivation, even if substrate cultivation growing systems in ditch, trough or bag are adopted in a few areas by some growers. Procedure for normal ways of soil cultivation is: making plots after applying some manure, covering plastic mulch on the plot, irrigating seedlings after transplanting, reviving water for seedlings after one week; then, no watering for about 20~30 days in order to enhance flowering; watering and fertilizing plants after fruit growing to 2~3cm size of first fruit truss.

For tomato growing in the soil in the greenhouse, furrow irrigation is normal watering pattern. Some growers irrigate plants under plastic mulch in order for reducing humidity in the greenhouse (Zhu and Li, 2005). Furrow irrigation is one of the traditional ways in agricultural production since it is no need other equipment, and it is simple, convenient and fast. After transferring from open field to protected cultivation, most growers are still used to furrow irrigation, limited by growing condition. Furrow irrigation in greenhouse is easily becoming high humidity, which would form water drop or dew on the leaves thereby inducing diseases afterwards. Meanwhile, furrow irrigation could spread some diseases; contaminate soil or ground water if there are some pesticides, chemicals or heavy metals. The main disadvantage of furrow irrigation is waste fresh water resources and decrease water use efficiency.

3 Water use efficiency in tomato cultivation

In recent years greenhouse tomato yield and water use efficiency are studied by some researchers under different irrigation pattern (Table 1). According to table 1, we could see that average yield of tomato in the greenhouse was about $7kg/m^2$ for furrow irrigation, $9\sim10kg/m^2$ for drip irrigation. And litres of water used for producing 1 kg fresh tomato were 95 and 40~45 for furrow and drip irrigation respectively. Comparing these results we found that tomato yield under furrow irrigation is only 70%~80% of that under drip irrigation, but drip irrigation could save water about 60% compared with furrow irrigation.

In open field, Sun et al. (2006) harvested tomato $2\sim2.6kg/m^2$ by using drip irrigation, which means that litres of water used for producing 1 kg fresh tomato were more than 200. Compared with this result, water production efficiency in greenhouse is about 6~8 time of that in open field.

4 Irrigation ways for saving water——Dutch ways

From the view in world level, water production efficiency of tomato crops is 30~40 litres per kg fresh tomato product in Spain and Israel under unheated greenhouse or plastic tunnel compared with those 60 litres in open field. In the Netherlands water used for producing 1 kg fresh tomato are 22 litres under the condition of glasshouse with climate control and CO_2 application, only 15 litres when closed growing system are adopted (Table 2). From these analyses, we could also say that the increase of production level means the increase water use efficiency. According to the analysis of Stanghellini et al. (2003), that greenhouse production has higher water use efficiency is the consequence of at least three factors:

- Reduced potential evaporation (less sun radiation; less wind and higher humidity)

- Increased production (better control of pathologies; better control of climate parameters)
- Application of advanced irrigation techniques (drip irrigation; re-use of drain water)

Compared with advanced level in the world, water use efficiency of tomato growing in China is still lower. The reasons of this are related with yield, cultivation pattern and irrigation ways. Zhang *et al*, (2006) established a quadratic equation between water use and tomato yield, and indicated that when the yield was 9.69kg/m^2 water use efficiency is lowest, after that with increasing yield litres water used for per kg tomato would be decreased. After analyzing Dutch greenhouse production Stanghellini *et al.* (2003) confirmed that present about 2/3 of the greenhouse area in Holland is substrate cultivation. The reason for growing on substrate is purely economical: better return. For example, return of the sweet pepper growing on substrate was 3 times more than that on soil-grown. Substrate cultivation of course makes it rather easy to collect drain-water and re-use it (closed system), which allows for some saving on the costs of fertilizer and water.

Table 2 Litres of irrigation water used for one kg of product, in various places and various growing systems

Place and growing condition	Water used per kg tomato
Israel and Almeria, field	60
Almeria, Unheated plastic (1990)	40
Israel, unheated glass	30
Almeria, improved unheated plastic (2000)	27
Holland, climatic-controlled glass with CO_2 injection	22
Holland, as above, with re-use of drain water	15

* From: Stanghellini *et al.* 2003.

5 Summery and suggestion

There is huge area of tomato growing in China; and therefore there is big room for saving water resource and increasing water use efficiency. Some suggestions are giving as following:

(1) Growing tomato under protected cultivation year-round.

(2) Covering plastic mulch and applying manure during tomato growing if adopted soil as cultivating substrate.

(3) Watering tomato crops with drip irrigation system.

(4) Adopted soilless cultivation system, especially closed growing system.

It is suggested that the development of sustainable agriculture must rely on the progress of scientific technology: increase production level, make the best use of fresh water or save water resource.

设施番茄砂培营养液配方筛选试验*

高艳明,李建设**,卜燕燕

(宁夏大学农学院,银川,750021)

摘要:采用三因素五水平通用旋转组合设计,在日光温室内研究了营养液配方中不同摩尔浓度的 NO_3-N、P、K 对番茄产量的影响,得到产量回归方程,并依此进行分析。结果表明:三个元素对番茄产量的影响大小顺序是:氮>磷>钾;通过计算机模拟寻优,番茄达到单株最高产量 2 123.13g 时,最优配方营养液中 NO_3-N、P、K/浓度分别为 8mmol/L、0.9mmol/L 和 5mmol/L。经过验证,最优配方处理 A、B 的株高、茎粗、生长势、产量、品质表现相对优于对照,模型准确可靠。

关键词:设施番茄;营养液配方;回归模型

Screening Experent of Nutrient Solution Famula on Sand Culture *Lycopersicon esculentum*

Gao Yanming, Li Jianshe, Bu Yanyan

(Agricultural College of Ningxia University, Yinchuan 750021, P. R. China)

Abstrat: The comprehensive effects of NO_3-N、P and K concentrations on *Lycopersicon esculentum* were studied by quadratic general spinning design under modern greenhouse. The results calculated from regresion equation between yields and there elements concentrations factors of N、P、K showed that: the effects sequence of there nutrients on the yield of *Lycopersicon esculentum* is nitric-N (NO_3-N) > potassium (K) > phosphorus (P); After simulation with computer, the optimal ration of N, P and K to achieve highest yield of individual *Lycopersicon esculentum* (2123.13g) is 8mmol/L, 0.9mmol/L and 5.0mmol/L respetibely. This ration A and B were verified because the height、stem diameter、growth potential、Yield and quality of Tomato were superior to CK.

Key words: Protected tomato; Nutrient solution famula; Regression model

1 实验材料与方法

1.1 试验概况

本试验于 2009 年 9 月至 2010 年 4 月在宁夏永宁县纳家户领鲜果蔬产业发展有限公司设施农业基地进行。温室长 87m,宽 7.5m,脊高 4.5m。

1.2 试验材料[1]

1.2.1 试验地供试水

试验用沙取自永宁县征沙渠腹部沙地;水为地下水,井深 20m。用沙理化性质及地下水

---

* 基金项目:"十一五"国家科技支撑计划项目(2007BAD57B03)。

** 作者简介:高艳明(1963—),女,宁夏平罗人,教授,主要从事设施蔬菜栽培和生理研究。
E-mail: jslinxcn@yahoo.com.cn。

水质理化性质分别见表1和表2。

表1 试验用沙理化性质
Table 1 The physicochemical properties and structures of sand

沙	pH	全盐 g/kg	有机质 g/kg	全量氮 g/kg	全量磷 g/kg	全量钾 g/kg	速效氮 mg/kg	速效磷 mg/kg	速效钾 mg/kg
	7.39	0.55	0.57	0.13	0.14	18	14	28	48

表2 试验地水质理化性状
Table 2 the physicochemical properties and structures of experiment area

水	pH	EC ms/cm	全盐 g/L	Ca^{2+} mg/L	Mg^{2+} mg/L	HCO_3^- mg/L	CO_3^{2-} mg/L	Na^+ mg/L	Cl^- mg/L	SO_4^{2-} mg/L
	7.14	1.32	1.015	103	75	477	0	116	117	232

1.2.2 供试蔬菜

番茄（*Lycopersicon esculentum*），品种：合作918。

1.3 试验设计

本试验以 NO_3-N、P、K 摩尔浓度为试验因素，采用三因素五水平通用旋转组合设计，根据日本园式通用配方[1,2]等权威配方，参照茄果类蔬菜番茄对N、P、K的需求范围设计出本试验的上下限及零水平（表4），共20个处理，3次重复，进行盆栽砂培试验。试验中微量元素采用通用配方（表3）。根据番茄的具体生长情况，采取人工每天定量浇液（500~1 000ml不等），以保证植株的正常生长。

表3 营养液微量元素配方浓度表
Table 3 The nutrition formula of microelement

元素	Fe	B	Mn	Zn	Cu	Mo
浓度（mg/L）	3	0.5	0.5	0.05	0.02	0.01

表4 番茄试验因子及水平编码值
Table 4 Encoding values of experimental factors and level of Tomato

因素	零水平	变化间距 $\triangle i$	无量纲编码/mmol/L				
			+r	1	0	-1	-r
NO_3-N	8	4	16	12	8	4	0
P	0.9	0.45	1.8	1.35	0.9	0.45	0
K	5	2.5	10	7.5	5	2.5	0

2 产量方程构建

根据设计原理[2,7]，试验的期望回归数学模型为（表5）：

$$Y = b_0 + \sum_{j=1}^{3} b_j x_j + \sum_{i<j}^{3} b_{ij} x_i x_j + \sum_{j=1}^{3} b_{jj} x_j^2$$

表5 番茄试验处理方案及对应产量表
Table 5 Effects of nutrients rations on yield of Tomato

处理	X_1	X_2	X_3	产量（g/株）
Tr1	1 (12)	1 (1.35)	1 (7.5)	1 956.05
Tr2	1 (12)	1 (1.35)	−1 (2.5)	1 849.75
Tr3	1 (12)	−1 (0.45)	1 (7.5)	1 586.32
Tr4	1 (12)	−1 (0.45)	−1 (2.5)	1 284.60
Tr5	−1 (4)	1 (1.35)	1 (7.5)	1 472.91
Tr6	−1 (4)	1 (1.35)	−1 (2.5)	1 341.92
Tr7	−1 (4)	−1 (0.45)	1 (7.5)	1 327.28
Tr8	−1 (4)	−1 (0.45)	−1 (2.5)	1 597.85
Tr9	−2 (0)	0 (0.9)	0 (5)	924.89
Tr10	2 (16)	0 (0.9)	0 (5)	1 428.45
Tr11	0 (8)	−2 (0)	0 (5)	1 592.28
Tr12	0 (8)	2 (1.8)	0 (5)	1 897.41
Tr13	0 (8)	0 (0.9)	−2 (0)	1 173.81
Tr14	0 (8)	0 (0.9)	2 (10)	1 698.22
Tr15	0 (8)	0 (0.9)	0 (5)	2 194.37
Tr16	0 (8)	0 (0.9)	0 (5)	2 202.72
Tr17	0 (8)	0 (0.9)	0 (5)	2 021.07
Tr18	0 (8)	0 (0.9)	0 (5)	2 324.89
Tr19	0 (8)	0 (0.9)	0 (5)	2 110.54
Tr20	0 (8)	0 (0.9)	0 (5)	1 912.42

以产量为目标函数（Y），营养液中 NO_3-N、P、K 三种元素的摩尔浓度为决策变量（X_1、X_2、X_3），建立番茄产量与三个肥料因子的回归方程（表6），砂培番茄营养液配方的回归方程为：

$$Y_{(番茄)} = 2\ 123.13 + 130.60X_1 + 97.95X_2 + 84.24X_3 - 306.57X_1^2 - 105.69X_2^2 - 214.88X_3^2 + 130.65X_1X_2 + 68.45X_1X_3 + 25.77X_2X_3 \tag{1}$$

表6 番茄回归方程的显著性检验表
Table 6 Significance test of regression equation on Tomato

试验作物	变异来源	平方和	自由度	均方	偏相关	F值
番茄	回归	2 540 222	9	282 246.8	$F_2 = 10.52277$	$F_{0.05} = 3.02$
	剩余	268 224.8	10	26 822.48		$F_{0.01} = 4.94$
	失拟	161 258.1	5	32 251.63	$F_1 = 1.50755$	$F_{0.05} = 5.05$
	误差	106 966.7	5	21 393.34		$F_{0.01} = 10.97$
	总和	2 808 446	19			

表6的方差分析表明，$F_2 > F_{0.01}$，番茄产量方程的回归关系极显著；$F_1 < F_{0.05}$，失拟不显著，说明产量的回归方程与实际拟合很好，可用于决策分析。

剔除 $a = 0.10$ 显著水平不显著项后，简化后的回归方程为：

$$Y_{(番茄)} = 2123.13 + 130.60X_1 + 97.95X_2 + 84.24X_3 - 306.57X_1^2 \\ - 105.69X_2^2 - 214.88X_3^2 + 130.65X_1X_2 \tag{2}$$

由方程（2）可以看出，番茄在 0.10 水平上氮与磷互作效应显著。

2.1 主效应分析

回归方程本身就已经过无量纲编码代换，其偏回归系数已经标准化，所以可以直接从一次项系数绝对值的大小来判断各因素对目标函数的相对重要性。NO_3-N、P、K 三因素对番茄产量的影响主次的线性项为：$X_1 > X_2 > X_3$，即 NO_3-N > P > K。

2.2 单因子效应分析

采用降维法固定其他两因素在零水平，肥料三因子与番茄产量的偏回归降维子方程为：

$$Y_{1(番茄)} = 2123.13 + 130.60X_1 - 306.57X_1^2 \tag{3}$$

$$Y_{2(番茄)} = 2123.13 + 97.95X_2 - 105.69X_2^2 \tag{4}$$

$$Y_{3(番茄)} = 2123.13 + 84.24X_3 - 214.88X_3^2 \tag{5}$$

根据上述偏回归子方程，得到 NO_3-N、P、K 三因子的单效应图（图1）：

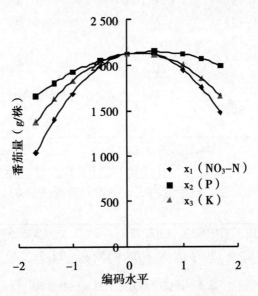

图1 番茄 NO_3-N、P、K 单因素效应曲线图

Figure 1 The single-factor effect diagram of Tomato

由图1可知，方程（3）、（4）和（5）均为一元二次抛物线方程，且一次项系数为正值，二次项系数为负值，说明番茄产量与 NO_3-N、P、K 三元素均呈开口向下的抛物线型关系，即3个方程均有最高产量对应的编码值存在[3]。

2.2.1 营养液中 NO_3-N 浓度对番茄产量的影响

由图1可以看出，随着 NO_3-N 浓度由 -2 增大到 0.213 时，番茄产量由 1 036.36g/株增大到 2 137.05g/株，增加单位浓度硝态氮番茄产量的平均增量为 132.63g；但随着硝态氮浓度的继续增加，由 0.213 增加至 2 水平时番茄产量有明显的下降趋势，产量降至 1 475.66g/株，产量平均增量为 -96.70g，也就是说番茄对氮素（尤其是硝态氮）的需求有一定的浓度范围要求，并不是越多越好，浓度太大反而抑制番茄对磷钾元素的有效吸收，导致产量减少。沙培番茄营养液中氮素取 0.213 水平最佳，即 8.80mmol/L。

2.2.2 营养液中 P 浓度对番茄产量的影响

图 1 表明，随着浓度由 -2 水平增大到 0.463 时，番茄产量由 1 659.447g/株增大到 2 145.82g/株，增加单位浓度 P 番茄产量的平均增量为 438.83g；但随着 P 浓度由 0.463 增加至 2 水平时番茄产量有下降趋势，产量降至 1 988.924g/株，产量平均增量为 -226.84g，也就是说番茄对磷素的吸收有一定的限度，沙培番茄营养液中磷取 0.463 水平为宜，即 1.11mmol/L。

2.2.3 营养液中 K 浓度对番茄产量的影响

由图 1 知，随着 K 浓度由 -2 增大到 0.196 时，番茄产量由 1 373.689g/株增加至 2 131.39g/株，增加单位浓度 K 素番茄产量的平均增量为 172.52g；但随着 K 浓度的继续增加，由 0.196 增加至 2 水平时番茄产量有明显的下降趋势，产量降至 1 657.02g/株，产量平均增量为 -131.48g，即番茄对钾素的吸收有限度，浓度超过一定的界限反而会抑制番茄的生长而造成减产，沙培番茄营养液中氮素最好取 0.196 水平，即 4.392mmol/L。

再令偏回归子方程（5）（6）（7）的偏导数为零，即 $\frac{\partial y_1}{\partial x_1} = 0$，$\frac{\partial y_2}{\partial x_2} = 0$，$\frac{\partial y_3}{\partial x_3} = 0$，可得 $X_1 = 0.213$，$X_2 = 0.463$，$X_3 = 0.196$。由于 $\frac{\partial^2 y_1}{\partial x_1^2} < 0$，$\frac{\partial^2 y_2}{\partial x_2^2} < 0$，$\frac{\partial^2 y_3}{\partial x_3^2} < 0$，故当 $X_1 = 0.213$，$X_2 = 0.463$，$X_3 = 0.196$，即氮磷钾的浓度分别为 8.80mmol/L、1.11mmol/L 和 4.39mmol/L 时，番茄产量 y_1、y_2、y_3 有最大值，分别为 2 137.05g/株、2 145.82g/株 和 2 131.39g/株。

2.3 单因素边际产量效应分析

边际产量就是每增施单位量肥料的增产量，实际上它是一定肥料水平时产量的曲线斜率[7]。由于氮磷钾各个因素对番茄产量的变化速率各异，因此有必要对边际产量效应作进一步分析，我们可以通过对之前所得回归方程求一阶偏导数而得到产量对各个因素水平值变化增减的速率，即：

$$\frac{\partial y}{\partial x_j} = b_j + \sum b_{ij} + 2 \sum b_{jj} x_j$$

将各变量固定在零水平时，其边际产量效应方程分别为：

$\frac{\partial y}{\partial x_1} = 130.60 - 613.14 X_1$ $\frac{\partial y}{\partial x_2} = 97.95 - 211.38 X_2$ $\frac{\partial y}{\partial x_3} = 84.24 - 429.76 X_3$

由以上的方程得氮磷钾三因素的平均变化速率分别为 $X_1 = 130.60$、$X_2 = 97.95$ 和 $X_3 = 84.24$。

图 2 中斜率反映了单位肥料对番茄产量的影响，在试验设定的编码水平范围内，在（-2，0）这个区间附近氮肥对产量的影响最明显，其次是钾肥，磷肥对产量的影响最小；

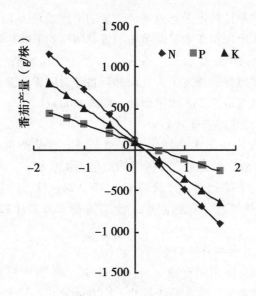

图 2 番茄边际产量效应图
Figure 2 The marginal product diagram of Tomato

而在（0，2）这个区间段，三个肥料因子对番茄产量的影响顺序为 P > K > NO_3-N。各因子在不同水平时对产量的影响不同，这为在不同条件下选择增产因素和决定施肥量大小提供了参数。随着三种肥料的增加，边际产量均呈下降趋势，边际产量为零时，产量达最大值，超过此水平，呈负效应，这符合肥料效应报酬递减定律。

综上所述，由于五个水平的编码值的变化间距相等，平均数值就相当于零水平，因此将各个因素固定在零水平时，所得到的产量就可以代表各个因素对产量的影响程度，所以可以用中值来很衡量各个因素的重要性，即氮磷钾对番茄产量的作用主次关系为 $x_1 > x_2 > x_3$，这与用一次项系数排序顺序一致。

2.4 互作效应

由方程（2）知番茄产量回归方程中氮磷钾三因子在 0.10 水平上存在显著交互作用的为 X_1、X_2，即氮与磷有互作效应（表7）。交互项系数为正，表明氮与磷配施要比单独施用效果更好，表现为正交互效应。将试验回归方程中的一个因素固定在零水平，用降维法可得其他两因素与产量的子方程，即为交互项对应的回归方程。

番茄栽培中 X_1 与 X_2 交互效应方程为：

$$Y_{(1,2)} = 2123.13 + 130.60X_1 + 97.95X_2 - 306.57X_1^2 - 105.69X_2^2 + 130.65X_1X_2 \quad (6)$$

由图3和表8可知，低氮水平时，随着磷元素的增加番茄的产量呈上升趋势，当磷素增加到中、高水平时产量反而下降；中、高氮水平时，磷元素的增加有利于番茄产量的形成，可见氮磷之间存在正的交互效应，两者共同作用促进高产，这与王合理等的研究结果一致[8]，P营养对N的吸收有促进作用，能提高了N的营养水平，两者共同促进肥水的番茄高产。从三维图中可以看出，番茄最高产量出现在氮水平取 0.21，磷水平取 0.46 时，而最低产量出现在氮取最高或磷取最高水平时。经计算，$X_1 = 0.21$，$X_2 = 0.46$ 时，产量最高。氮、磷浓度变化对番茄产量的影响见表8。

表7 番茄与氮磷交互效应和对应产量
Table 7 The yields of NO$_3$-N and P effects of interaction on Tomato (g)

	-2	-1	0	1	2
-2	539.71	693.31	635.65	366.61	113.81
-1	1 328.6	1 613	1 685.96	1 547.57	1 197.8
0	1 504.5	1 919.5	2 253.73	2 115.39	1 896.27
1	1 067.2	1 514.9	1 947.16	2 070.07	1 981.6
2	16.79	693.11	1 158.05	1 411.61	1 453.79

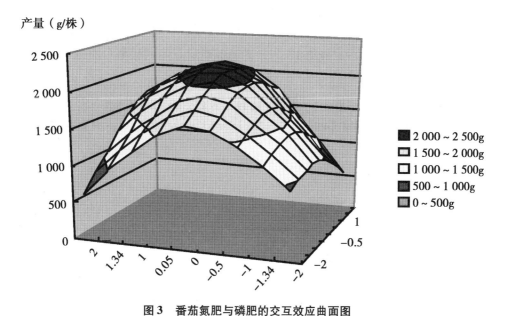

图3 番茄氮肥与磷肥的交互效应曲面图
Figure 3 The effects of interaction of NO$_3$-N and P on Tomato

表8 番茄氮、磷浓度变化对产量的影响
Table 8 The Effects of variety of NO$_3$-N and P concentration to Tomato yields

区域	因素变化	产量变化
当 $X_1 < 0.21$, $X_2 < 0.46$	增加氮、磷时	产量提高
当 $X_1 > 0.21$, $X_2 < 0.46$	增加磷、氮不变时	产量提高
	增加氮、磷不变时	产量提高
当 $X_1 < 0.21$, $X_2 > 0.46$	增加磷、氮不变时	产量下降
	增加氮、磷不变时	产量提高
当 $X_1 > 0.21$, $X_2 > 0.46$	增加磷、氮不变时	产量降低

2.5 模型寻优

根据回归方程（1）在计算机上的优选结果，番茄最高产量达2123.13g/株，此时影响产量的三个因素 NO$_3$-N、P、K 在营养液中的浓度分别为 8.0mmol/L、0.9mmol/L、

5.0mmol/L，如表9所示：

表9　番茄9产量达最大时各因素的编码值组合
Table 9 The combination of different encoding values when yields to be maximized output

类别	X_1	X_2	X_3	Y_{max}（g/株）
番茄	0（8）	0（0.9）	0（5.0）	2 123.13

这里我们得出的结论是纯理论值，大田农业生产试验不可避免的会受到环境因素、人为误差因素等的影响，实际值和我们得出的理论值会有些许的出入，因此这些理论值只能作为我们学习工作的宏观指导，为了确定试验数值和理论值的无限靠近，有必要进行频数分析。

2.6　频数分析

将番茄产量大于1 694.89g/株的19个方案中各变量取值的频率分布列于下表（表10、表11）：

表10　产量大于1 694.89的19个方案取值的频率分布表
Table 10 The frequency distribution table of 19 schemes of the yield per plant above 1694.89g

水平	X_1	频率	X_2	频率	X_3	频率
-2	0	0	0	0	0	0
-1	0	0	2	0.1053	4	0.2105
0	10	0.5263	5	0.2632	8	0.4211
1	8	0.4211	6	0.3158	7	0.3684
2	1	0.0526	6	0.3158	0	0

表11　19个方案中各因子在95%置信区间的水平值
Table 11 The level of 95% confidence interval in 19 schemes

	加权均数	标准误	95%的分布区间
X_1	0.51	0.128	0.259...0.760
X_2	0.742	0.201	0.349...1.135
X_3	0.158	0.171	-0.177...0.493

由表10和表11可以看出，当X_1的投入水平为0.259~0.760，X_2的投入水平为0.349~1.135，X_3的投入水平为-0.177~0.493时，有95%的把握使番茄产量达到最高，为2 123.13g/株，即NO_3-N、P、K浓度分别是8.30~9.04mmol/L、1.06~1.41mmol/L和4.56~6.23mmol/L。

3　验证试验

为了确保数据模型的准确和可靠，根据前期筛选试验的优选结果筛选出两个最优组合（A和B）和两个最差组合（C和D），再以日本园试配方1/2单位为CK进行验证试验（共

5个处理)(表12)。其中,分不同的生育期测定相应的形态指标和生理指标,包括株高、茎粗、叶绿素含量、叶片数、根系活力、地上部干鲜重、地下部干鲜重、果实硝酸盐含量和维生素C含量[4~6,9]。

表12 番茄 NO_3-N、P、K 三因素验证试验营养液配方

Table 12 Validative ration of NO_3-N、P、K for Tomato

处理	NO_3-N	P	K
A	8	0.9	5
B	12	1.35	7.5
CK	8	0.665	4
C	8	0	5
D	8	0.9	0

3.1 不同处理番茄株高、茎粗的动态变化

由图4和图5可知,不同处理植株在不同生育时期其株高、茎粗均有增长,但增幅差异较大。图4中,番茄株高大小排序为:B>CK>A>C>D。处理CK与B的株高增长最快,拉秧时株高分别为118.67cm和117.33cm;处理D株高变化最小。图5表明各个处理茎粗的变化趋于一致,处理B茎粗增长最快,处理B、CK与A的茎粗均大于处理C、D。综上所述,两个最优组合处理A与处理B的株高、茎粗明显优于最差组合C、D的表现,初步说明筛选结果是正确科学的。

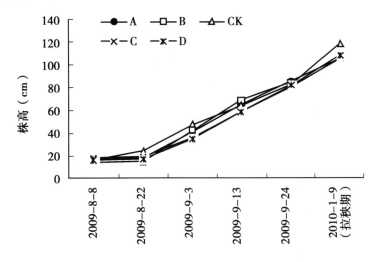

图4 不同处理对番茄株高的影响

Figure 4 The effects of different treatment on Tomato's height

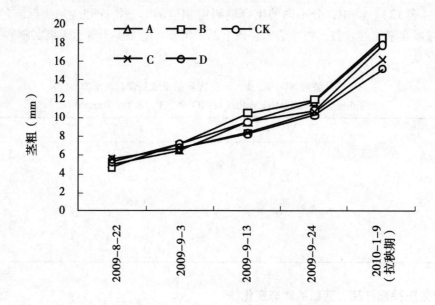

图 5 不同处理对番茄茎粗的影响
Figure 5 The effects of different treatment on Tomato's stem diameter

3.2 不同处理对番茄生长发育的影响

表 13 不同处理对番茄生长的影响
Table 13 The effects of different treatment on growth potential of Tomato

处理	番茄叶片数（片）	平均节间长（cm）	叶绿素含量（μg/cm²）	地上部鲜重（g）	地上部干重（g）	地下部鲜重（g）	地下部干重（g）	番茄根系活力（μg/g·h）
A	17.50aA	5.99bB	51.27a	1 092aA	147.25bB	108.67abAB	13.87aA	9.56aA
B	16.00abAB	6.32bAB	51.78a	1 144aA	180.27aA	122.50aA	13.54aA	6.01abAB
CK	16.25abAB	6.59abAB	51.29a	1 070aA	156.20bB	113.67abAB	13.06aA	4.01bB
C	14.67bcB	6.50abAB	50.02a	516b	71.50dD	67.33cB	7.95cC	3.44bB
D	14.00cB	7.01aA	53.07a	890aAB	102.14cC	88.00bcAB	9.34bB	3.99bB

由表 13 生长势的各项指标知，处理 C、D 栽植的番茄表现相对较差。节间长是一定生长时期生长量积累的反映，各处理节间长大小排序为 D＞CK＞C＞B＞A，处理 D 的平均节间较其他处理的长，这可能是因为营养液内缺少 K 元素，造成番茄对氮肥和磷肥的奢侈吸收，最终使得植株在营养生长时期出现徒长现象，节间较长。

地上部干鲜重是养分在植株体吸收、利用、积累后的表现。处理 B 地上部干重极显著的大于其他各个处理，CK 与处理 A 无显著性差异，其他处理两两间均达到极显著差异；处理 C 地上部鲜重极显著的小于处理 A、B、CK，而处理 A、B、CK 之间无显著性差异，这表明营养液被植株吸收利用后在植株形态上的表现和试验筛选的结果一致，证明了筛选结果的客观正确性。

根系活力是衡量根系功能的主要指标之一，试验结果表明：处理A、B的根系活力极显著地优于C、D，这也解释说明了处理A、B地上部干鲜重大于处理C、D，是营养元素被植株根系吸收后经过一系列的生理生化反应后的直观表现，也是试验筛选结果中营养液最优配方有利于根系吸收运输矿质营养的表现。

3.3 不同处理对番茄产量的影响

表14的产量表明，处理C产量极显著的低于处理A、B和CK，比CK单株产量减少981.44g；处理A与CK产量差异显著，与处理C、D差异极显著。筛选出的最优处理（A、B），其产量较CK均有增产作用，分别增加570.03g/株和186.71g/株；而筛选出的最差处理（C、D），其产量较CK均有减产作用，分别减少981.44g/株和463.61g/株。验证试验结果证明了筛选出的最优配方确实有使番茄产量增加的优势存在。

表14 不同处理对番茄产量的影响
Table 14 The Effects of different treatment on yield of Tomato

处理	产量（g/株）				平均 g/株	比 CK ±	显著水平	
	重复1	重复2	重复3	重复4			5%	1%
A	4 012.54	4 448.1	4 885.61	4 105.88	4 363.03	+570.03	a	A
B	4 190.89	3 925.32	4 480.8	3 321.82	3 979.71	+186.71	ab	AB
CK	4 071.31	3 956.36	3 683.25	3 461.08	3 793.00	0	bc	AB
C	2 697.94	2 657.91	2 840.35	3 050.05	2 811.56	-981.44	d	C
D	3 675.11	3 146.77	3 166.28	3 329.39	3 329.39	-463.61	c	BC

3.4 不同处理对番茄品质的影响

表15 不同处理对番茄品质的影响
Table 15 The effects of different treatment on quality of Tomato

处理	维生素C含量（mg/100gFW）	硝酸盐含量（μg/g）	可溶性固形物（%）	可溶性糖含量（%）	有机酸（%）
A	11.15 bAB	164.96 aA	5.6 a	2.54 aA	0.0032 aAB
B	12.37 aA	349.93 aA	5.7 a	2.16 bA	0.0032 aAB
CK	10.64 bB	156.99 bB	5.7 a	2.11 bA	0.0028 bBC
C	7.82 cC	144.98 bB	5.6 a	2.21 bA	0.0035 aA
D	4.80 dD	209.95 bAB	5.6 a	2.20 bA	0.0027 bC

维生素C是衡量蔬菜品质的重要指标。由表15知，番茄维生素C含量大小为：B>A>CK>C>D，处理B与除处理A以外的其他处理均有极显著差异，处理C、D之间差异极显著，并极显著的低于最优营养液处理A、B和对照的维生素C含量；试验结果表明，各处理番茄果实可溶性固形物差异不大，不具有统计学上的显著性，有待于进一步的研究；硝酸盐在人体内可以转化为致癌物质亚硝酸盐，它在蔬菜中的标准含量是不高于432mg/kg。由试

验结果知番茄果实硝酸盐含量顺序为：B>D>A>CK>C。处理B硝酸盐含量之所以最高，是因为处理B中添加的NO_3-N肥最多，而NO_3-N易使蔬菜中的硝酸盐含量升高；处理D中硝酸盐含量较高，可能是营养液中缺钾造成植株对氮磷的奢侈吸收导致的结果；处理C硝酸盐含量低于CK，尚需进一步的试验研究求证其缘由。可溶性糖含量顺序为：A>C>D>B>CK，处理A与其他4个处理间差异显著，处理B、C、D与CK两两间均无显著差异性；有机酸是评价果蔬风味的主要指标，番茄有机酸含量顺序为：C>A=B>CK>D，处理A、B、C与处理D、CK之间差异极显著，处理A、B、C与处理D、CK之间差异极显著，处理C糖酸比最低。

4 结论与讨论

（1）银川沙培番茄无土栽培营养液中，NO_3-N、P、K三种元素对其产量影响的顺序为：氮>磷>钾。

（2）三种营养元素对番茄产量的影响表现为典型的抛物线型，即在一定范围内随NO_3-N、P、K浓度的提高，番茄产量增加；而过高的浓度则造成减产，符合报酬递减规律。单施三种营养元素浓度分别达到NO_3-N 8.80mmol/L、P 1.11mmol/L 和 K 4.39mmol/L 时，番茄单株最大产量分别可达到 2 137.05g/株、2 145.82g/株和2 131.39g/株。

（3）在0.10水平，氮磷对番茄产量存在正交互效应，即氮与磷肥的配施效果好于单施，能显著的提高番茄产量。

（4）在试验设定范围内，营养液中NO_3-N、P、K的浓度分别为8.0mmol/L、0.9mmol/L和5.0mmol/L时，番茄单株产量达到最高，为 2 123.13g。

（5）经过验证，结果说明筛选出的营养液配方对番茄生长科学、合理。

（6）该试验结果是针对银川市的水质筛选出来的，有一定的地域特殊性；本试验中以番茄全生育期为研究范围，建议接下来的试验最好能分不同的生育期探讨各个生长发育阶段对氮磷钾肥的需要量，进而为精确施肥、高效优质服务。

参考文献

[1] 高祥照，郑义．肥料使用手册．北京：中国农业出版社，2002
[2] 程慧娟，白大鹏．早熟高粱综合农艺栽培措施与产量关系模型的研究．内蒙古农牧学院学报，1999：20（2）
[3] 王桂良．稻田套播油菜氮磷钾肥三因子优化组合试验研究．耕作与栽培学，2007（6）
[4] 高俊凤主编．植物生理学实验技术．西安：世界图书出版公司，2000
[5] 李合生主编．植物生理生化实验原理和技术．北京：高等教育出版社，2000
[6] 张志良主编．植物生理生化实验指导．北京：高等教育出版社，2002
[7] 徐中儒．回归分析与实验设计．北京：中国农业出版社，1998
[8] 王合理．无机及有机栽培对黄瓜生育的影响．塔里木大学学报，2005：17（3）
[9] 邹琦．植物生理学实验指导．北京：中国农业出版社，2004

Reducing Nitrate Concentration in Lettuce by Continuous Light Emitted by Red and Blue LEDs before Harvest

Zhou Wanlai, Liu Wenke, Yang Qichang*

(*Institute of Environment and Sustainable Development in Agriculture, Chinese Academy of Agricultural Sciences, Key Lab. for Agro-Environment and Climate Change, Ministry of Agriculture, Beijing 100081, P. R. China*)

Abstract: Influence of continuous light with different red/blue ratio on photon flux density (R/B ratio) on nitrate and soluble sugar content was studied by growing lettuce (*Lactuca sativa* L.) under continuous illumination delivered by light-emitting diodes (LEDs). Results showed that compared with the initial values, nitrate concentrations in leaf blade and petiole of lettuce were significantly decreased while soluble sugar contents were dramatically increased after 48h light. In addition, significant negative correlation between nitrate concentration and soluble sugar content was found both in leaf blade ($r = 0.89$) and petiole ($r = 0.9$). Compared with single red light treatment, the decrease in nitrate concentration and increase in soluble sugar content in lettuce under mixed red and blue light was more pronounced. The optimal R/B ratio of light for short-term measures to reduce nitrate accumulation in lettuce is 4, because the lowest nitrate concentration and the highest soluble sugar content were obtained under this light condition. It is concluded that pre-harvest short-term exposure to continuous light is an effective method for producing lettuce with low nitrate accumulation and high soluble sugar content.

Key words: Continuous light; LEDs; Nitrate; R/B ratio; Soluble sugar

1 Introduction

Vegetables contribute about 80% of human dietary intake of nitrate (Eichholzer and Gutzwiller, 1998). Many vegetables, especially leaf vegetables had exorbitant nitrate accumulation that could easily lead to excessive ingestion of nitrate in human body (Feng et al., 2006; Umar et al., 2007). Excessive ingested nitrate is considered as a serious potential threat to human health, as it could be converted in human body to nitrite causing methaemoglobinaemia (Mensinga et al., 2003) or carcinogenic nitrosamines (Gangolli et al., 1994).

The accumulation of nitrate in vegetable is genetically controlled (Chen et al., 2000; Du et al., 2008), and is intensely affected by exogenous factors (e.g., nutrition, light, temperature, water supply) (Santamaria, 2006; Ma et al., 2008). Current effective means to minimize the accumulation of nitrate are mainly through control on nutrition, including pre-harvest transfer to Nitrogen-free or Nitrogen-reduced solutions (Mozafar, 1996; Dong and Li, 2003), partial replacement of nitrate in nutrient solution by amino acid or amide-nitrogen (Gunes et al., 1994; Liu et al., 2009), proportioning the nutrient elements (Sun et al., 2003) and utilizing nitrification in-

---

* Correspondence: Yang Qichang, Institute of Environment and Sustainable Development in Agriculture, Chinese Academy of Agricultural Sciences, Beijing, 100081, China. E-mail: yangqichang1@vip.sina.com.

hibitor (Montemurro et al., 1998), etc.

On the other hand, nitrate accumulation is strongly influenced by light (Travis et al., 1970). Light controls the nitrate accumulation by modulating its uptake, or its reduction either through synthesis or regulation of nitrate reductase activity (Lillo, 2004). Nitrate reductase is rapidly deactivated in dark (Kaiser and Brendle-Behnisch, 1991; Huber et al., 1992), while nitrate uptake in dark doesn't significantly decrease (Scaife and Schloemer, 1994; Aslam et al., 2001) even has an absorption peak (Carrasco and Burrage, 1993). Such discrepancy in dark period could lead massive nitrate accumulation in plant, thus if we eliminate dark period with artificial light and expose plants to continuous light for a period (e.g. 2~3days) before harvest, the nitrate content may be significantly reduced. The red/blue ratio on photon flux density (R/B ratio) was found to have a great impact on growth and development of lettuce (Wen et al., 2009), tomato (Ohashi et al., 2004) and chrysanthemum (Di et al., 2008), but quite few studies on the influence of R/B ratio on nitrate content and nutritional quality of vegetable were reported.

In this study, we exposed lettuces to 48h LED-based continuous lights with different R/B ratio, Which was done with two objects: ①to investigate whether the continuous light can reduce the nitrate accumulation and its effect on soluble sugar content; ②to study the effect of R/B ratio on nitrate and soluble sugar content in vegetables under continuous light.

2 Materials and methods

2.1 Experimental design

Lettuce (*Lactica Sativa* L.) were transplanted into a recirculating-solution hydroponic culture system at the third true leaf stage. The light regime was 10/14 h (light/dark), a temperature regime of 21/15℃ (day/night) and a photosynthetic photon flux density of 150 $\mu mol \cdot m^{-2} \cdot s^{-1}$ provided by fluorescent lamps were maintained throughout lettuce growth period. 25 days after transplantation, uniform lettuce plants in weight were selected to grow under continues light delivered by light-emitting diodes (LEDs). The same temperature regime (21/15℃ day/night) and nutrient solution were used. The composition of nutrient solution was as follows: KNO_3, 5.94; $MgSO_4$, 1.42; $NH_4H_2PO_4$, 1.00; KH_2PO_4, 0.44; $Ca(NO_3)_2$, 2.12; EDTA-Fe, 4.29×10^{-2}; H_3BO_3, 4.839×10^{-2}; $MnSO_4$, 1.325×10^{-2}; $ZnSO_4$, 1.35×10^{-3}; $CuSO_4$, 5×10^{-4}; $(NH4)_6Mo_7O_{24}$, 4×10^{-4} mmol $\cdot L^{-1}$.

Table 1 Photon flux densities in each spectral component in different treatments

Treatments	R/B ratio	Photon flux density of red and blue component ($\mu mol \cdot m^{-2} \cdot s^{-1}$)	
		Red	Blue
LED1	—	150	0
LED2	2	100	50
LED3	4	120	30
LED4	8	133	17

Four treatments under continuous illumination with different R/B ratio were performed, as specified in table 1. Spectra in these treatments contained a red component with peak emission at 630nm delivered by red LEDs and a blue component with peak emission at 460nm delivered by blue LEDs. The photosynthetic photon flux density at plant height was maintain at $150 \mu mol \cdot m^{-2} \cdot s^{-1}$. A spectrometer (AvaSpec-2048-USB2, *Avantes Inc.*, Netherlands) was used to measure photon flux density in each spectral component. The continuous light began at the end of the dark period of last 10/14h light regime and lasted for 48h.

2.2 Harvest and determination

Lettuce plants were randomly sampled at the beginning and the end of 48h continuous illumination for determination of initial and terminal value respectively. Samples for determination of nitrate concentration and soluble sugar content were prepared as follows. Fresh samples were cut up and mixed carefully, then 3~4g of pieces were placed into 50ml tubes with 20 ml of hot distilled water, after being boiled in a water-bath for 30 minutes, the extract was filtered. Nitrate concentration was determined using the method described by Liu *et al.* (2009) and soluble sugar content was determined by the method described by Li (2005).

3 Results

Figure 1 shows the initial and terminal nitrate concentrations in lettuce. In contrast to the initial, the nitrate concentration in lettuce significantly decreased in 48h continuous illumination, specifically, the decrement in leaf blade ranged from 1648.0 to 2061.1 mg/kg, and in petiole ranged from 962.9-2090.3mg/kg. Decrement in leaf blade was greater than in petiole except in the light of LED3, however, difference in decrement in petiole between treatments was greater than in leaf blade. Nitrate concentrations in petiole of lettuce under red and blue light were lower than under red light alone but it was not very significant in leaf blade, and in LED3 (R/B ratio = 4), the nitrate concentration was the lowest, compared to the initial value, decrement of 65.9% in leaf blade and 41.0% in petiole were recorded respectively.

The soluble sugar contents in different treatments were presented in figure 2. The soluble sugar content increased dramatically under continuous illumination, but varied between treatments. The data showed that soluble sugar contents in lettuce under red and blue light were significantly higher than under red light alone. In the treatment LED3, where the lowest nitrate concentration was observed, the soluble sugar content in leaves was 17 times as high as the initial and in petiole it was 4.8 times. However, soluble sugar content in this treatment was not significantly higher than in the other two mixed light treatments (LED2 and LED4).

4 Discussion

Our study evidenced that the nitrate concentration in lettuce substantially decreased under continuous light. It is estimated that the nitrate content in lettuce plant decreased by about 55% in the treatment LED3. It's in line with the previous observation that nitrate concentration continuously de-

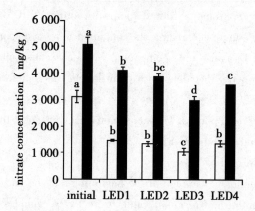

Figure 1 Initial (initial) and terminal (LED1, LED2, LED3, LED4) nitrate concentration in lamina (□) and in petiole (■). Different letters on the column in the same series show significant difference at $P < 0.05$, by LSD.

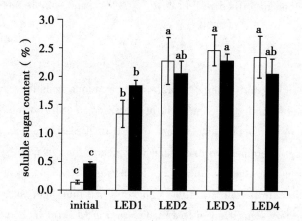

Figure 2 Initial (initial) and terminal (LED1, LED2, LED3, LED4) soluble sugar content in lamina (□) and in petiole (■). Different letters on the column in the same series show significant difference at $P < 0.05$, by LSD.

creased in 24h light (Scaife and Schloemer, 1994). Light promotes nitrate uptake of plant (Cardenas-Navarro et al., 1998; Lillo, 2004). However, it was revealed that under light nitrate uptake only slightly increased while assimilation rate substantially increased (Aslam et al., 1979; Scaife and Schloemer,1994), which could lead to a decline in plant nitrate accumulation. Moreover, it was reported that nitrate accumulated in dark for osmotic purposes because of a lack of organic solutes in cells (Steingrover et al., 1986). it implies that in the light additional accumulation of nitrate is unnecessary since there is sufficient photosynthetic products to maintain cell osmotic pressure, which may lend a further explanation for the nitrate decline in light. Compared to the decrement in 24h continuous light (measured by us, unpublished), a further decrease occurred in 48h continuous light, especially in petiole, where quite few decline occurred in 24h light, and this fact suggests a possibility for further reducing nitrate content in vegetable through further elongating duration of continuous illumination.

The changes under continuous light of soluble sugar and nitrate level are reverse. Significant negative correlations between nitrate concentration and soluble sugar content were found both in leaf blade ($r = -0.89$) and in petiole ($r = -0.90$). It's in agreement with the observation of Hanafy Ahmed et al. (2000). Carbohydrates provide energy and carbon skeleton for nitrogen metabolism (Vanlerberghe et al., 1992). It was found that nitrogen metabolism ceased when intracellular carbohydrate stores were depleted during dark (Huppe and Turpin, 1994). Sucrose-enhanced expressions of both NR and NiR were reported (Cheng et al., 1992; Sivasankar et al., 1997). Therefore, the high soluble sugar content maybe the key to support the continuous nitrogen metabolism leading decline in nitrate accumulation. On the other hand, dramatic increase in soluble sugar content also greatly improves the nutritional quality of vegetables.

Our results showed that mixed red and blue light is more conducive to reduce nitrate accumulation while increase soluble sugar content in lettuce. It's in correspondence with the conclusion that supplementation of blue component to red light promoted assimilation of nitrogen (Ohashi et al., 2006), and that a combination of red and blue is more favourable for plant growth (Yorio et al., 2001; Matsuda et al., 2004). It was reported that the nitrate concentration in lettuce was lowest at R/B ratio of 8 (Wen et al., 2009), Urbonaviciut et al. (2007) also showed that lettuce grown under a combination of 86% red light and 14% blue light had relatively low nitrate accumulation. However, according to our results, in order to reduce the nitrate level in lettuce, the best R/B ratio should be 4. Reason for this difference may lie in the fact that lights in previous study were throughout the entire growth period of plants, and hence the optimal R/B ratio reflected the long-term effect on promoting both carbon and nitrogen metabolism. Short-term control measures to reduce nitrate accumulation seem to be more focused on the promotion of plant nitrogen metabolism, which, reportedly, is effectively enhanced by blue light (Shi et al., 1999; Deng et al., 2000).

In conclusion, nitrate concentration in lettuce substantively decreased while soluble sugar content increased dramatically, it's suggested that short-term exposing plants to continuous illumination before harvest is suited to produce lettuce or other leafy vegetables with low nitrate accumulation and high soluble sugar content. Unlike in long-term culture, the best R/B ratio for short-term regulation on reducing nitrate level is 4, this feature might be of practical importance for lettuce commercially grown under artificial lighting, for perfect unity of vegetable yield and quality could be achieved by adopting distinguished lights in plant growth period and harvest stage respectively. However, it's still not clear whether the decrement would be proportional with the duration in a longer continuous light, and the optimal duration, under which sufficient decrement could be obtained at as low cost of light energy as possible, is worth of further study.

5 Acknowledgements

We want to thank the financial support of 948 projects of Agriculture Ministry (2011-Z7).

References

[1] Aslam, M., Huffaker, R. C., Rains, D. W., & Rao, K. P. (1979). Influence of light and ambient carbon dioxide concentration on nitrate assimilation by intact barley seedlings. *Plant Physiol.*, 63, 1 205~1 209

[2] Aslam, M., Travis, R. L., & Rains, D. W. (2001). Diurnal fluctuations of nitrate uptake and in vivo nitrate reductase activity in pima and acala cotton. *Crop Sci.*, 41, 372~378

[3] Cardenas-Navarro, R., Adamowicz, S., & Robin, P. (1998). Diurnal nitrate uptake in young tomato (Lycopersicon esculentum Mill.) plants: test of a feedback-based model. *Journal of Experimental Botany*, 49, 721~730

[4] Carrasco, G. A., & Burrage, S. W. (1993). Diurnal fluctuations in nitrate uptake and nitrate accumulation in lettuce (*Lactuca sativa* L.). *Acta Horticultura*, 339, 137~147

[5] Chen, X. P., Zou, C. Q., Liu, Y. P., & Zhang, F. S. (2000). The nitrate content difference and the reason among four spinach varieties. *Plant Nutrition and Fertilizer Science*, 6, 30~34

[6] Cheng, C. L., Acedo, G. N., Christinsin, M., & Cronkling, M. A. (1992). Sucrose mimics the light induction of Arabidopsis nitrate reductase gene transcription. *Proc Natl Acad Sci USA*, 89, 1 861~1 864

[7] Deng, J. M., Bin, J. H., Pan, R. C. (2000). Effects of light quality on the primary nitrogen assimilation of rice (*Oryza sativa* L.) seedlings. *Acta Botanica Sinica*, 42, 234~238

[8] Di, X. R., Jiao, X. L., Cui, J., Liu, X. Y., Kong, Y., & Xu, Z. G. (2008). Effects of different light quality ratios of LED on growth of chrysanthemum plantlets in vitro. *Plant Physiology Communications*, 44, 661~664

[9] Dong, X. Y., & Li, S. J. (2003). Effect of nutrient management on nitrate accumulation of pakchoi under solution culture. *Plant Nutrition and Fertilizer Science*, 9, 447~451

[10] Du, S. T., Li, L. L., Zhang, Y. S., & Lin, X. Y. (2008). Nitrate accumulation discrepancies and variety selection in different Chinese cabbage (*Brassica chinensis* L.) genotypes. *Plant Nutrition and Fertilizer Science*, 14, 969~975

[11] Eichholzer, M., & Gutzwiller, F. (1998). Dietary nitrates, nitrites, and N-Nitroso compounds and cancer risk: a review of the epidemiologic evidence. *Nutrition Review*, 56, 95~105

[12] Feng, J. F., Shi, Z. X., Wu, Y. N., Wu, H. H., & Zhao, Y. F. (2006). Assessment of nitrate exposure in Beijing residents via consumption of vegetables. *Chinese Journal of Food Hygiene*, 18, 514~517

[13] Gangolli, S. D., van den Brandt, P., Feron, V., Janzowsky, C., Koeman, J., Speijers, G., Speigelhalder, B., Walker, R., & Winshnok, J. (1994). Assessment of nitrate, nitrite, and N-nitroso compounds. *Eur J Pharmacol Environ Toxicol Pharmacol Sect*, 292, 1~38

[14] Gunes, E., Post, W. H. K, Kirkby, E. A., & Aktas, M. (1994). Influence of partial replacement of nitrate by amino acid nitrogen or urea in the nutrient medium on nitrate accumulation in NFT grown winter lettuce. *Journal of Plant Nutrition*, 17, 1 929~1 938

[15] Hanafy Ahmed, A. H., Mishriky, J. F., & Khalil, M. K. (2000). Reducing nitrate accumulation in lettuce (*Lactica Sativa* L.) plants by using different biofertilizers. ICEHM2000, Cairo University, Egypt, 509~517

[16] Huber, J. L., Huber, S. C., Campbell, W. H., & Redinbaugh, M. G. (1992). Reversible light/dark modulation of spinach leaf nitrate reductase activity involves protein phosphorylation. *Arch. Biochem. Biophys*, 296, 58~65

[17] Huppe, H. C., & Turpin, D. H. (1994). Integration of carbon and nitrogen metabolism in plant and algal cells. *Annu Rev Plant Phys.*, 45, 577~607

[18] Kaiser, W. M. & Brendle-Behnisch, E. (1991). Rapid modulation of spinach leaf nitrate reductase activity by photosynthesis. *Plant Physiol.*, 96, 363~367

[19] Li, H. S. (2005). Principles and techniques of plant physiological biochemical experiment, Higher Education Press

[20] Lillo, C. (2004). Light regulation of nitrate uptake, assimilation and metabolism. In S. Amancio, & I. Stulen (eds), . Plant Ecophysiology (pp. 149~184), Dordrecht: Kluwer Academic Publisher

[21] Liu, W. K., Du, L. F., & Yang, Q. C. (2009). Biogas slurry added amino acids decreased nitrate concentrations of lettuce in sandculture. *Acta Agriculturae Scandinavica Section B-Soil and Plant Science*, 59, 260~264

[22] Ma, M. T., An, Z. Z., Zhao, T. K., Du, L. F., Li, P., Liu, B. C., & Wu, Q. (2008). Progress in vegetable nitrate accumulation control measurements. *Chinese Agricultural Science Bulletin*, 24, 208~212

[23] Matsuda, R., Ohashi, K. K., Fujiwara, K., Goto, E., & Kurata, K. (2004). Photosynthetic characteristics of rice leaves grown under red light with or without supplemental blue light. *Plant and Cell Physiology*, 45, 1 870~1 874

[24] Mensinga, T. T., Speijers, G. J. A., & Meulenbelt, J. (2003). Health implications of exposure to environmental nitrogenous compounds. *Toxicol Rev*, 22, 41~51

[25] Montemurro, F., Capotorti, G., Lacertosa, G., & Palazzo, D. (1998). Effects of urease and nitrification inhibitors application on urea fate in soil and nitrate accumulation in lettuce. *Journal of Plant Nutrition*, 21, 245~252

[26] Mozafar, A. (1996). Decreasing the NO_3^- and increasing the vitamin C contents in spinach by a nitrogen deprivation method. *Plant Foods for Human Nutrition*, 49, 155~162

[27] Ohashi, K. K., Fujiwara, K., Kimura, Y., Matsuda, R., & Kurata, K. (2004). Effects of red and blue LEDs low light irradiation during low temperature storage on growth, ribulose-1, 5-bisphosphate carboxylase/oxygenase content, chlorophyll content and carbohydrate content of grafted tomato plug seedlings. *Environment Control in Biology*, 42, 65~73

[28] Ohashi, K. K., Goji, K., Matsuda, R., Fujiwara, K., Goto, E., & Kurata, K. (2006). effects of blue light supplementation to red light on nitrate reductase activity in leaves of rice seedings. *Acta Horticulturae*, 711, 351~356

[29] Santamaria, P. (2006). Review nitrate in vegetables: toxicity, content, intake and EC regulation. *Journal of the Science of Food and Agriculture*, 86, 10~17

[30] Scaife, A., & Schloemer, S. (1994). The diurnal pattern of nitrate uptake and reduction by spinach (*Spinacia oleracea* L.). *Annals of Botany*, 73, 337~343

[31] Shi, H. Z., Han, J. F., Guan, C. Y., & Yuan, T. (1999). Effects of red and blue light proportion on leaf growth, carbon-nitrogen metabolism and quality in tobacco. *Acta Agronomica Sinica*, 25, 215~220

[32] Sivasankar, S., Rothstein, S., & Oaks, A. (1997). Regulation of the accumulation and reduction of nitrate by nitrogen and carbon metabolites in maize seedlings. *Plant Physiol.*, 114, 583~589

[33] Steingrover, E., Ratering, P., & Siesling, J. (1986). Daily changes in uptake, reduction and storage of nitrate in spinach grown at low light intensity. *Physiologia Plantarum*, 66, 550~556

[34] Sun, Q., Gao, Y. M., & Li, J. S. (2003). Effects of different proportions of N, P, and K Oil nitrate contents in rape. *Chinese Journal of Eco-Agriculture*, 11, 84~86

[35] Travis, R. L., Jordan, W. R., & Huffaker, R. C. (1970). Light and nitrate requirements for induction of nitrate reductase activity in Hordeum vulgare. *Physiol. Plant*, 23, 678~685

[36] Umar, S. A., Iqbal, M., & Abrol, Y. P. (2007). Are nitrate concentrations in leafy vegetables within safe limits? *Current Science*, 92, 355~360

[37] Urbonaviciut, A., Pinho, P., Samuolien, G., Duchovskis, P., Vitta, P., Stonkus, A., Tamulaitis, G., Zukauskas, A., & Halonen, L. (2007). Effect of short-wavelength light on lettuce growth and nutritional quality. *Scientific Works of the Lithuanian Institute of Horticulture and Lithuanian University of Agriculture*, 26, 157~165

[38] Vanlerberghe, G. C., Huppe, H. C., Viossak, K. D. M., & Turpin, D. H. (1992). Activation of respiration to support dark NO_3^- and NH_4^+ assimilation in the green alga selenastrum minutum. *Plant Physiol.*, 99, 495~500

[39] Wen, J., Bao, S. H., Yang, Q. C., & Cui, H. X. (2009). Influence of R/B ratio in LED lighting on physiology and quality of lettuce. *Chinese Journal of Agrometeorology*, 30, 413~416

[40] Yorio, N. C., Goins, G. D., Kagie, H. R., Wheeler, R. M., & Sager, J. C. (2001). Improving spinach, radish and letuuce growth under red light emitting diodes (LEDs) with blue light supplementation. *HortScience*, 36, 380~383

项目支持：科技部国际合作重点项目（2010DFB30550）；农业部948项目（2011-Z7）。

不同红蓝 LED 对生菜形态与生理品质的影响*

闻 婧[1,2]**，杨其长[1]***，魏灵玲[2]，程瑞锋[2]，刘文科[2]，孟力力[1]，鲍顺淑[2]，周晚来[2]

（[1] 江苏省农业科学院蔬菜研究所，南京 210014；
[2] 中国农业科学院农业环境与可持续发展研究所/农业部农业环境与
气候变化开放试验室，北京 100081）

摘要：利用波长为 660nm 红光 LED 与 450nm 蓝光 LED 作为光源，研究不同红/蓝配比（R/B）下生菜（*Lactuca sativa* L.）的形态、生理及品质差异，探索人工可控条件下栽培高品质生菜的红蓝光比例组合。试验以水耕栽培为模式种植生菜，环境条件为：昼温 25 ± 1℃、夜温 15 ± 1℃、湿度 60%～80%、CO_2 浓度 $1\,500 \pm 30\mu molCO_2/mol$、光照强度 $150 \pm 11\mu mol/m^2 \cdot s$、光周期 12h/d。试验以设施栽培中普遍使用的荧光灯为对照，设置 3 个 R/B 试验区，分别为 LED1（R/B = 10）、LED2（R/B = 8）、LED3（R/B = 6）。在该可控环境下定植 30d 后，试验区 LED2（R/B = 8）的生菜净光合速率、维生素 C 含量及鲜重增加量均明显高于其他试验区，而硝酸盐含量最低，表现出较快的生长速度和良好的品质特性，所以 R/B = 8 时较其他试验区更适宜生菜的生长发育。试验还表明：红蓝光可以有效提高植物的光能利用率，促进光合作用；叶绿素 A 和 B 会随着红光的减少与蓝光的增加而减少；适宜的红蓝光比例还能有效增加植物维生素 C 含量，并降低硝酸盐含量。

关键词：LED；生菜红/蓝配比

Influence of R/B Ratio in LED Light Quality on Morphology and Physiology of Lettuce (*Lactuca sativa* L.)

Wen Jing[1,2], Yang Qichang[2], Wei Lingling[2], Cheng Ruifeng[2],
Liu Wenke[2], MengLili[2], Bao Shunshu[2] and Zhou Wanlai[2]

(1. *Institute of Vegetable Crops, Jiangsu Academy of Agricultural Sciences, Nanjing 210014, P. R. China*;
2. *Institute of Environment and Sustainable Development in Agriculture, CAAS/Key Open Laboratory for Agro-environment and Climate Change, Ministry of Agriculture, Beijing 100081, P. R. China*)

Abstract: The ratio of red (R) to blue (B) was studied to investigate the influence of R/B ratio on the growth and development of lettuce (*Lactuca sativa* L.). The experiment was conducted under controlled daytime temperature by 25 ± 1℃, night temperature by 15 ± 1℃, relative humidity by 60%～80%, CO_2 concentration by $1\,500 \pm 50\mu mol/mol$, light intensity by $150 \pm 11\mu mol/(m^2 \cdot s)$ and photoperiod by 12h/d. The lettuce was planted by the hydroponic cultivation. The light emitting diode (LED) composed by red light LED whose wavelength was 660 nm and blue light LED whose wavelength was 450 nm were used as lighting source. Three treatments were designed with LED1

---

* 基金项目：科技部国际合作重点项目（2010DFB30550）；农业部 948 项目（2011 - Z7）；北京市科委重点项目（D111100000811001）。

** 作者简介：闻 婧（1983— ），女，内蒙古呼和浩特人，硕士研究生，从事设施栽培方面研究。

*** 通讯作者 Author for correspondence：E-mail：yangq@cjac.org.cn。

（R/B = 10），LED2（R/B = 8）and LED3（R/B = 6）using LED light quality respectively, compared to treatment of fluorescent lamp for 30 days. The net photosynthetic rate, vitamin C contents and fresh weight were the highest, and the nitrate contents was the least in LED2. The quality and growth of lettuce in LED2 were better than other treatments. Moreover, the result showed that red light and blue light could improve the utilization of light energy and promote photosynthesis；Chlorophyll A and B would decrease with the increase of red light and the decrease of blue light；Vitamin C contents would enhance and nitrate contents would reduce effectively though adjusting of R/B.
Key words：LED；Lettuce；R/B ratio

生菜的营养丰富、市场需求量大，传统露地或日光温室栽培的生菜无论从数量上还是品质上都不能满足人们日益增长的需求[1~3]，它的工厂化栽培迫在眉睫。在工厂化栽培中，我们采用多层立体栽培方式，充分发挥生菜生长空间小的特点，在提高空间利用率的同时极大提高生菜单位面积产量[4]；通过高精度的环境调控与规范化操作，为生菜生长发育创造良好的生长条件，有效减少或避免病虫害的发生，缩短生长周期，提高品质[5,6]，最大限度地满足人们对生菜质量与品质需求。

在工厂化栽培中，人工光源是植物生长的唯一能量来源。目前，在工厂化栽培系统中多以荧光灯为主要人工光源，但是荧光灯存在耗电量大、产热量大、电能利用率低等缺点，而作为21世纪新型绿色光源的LED则弥补了荧光灯存在的不足。LED光源除具有节能、环保、使用寿命长、体积小等特点外，它发出的光波窄（±20nm），能够实现精确的光质配比[7,8]。研究表明，在光源的可见光光谱（380~760nm）中，植物吸收的光能约占生理辐射光能的60%~65%。其中主要是波长为610~720nm（波峰为660nm）的红、橙光辐射，其次是波长为400~510nm（波峰为450nm）的蓝、紫光辐射。如果用红蓝两种波长的光照射作物可以极大地提高植物对光能的利用率[9~11]。因此，根据作物的这一生理特征与LED光源的物理特性，利用红蓝LED光源作为植物生长光源已经成为国内外研究的热点。

目前，国内外有关红蓝LED光源栽培的研究主要集中于单色光对植物生长发育的影响，研究表明，植物在纯红光或纯蓝光下均能完成生长过程。在纯红光照射下，植株干物质积累多，节间较长而茎较细且叶片细小，总糖含量高；在纯蓝光照射下，植株干物重小，节间较短而茎较粗，伸长生长受到一定抑制作用[12~14]。1994年，日本采用660nm红光LD加5%蓝光LED的组合光源栽培生菜，证明与荧光灯相比，组合光源下生菜的地上部鲜重增加量最多，但对其他生理及品质指标未做详细的分析研究[15]。1999年，孟继武等人利用光生态膜进行了莴苣的光形态研究，结果表明440nm和680nm两个波段范围的光可以促进光合作用。但是由于试验条件的限制，不能精确的控制光照的波长，只从定性的角度说明增加440nm和680nm两个波段范围的光可以增加产量。本文在前人的研究基础上，利用新型LED光源装置，精确调控光照波长，并以应用范围最广的荧光灯为对照，研究不同R/B下生菜的生长形态、生理特性及营养品质之间存在的差异。

1 试验材料与方法

1.1 光源设备

试验使用由中国农业科学院农业环境与可持续发展研究所与中国科学院半导体研究所合

作开发研制的新型 LED 植物生长光源。它由 660nm（±20nm）的红光 LED 与 450nm（±20nm）的蓝光 LED 组成，可以根据试验需要调节不同光质、光强、光周期及灯板距作物的高度。试验中将其置于作物顶部 20cm 处，使其尽量近距离照射植物，减少不必要的光能损失。试验以荧光灯为对照，选用松下 36W 三基色荧光灯（YZ36RR6500K，北京松下照明光源有限公司），根据灯管的开启个数和高度的不同调节其光照强度。

1.2 试验材料

试验选用"奶油生菜"为材料。2007 年 11 月 19 日在中国农业科学院作物科学研究所重大科学工程试验温室内育苗，2007 年 12 月 10 日，幼苗达到两叶一心时定植于中国农业科学院农业环境与可持续发展研究所研制的植物工厂内。采用 DFT 水耕栽培方式培育生菜，营养液 pH = 6.5 ± 1、EC = 1.3 ± 1、营养液温度 20 ± 1℃，每小时循环供液 10min。

1.3 试验环境条件与试验设置

试验在中国农业科学院农业环境与可持续发展研究所设施农业中心植物工厂内完成。该植物工厂通过计算机对植物生育过程的温度、湿度、CO_2 浓度以及营养液循环等环境条件进行自动监控，不受或很少受自然条件的影响。试验过程中各项环境指标为：昼温 25 ± 1℃、夜温 15 ± 1℃、湿度 60% ~ 80%、CO_2 浓度 1 500 ± 30$\mu molCO_2/mol$，均达到预期要求。试验设置 3 个光质处理区，分别为 LED1（R/B = 10）、LED2（R/B = 8）、LED3（R/B = 6）；一个荧光灯对照区，即 PPF150；各试验区光强均设置为 150μmol/（m·s），光周期为 12h/d。

1.4 参数测量

在生菜定植 30d 后，观察其形态指标，并使用 LI – 6400 光合仪测量各试验区材料的净光合速率、气孔导度和蒸腾速率等生理指标；采用比色法测量叶绿素和类胡萝卜素含量，首先将叶片剪碎并置于 1∶1 的乙醇—丙酮混合液，在常温黑暗中浸提 16h，直至叶片完全变白后用 UNIC—7200 分光光度计测量其吸光度值，通过计算得到叶绿素和类胡萝卜素含量；分别采用 2.6 – 二氯酚靛酚钠法、铜还原碘量法、凯氏定氮法和紫外分光光度法测量生菜维生素 C、总糖、粗蛋白和硝态氮的含量。

2 结果与分析

2.1 形态特征

在生菜定植 30d 后，不同红/蓝配比的试验区中，植株形态存在较大差异。从图 1 中可以看到 PPF150 与 LED1（即 R/B = 10）两个试验区的植株形态差异不明显，叶片宽大、松散舒展，但是没有结球，商品价值较低。LED2（即 R/B = 8）植株形体最大，已经形成叶球，叶片肥硕、舒展度适中，生长发育状态良好，具有较佳的商品价值。LED3（即 R/B = 6）植株形体较小，未结球，叶片紧实、卷曲程度大，植株生长受到明显的抑制作用。实验结果表明增加蓝光的比例可以抑制植物的伸长，促使植物矮壮，但是蓝光比例过大会明显抑制植株的正常生长发育，影响产量。而适宜的 R/B 配比可以促进生菜的生长发育，增加经济产值，而 R/B = 8 比 R/B = 10 和 R/B = 6 更适合生菜的生长发育。

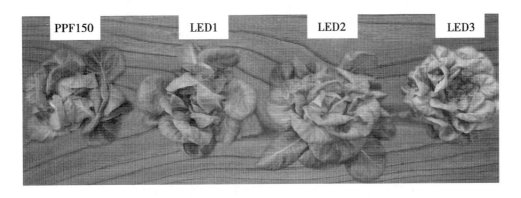

图1 各试验区定植30d的形态外观对照

Figure 1 Morphology compares of lettuce at different treatments after planting 30d

2.2 光合特性分析

表1 光合特征指标

Table 1 Indicators of photosynthetic characteristics

试验区	光合速率 ($\mu molCO_2/m^2 \cdot s$)	蒸腾速率 ($molH_2O/m^2 \cdot s$)	气孔导度 ($molH_2O/m^{-2} \cdot s^{-1}$)	胞间CO_2浓度 ($\mu molCO_2/mol$)
LED1	7.30±0.177c	2.42±0.046b	0.10±0.002b	294.8±4.66b
LED2	7.88±0.073a	3.86±0.613a	0.16±0.032a	332.8±16.98a
LED3	7.63±0.031b	2.11±0.114b	0.08±0.004b	247.8±7.58c
PPF150	5.70±0.138d	2.52±0.025b	0.09±0.003b	306.8±1.40b

注：同列中的不同字母代表$P<0.01$极显著性差异。

Note: Different letters in the same column indicate statistically significant difference at $P<0.01$.

从表1可以看出，LED2各项指标均显著高于其他试验区，其中光合速率达到7.88$\mu molCO_2/$（$m^2 \cdot s$），蒸腾速率达到3.86$molH_2O/$（$m^2 \cdot s$），气孔导度达到0.16$molH_2O/$（$m^2 \cdot s$），胞间CO_2浓度达到332.8$\mu molCO_2/mol$，说明R/B=8时，表现出明显的生长优势，与图1所示结果相符。另外，表1中各试验区的光合速率均存在极显著差异，LED试验区均显著高于PPF150试验区，表明660nm红光和450nm蓝光可以有效促进生菜的光合作用，适宜红/蓝配比的LED光源则能够更有效的提高生菜对光的利用率，从而进一步提高生菜的光合速率。表1中蒸腾速率与气孔导度两项指标在LED1、LED3和PPF150三个试验区内没有显著差异，胞间CO_2浓度在各试验区内也没有呈现规律变化，而LED2显著高于其他实验区，表明只有适宜的红/蓝光比例才能有效地提高生菜的蒸腾速率、气孔导度和胞间CO_2浓度。结果表明：在光照强度一致的情况下，红蓝光配比为R/B=8时，生菜的各项光合特征指标均显著高于其他试验区，表现出良好的生长发育优势，能有效促进生菜生长发育。

2.3 叶绿素含量

从图2可以看出，LED1试验区的叶绿素A和B含量最高，分别达到1.07mg/g和

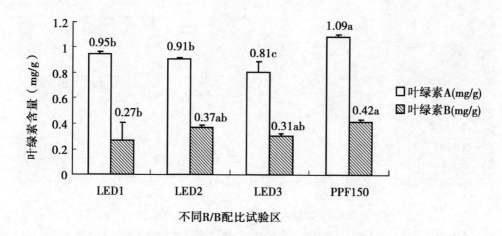

图 2 不同 R/B 配比试验区内生菜叶绿素含量
Figure 2 Chlorophyl contents of lettuce at different treatments
注：图中的不同字母代表平均值的显著性差异（P<0.05）。
Note: Different letters in the figure indicate statistically significant difference at P<0.05.

0.42mg/g，但与 LED2 和 PPF150 试验区均无有显著差异，LED3 试验区的叶绿素 A 和 B 含显著低于其他实验区。试验中 PPF150 试验区的叶绿素 A 和 B 的含量均较高，推测是荧光灯中含有远红光的原因，但尚未见有关远红光对叶绿素含量影响的详细报告，有待进一步研究。尽管如此，试验结果仍表明：随着 R/B 的减少，叶绿素 A 和叶绿素 B 的含量逐渐降低，因此红光和蓝光对叶绿素具有调节作用，而适宜的叶绿素 A 和 B 才能更有效的促进光合作用。

通过对植株光合特性与叶绿素含量的对比分析发现：LED3 试验区的叶绿素含量较低但光合速率却显著高于荧光灯试验区，推测 LED 单色光更有利于叶绿素的的吸收和光能的转化，但还未见相关生理方面的研究，有待于进一步探索。

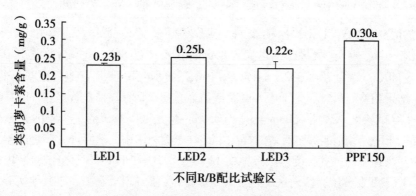

图 3 不同 R/B 配比试验区内生菜类胡萝卜素含量
Figure 3 Carotenoid contents of lettuce at different treatments
注：图中的不同字母代表平均值的极显著性差异（P<0.01）。
Note: Different letters in the figure indicate statistically significant difference at P<0.01.

2.4 类胡萝卜素

类胡萝卜素在植物叶绿体光合作用中起着重要作用，是光合作用中光传导途径和光反应中心的重要结构成分，帮助叶绿体吸收光能；并保护其在高温、强光下免受破坏。从图3可以看出，PPF150的类胡萝卜素含量最高，达到0.30mg/g，显著高于3个LED试验区，结果表明：LED光源与荧光灯相比降低了类胡萝卜素的含量，却提高了光合速率。目前还未见有关不同光质对类胡萝卜素影响的研究，还有待于进一步探索。

2.5 营养品质分析

生菜定植30d后，每个试验区随机摘取500g可食用部分作为供试材料，送至农业部蔬菜品质监督检验测试中心进行品质检验，具体检测结果如下：

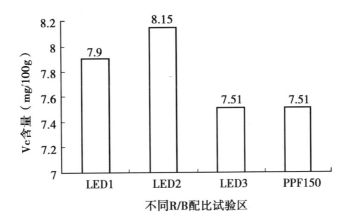

图4 不同R/B配比试验区内生菜Vc含量

Figure 4 Vitamin C contents of lettuce at different treatments

2.5.1 维生素C（Vc）含量

Vc是生菜中重要的营养成分，是评价生菜品质的重要指标。从图4可以看出LED2的Vc含量最高达到8.15mg/100g，而LED3与PPF150的含量最低，均为7.51mg/100g。其中LED2试验区的优势明显，分别比LED1和LED3高出0.25mg/100g和0.58mg/100g。结果表明：LED试验区的维生素C含量高于荧光灯试验区，并且适宜的R/B配比可以增加生菜Vc含量。

2.5.2 硝酸盐含量

硝酸盐是生菜中可能对人体造成不利的影响的一种化学物质，如果人体摄取过量可导致活动迟钝，工作能力减退，头晕、昏迷；一次用量过大甚至会导致死亡。而人体中摄取的硝酸盐81.5%来自蔬菜，因此硝酸盐含量是评价蔬菜质量的关键。从图5中可以看出，LED2的硝酸盐含量最低，为2.96g/kg，比最大值小0.65g/kg。因此，通过调节红蓝光比例可以降低蔬菜中硝酸盐含量。

2.5.3 总糖和粗蛋白含量

各试验区中总糖与粗蛋白含量总体差异不大，其中粗蛋白含量最大相差仅为0.07%，所以红蓝光比例对生菜的粗蛋白含量没有明显影响。PPF150的总糖含量最高为0.75%，LED3最低仅0.5%，LED2的含量居中为0.64%，所以LED2比LED1和LED3更有利于总

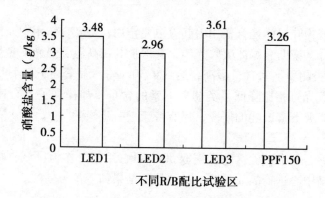

图 5　不同 R/B 配比试验区内生菜硝酸盐含量

Figure 5　Nitrate contents of lettuce at different treatments

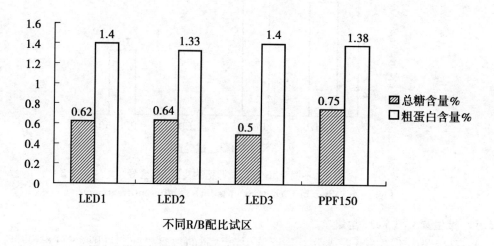

图 6　不同 R/B 配比试验区内生菜总糖和粗蛋白含量

Figure 6　Total sugar and crude protein contents of lettuce at different treatments

糖的积累（图 6）。

3　结论与讨论

试验表明，LED2 试验区中生菜植株的各项光合指标都显著高于其他试验区，表现出明显的生长优势，维生素 C 含量高，硝酸盐含量最低，具有很高的商品价值。R/B = 8（即 LED2）时，生菜在生长发育形态和营养品质上更有优势，在工厂化栽培时，可选用此 R/B 进行生产。

通过试验发现红蓝光对植株的生长发育存在以下 3 个作用：

第一，随着红光 LED 的减少与蓝光 LED 的增加，叶绿素 A 和 B 的含量逐渐降低。

第二，红蓝光可以有效地促进植物光合速率，并有助于维生素 C 的积累，表明 LED 红蓝光较荧光灯更有利于植物对光能的吸收和利用，提高了植物的光能利用率。

第三，适宜的 R/B 可以使植株体内硝酸盐含量的降低。

综上所述，LED 是优于荧光灯的新型植物生长灯，但是只有适宜的红蓝光比例才能充分发挥植物的生长潜力，而且不同植物对光的需求是不同的，以后的试验中应选用不同作物，进一步探索红蓝光对植株生长发育的作用规律。此外，试验过程中发现 LED 试验区中虽然叶绿素 A、B 和类胡萝卜素含量低于荧光灯试验区，但是光合速率却显著高于荧光灯，目前对其原因还不清楚，有待于进一步研究。

Reference

[1] 郭世荣. 无土栽培学. 北京：中国农业出版社，2003
[2] 王明明. 千金菜—生菜. 营养保健，2003
[3] 潘杰. 水培生菜技术研究. 硕士学位论文. 河南农业大学，2003
[4] 魏灵玲，杨其长，刘水丽. LED 在植物工厂中的研究现状与应用前景. 中国农学通报，2007，23（11）：408~410.
[5] 李程，冯志红，李丁仁. 2002. 蔬菜无土栽培发展现状及趋势. 北方园艺，(6)：9~11
[6] 宋亚英，陆生海. 温室人工补光技术及光源特性与应用研究. 温室园艺，2005，(1)：28~29
[7] 工海鸥，李广安. 认识照明 LED. 中国照明电器，2004，(2)：1~3
[8] 杨其长，张成波. 植物工厂概论. 北京：中国农业科学技术出版社，2005
[9] 廖祥儒，张蕾，徐景. 光在植物生长发育中的作用. 河北大学学报（自然科学版），2001，21（3）：341~346
[10] Hiroshi Shimizu, Shinji Tazawa, Zhiyu Ma, et al. The Application of Blue Light as a Growth Regulator. An ASAE Meeting Presentation, US：ASAE Tampa Convention Center, 2005, paper No：054152
[11] R. C. Jao, W. Fang. An adjustable light source for photo-phytorelated research and young plant production. American Society of Agricultural Engineers, 2003, 19 (5), 601~608
[12] Barta DJ, Tibbitts TW, Bula RJ, et al. Evaluation of lighting-emitting diodes characteristics for a space-based plant irradiation source. Advances Space Res, 1992, 12：141~149
[13] Critten, D. L. A review of the light transmission into greenhouse crops. Acta Hort, 1993. 248：101~108
[14] Fang, W., R. C. Jao and D. H. Lee. Artificial lighting apparatus for young plants using light emitting diodes as light source. 2002. US patent no：US 6474838 B2
[15] Gaudreau, L., Charbonneau, J., Veaina, L. P. and Gosselin, A. 1994. Photoperiod and PPF influence growth and quality of greenhouse-grown lettuce. HortScience, 29：1 285~1 289

Effects of Silicon on Plant Growth and Antioxidan Enzyme Activities in Leaves of Cucumber Seedlings under NO_3^- Stress

Song Yunpeng[1], Wang Xiufeng[1,2], Wei Min[1,2], Shi Qinghua[1,2], Yang Fengjuan[1,2]*

(1. College of Horticulture Science and Engineering, Shandong Agricultural University, Tai'an, 271018, P. R. China; 2. State Key Laboratory of Crop Biology, Tai'an, Shandong, 271018, P. R. China)

1 Purpose

Cucumber is one of the most important vegetables in the world. Recently, over-utilization of chemical fertilizer has caused secondary salinization in Chinese greenhouse. The excessively accumulated anion in soil of the greenhouse is NO_3^-. The large accumulation of salt and salt ions might induce other limiting factors of greenhouse cropping system, such as the nutritional disorders, acidification of soil and so on. Silicon (Si) is the second most abundant element on the surface of the earth, yet its role in plant biology has been poorly understood and the attempts to associate Si with metabolic or physiological activities have been inconclusive. At present, salt stress has been shown in some investigations to be mitigated by Si. But most of salt stresses regarded NaCl as means of research, while the effects of silicon on plant growth under excess NO_3^- was seldomly investigated.

2 Materials and Methods

Cucumber (*Cucumis. sativus* L. cv. Xintaimici) seeds were germinated on moisture lter paper in an incubator at 28 ℃ for 2 days. The germinated seeds were sown in plastic plugs (50holes) filled with nursery substrate (peat: vermiculite: perlite =2:1:1). When there was one completely expanded true leaf, batches of 8 seedlings were transferred to plastic tank with 10 L nutrient solution with pH 6.0 containing aerated full nutrient solution: Ca$(NO_3)_2 \cdot 4H_2O$ 3.5mmol/L, KNO_3 7mmol/L, KH_2PO_4 1mmol/L, $MgSO_4 \cdot 7H_2O$ 2mmol/L, H_3BO_3 46.3μmol/L, $MnSO_4 \cdot H_2O$ 10μmol/L, $ZnSO_4 \cdot 7H_2O$ 1.0μmol/L, $(NH_4)_6Mo_7O_2 \cdot 4H_2O$ 0.3 μmol/L, $CuSO_4 \cdot 5H_2O$ 0.75μmol/L, EDTA-FeNa 100μmol/L. The pH of the nutrient solution was adjusted to 6.0 daily using0.1 mol/L H_2SO_4.

Salinity and silicon treatments were started by adding NO_3^- (KNO_3 and Ca$(NO_3)_2 \cdot 4H_2O$ provide the same mol of NO_3^-) and sodium silicate (Na_2SiO_3) to the nutrient solution immediately while the cucumber seedlings were at three-true-leaf. The pH of the nutrient solution after the addition of sodium silicate was adjusted to 6.0 using H_2SO_4 before transplanting. The experimental design consisted of six treatments as follows: CK: 14mmol/L NO_3^-; Ⅰ: 140mmol/LNO_3^-; Ⅱ: 140mmol/LNO_3^- + 0.5mmol/L Na_2SiO_3; Ⅲ: 140mmol/LNO_3^- + 1mmol/L Na_2SiO_3; Ⅳ:

140mmol/L NO$_3^-$ + 1.5mmol/L Na$_2$SiO$_3$; V: 140mmol/L NO$_3^-$ + 2mmol/L Na$_2$SiO$_3$. One treatment has three replications. The young fully expanded leaves were sampled after 7 days of the treatment to determine malondialdehyde (MDA), superoxidase (SOD), peroxidase (POD), catalase (CAT) and ascorbic acid peroxidase (APX) activities.

3 Results and Discussion

In this study, cucumber seedlings were cultivated in nutrient solution added with different concentration (0, 0.5, 1, 1.5, 2 mmol/L) Na$_2$SiO$_3$ as Si donor to study the effects of exogenous Si on the growth of cucumber seedlings and the activities of antioxidant enzymes in cucumber leaves under NO$_3^-$ stress. Under the stress of 140mmol/L, treating with 1mmol/L of Na$_2$SiO$_3$ for 7d increased the leaf superoxidase (SOD), peroxidase (POD), catalase (CAT) and ascorbic acid peroxidase (APX) activities, and decreased the leaf malondialdehyde (MDA) content significantly, suggesting that exogenous Si could enhance the capacity of cucumber seedlings in scavenging active oxygen species, protect the seedlings from the peroxidation of membrane lipids, and promote the seedlings growth and increase resistance to high concentration NO$_3^-$ stress. After the cucumber seedlings grew in 2mmol/L of Na$_2$SiO$_3$ for 7d, the activities of SOD, POD, CAT and APX in leaves decreased, and the MDA content increased, resulting in the injury of cucumber seedlings. It was indicated that certain concentration (0.5 ~ 1 mmol/L) exogenous Si could alleviate the NO$_3^-$ stress to cucumber seedlings.

Table 1 Effects of exogenous silicon on cucumber seedlings growth under NO$_3^-$ stress

Treatments	Plant Height (cm)			Leaf area (cm^2)		
	0d	7d	Increment	0d	7d	Increment
CK	5.74 ±0.11a	23.94 ±1.16a	18.20 ±1.19a	160.54 ±7.89a	350.54 ±9.21a	190.02 ±15.75a
I	5.20 ±0.16a	13.14 ±0.76d	7.94 ±0.63d	151.95 ±6.64a	227.35 ±4.94d	75.41 ±7.89d
II	5.38 ±0.19a	14.98 ±0.99c	9.60 ±1.05c	159.37 ±3.89a	240.89 ±7.49c	81.51 ±9.83c
III	5.54 ±0.20a	18.22 ±0.62b	12.68 ±0.67b	158.55 ±11.14a	261.82 ±9.08b	103.27 ±11.43b
IV	5.50 ±0.17a	17.22 ±1.62b	11.72 ±1.74b	160.17 ±8.01a	251.47 ±9.98bc	91.30 ±13.91bc
V	5.32 ±0.13a	14.46 ±0.81cd	9.14 ±0.69cd	159.94 ±5.08a	241.12 ±7.76c	81.18 ±3.37c

Note: Values followed by different letters in the same column mean significant difference among treatments at 0.05 level.

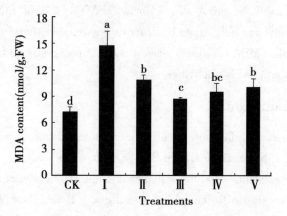

Figure 1 Effects of exogenous silicon on MDA content in leaves of cucumber seedlings under NO_3^- stress

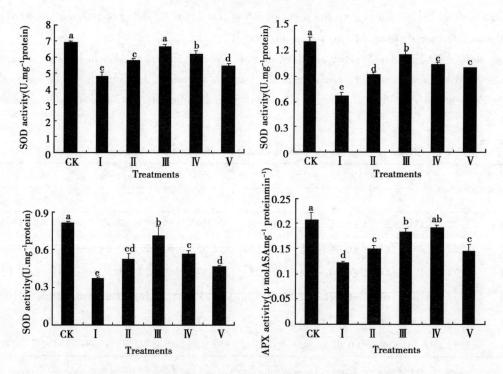

Figure 2 Effects of exogenous silicon on SOD, POD, CAT and APX activities in leaves of cucumber seedlings under NO_3^- stress

采前硝酸钙和氯化钾对水培生菜硝酸盐含量的影响[*]

刘文科[**]，杨其长

（中国农业科学院农业环境与可持续发展研究所，中国农业科学院设施农业环境工程研究中心，北京 100081）

摘要：采用温室盆栽试验的方法研究了采前单独或共同供应硝酸钙和氯化钾对水培生菜展开叶叶片和叶柄中硝酸盐含量的影响。结果表明，与单独供应硝酸钙和氯化钾处理相比，共同供应硝酸钙和氯化钾显著增加了水培生菜展开叶叶片中的硝酸盐含量。此外，共同供应硝酸钙和氯化钾处理中，10mmol/L硝酸钙处理生菜展开叶叶片中的硝酸盐含量高于5mmol/L硝酸钙处理。供应5mmol/L硝酸钙和10mmol/L氯化钾处理的生菜叶柄中含有较高的硝酸盐含量。所以，硝酸钙和氯化钾在调控水培生菜硝酸盐累积上具有协同交互作用，促进了硝酸盐的累积。

关键词：生菜；硝酸钙；氯化钾；叶片；叶柄

Together or Solely Supplying Calcium Nitrate and Potassium Chloride Affect Nitrate Concentrations in Lettuce Differently Before Harvest

Liu Wenke, Yang Qichang

(*Environment and Sustainable Development in Agriculture, Chinese Academy of Agricultural Sciences, Research Center for Protected Agriculture & Environment Engineering of CAAS, Beijing 100081, P. R. China*)

Corresponding author: liuwke@163.com

Abstract: A glasshouse experiment was conducted to investigate the effects of together or solely supplying potassium chloride and calcium nitrate on nitrate concentrations in expanded leaf blades and petioles of lettuce cultivated hydroponically before harvest. The results showed that together supplying calcium nitrate plus potassium chloride significantly increased the nitrate concentration in expanded leaf blades of lettuce compared with solely supplying calcium nitrate and potassium chloride. In addition, the nitrate concentration in expanded leaf blades of lettuce treated by 10mmol/L calcium nitrate plus potassium chloride contained higher nitrate concentration than that of 5mmol/L calcium nitrate plus potassium chloride. Together supplying 5mmol/L calcium nitrate plus 10mmol/L potassium chloride had higher nitrate concentration in expanded leaf petioles of lettuce than other treatments. Furthermore, treatments, 10mmol/L calcium nitrate plus 10mmol/L potassium chloride, 10mmol/L calcium nitrate, and 10mmol/L potassium chloride made no difference in nitrate concentrations in expanded leaf petioles of lettuce. To conclude, together use of calcium nitrate and potassium chloride before harvest increased nitrate accumulation of hydroponic lettuce, and the outcomes depend on the calcium nitrate concentrations used.

---

[*] 中央级科研院所基本科研业务费项目（BSRF201004）；"十二五" 863项目（2011AA03A1）资助。

[**] 作者简介：刘文科，男，博士，副研究员，硕士生导师，研究方向为设施蔬菜营养与品质生理。Tel：010-8106015，E-mail：liuwke@163.com。

Key words: Lettuce; Calcium nitrate; Nitrate; Potassium chloride; Leaf blade; Petiole

Introduction

Vegetables, particularly the off-season leafy vegetables, are readily contaminated by nitrate due to overuse of nitrogen fertilizer (Müller & David, 1987; Wang & Li, 2003; Chen et al., 2004; Boroujerdnia et al., 2007; Konstantopoulou et al., 2010) and weak light (Müller & David, 1987; Riens & Heldt, 1992; Ysart et al., 1999) inner protected facilities in winter no matter soilless or soil cultivation. Nowadays, off-season leafy vegetables are generally characterized by high nitrate accumulation and low Vc content, which determines to some extent the nutritional quality at harvest and that in subsequent storage (Poulsen et al., 1995; Konstantopoulou et al., 2010). It was reported that more than 80% nitrate intake by human was from vegetables (Gangolli et al., 1994; Eichholzer et al., 1998). Moreover, excessive intake of nitrate will pose a potential hazard to human health, especially for infants. The toxic effects of nitrate are due to its endogenous conversion to nitrite, which is implicated in the occurrence of methaemoglobinaemia, gastric cancer and many other diseases (Wolff & Wasserman, 1972; Santamaria, 2006). Therefore, decreasing the content of nitrate is preharvest is the main ways to improve efficiently vegetable quality.

Hydroponic leafy vegetables are proved to be easily regulated to control nitrate concentration by mediating nutrient element components in nutrient solution (Gunes et al., 1996) or by treating with nitrogen-free solutions (Mozafar, 1996). Generally, nitrate supplying level determined nitrate concentrations in shoot tissues, and some osmotic ions, e.g. Cl^-, SO_4^{2-} and acetic ion, supplying could decrease nitrate concentrations in shoot tissues (Dong & Li, 2003; Liu & Yang, 2011). Some literatures found that preharvest treatment with nitrogen-free solutions could decrease nitrate contents of lettuce (Liu & Yang, 2011), Pakchoi (Dong & Li, 2003) and spinach (Mozafar, 1996). Based on the advantages of low costs and high efficiency, preharvest nitrogen disruption treatment is supposed as a promising method to improve nutritional quality of leafy vegetables in hydroponics practically. However, no investigation has been carried out to examine the interactions between nitrate and osmotic ions in hydroponic solution on nitrate concentration in the edible parts (expanded leaf blades and petioles) before harvest. In this study, a glasshouse experiment was conducted to investigate the effects of together or solely supplying potassium chloride and calcium nitrate on nitrate concentrations in expanded leaf blades and petioles of lettuce cultivated hydroponically before harvest.

Materials and methods

Experimental materials

Lettuce (*Lactuca sativa* L), an Italian cultivar, was cultivated in a seedling tray (vermiculite used as substrate) supplied with full nutrient solution for 20 days. On 10 Nov., 2010, uniform lettuce seedlings were selected and transplanted into an experimental pot (length × width × height is 32cm × 24cm × 11cm). Two lines and seven holes were made on the lid at an equably distance, and the central one was used as a gasvent to aerate the nutrient solution by a air-pump (Atman EP-

9000) to ensure the supply of soluble oxygen around the roots. Six liters of full nutrient solution was filled in one pot, and six lettuce seedlings in one were planted. 15 pots were used, and 90 lettuce plants were cultivated. Nutrient solution was prepared with distilled water and analytical reagents. The full nutrient solution is composed of 5.0 $Ca(NO_3)_2$, 0.75 K_2SO_4, 0.5 KH_2PO_4, 0.65 $MgSO_4$, 0.1 KCl, 1.0×10^{-3} H_3BO_3, 1.0×10^{-3} $MnSO_4$, 1.0×10^{-4} $CuSO_4$, 5.0×10^{-6} $(NH_4)_6MO_7O_{24}$, 1.0×10^{-3} $ZnSO_4$, 0.1 EDTA-Fe (mmol/L).

Experimental design and determination

On 24 Dec., 2010, 48 uniform lettuce plants were selected from the ninety lettuce plants prepared, and they were transplanted into 12 pots the sizes of which are the same as previously used. There are four treatments in all (see table 1). All treatments replicated three times. After five days' growth, two plants for each replication were randomly harvested by removing root off the roots. After weighing the fresh weight, lettuce shoot was separated two parts, i.e. expanded leaf blades and petiole for nitrate concentration determination. The nitrate concentration of fresh samples was determined as the method described by Liu et al. (2009). The experiment was carried out in a glasshouse, with a temperature ranging from 15 to 30℃. During the study period, the weather during treatment days is clear.

Statistical analyses

Data were analyzed using SAS software 6.12. Significant differences between means were established by using the least significance difference (LSD) test.

Results

Nitrate concentration in expanded leaf blades of lettuce

After treatment, the biomass of different treatments was similar (Figure 1). Together supplying calcium nitrate and potassium chloride significantly increased the nitrate concentration in expanded leaf blades of lettuce compared with solely supplying calcium nitrate and potassium chloride (Figure 2). In addition, the nitrate concentration in expanded leaf blades of lettuce of treatment 2 (10mmol/L calcium nitrate plus potassium nitrate) contained higher nitrate concentration than that of treatment 3 (5mmol/L calcium nitrate plus potassium nitrate). Solely supplying 10mmol/L calcium nitrate and 10mmol/L potassium chloride made no difference in expanded leaf blades of lettuce.

Nitrate concentration in expanded leaf petioles of lettuce

Together supplying 5mmol/L calcium nitrate plus 10mmol/L potassium chloride (treatment 3) had higher nitrate concentration in expanded leaf petioles of lettuce than other treatments (Figure 3). Furthermore, treatments, 10mmol/L calcium nitrate plus 10mmol/L potassium chloride, 10mmol/L calcium nitrate, and 10mmol/L potassium chloride made no difference in nitrate concentrations in expanded leaf petioles of lettuce.

Discussions

The results showed that together supplying calcium nitrate plus potassium chloride significantly increased the nitrate concentration in expanded leaf blades of lettuce compared with solely supplying calcium nitrate and potassium chloride. Our data also found that the nitrate concentration in expanded leaf blades of lettuce treated by 10mmol/L calcium nitrate plus 10mmol/L potassium chloride contained higher nitrate concentration than that of 5mmol/L calcium nitrate plus 10mmol/L potassium chloride, which means that nitrate concentration in expanded leaf blades is related to the level nitrate supplied. On the contrary, together supplying 5mmol/L calcium nitrate plus 10mmol/L potassium chloride had higher nitrate concentration in expanded leaf petioles of lettuce than other treatments. That suggested that 10mmol/L calcium nitrate plus 10mmol/L potassium chloride did not contribute highest nitrate concentration in expanded leaf petioles although its nitrate level was twofold high. As previously reported, nitrate concentration in leaves could be decreased when supplied osmotic ions before harvest (Dong & Li, 2003; Liu and Yang, 2011). Various responses of nitrate concentrations in expanded leaf blades and petioles can be attributed to the osmotic function of chloride ion, also which relates to the nitrate level supplied together. Our results indicated that osmotic ions, i. e. KCl, should not be used together with nitrate ion by which nitrate accumulation could be inhibited for hydroponic lettuce. Ivashikina et al. (1998) suggested that some cations, e. g. K^+, Mg^{2+} > Na^+ > Ca^{2+} could improve nitrate uptake, but K^+ improved the xylem loading instead of uptake (Casadesus et al., 1995). As results presented in this study, nitrate uptake might be improved for together presence of K^+ and Ca^{2+} in hydroponic solutions. The physiological mechanisms need further investigations, particularly the interactions between chloride ion, nitrate ion and K^+ on nitrate accumulation in leaves before harvest.

To conclude, solely supplying calcium nitrate did not increase nitrate concentration in expanded leaf blades and petioles; together supplying calcium nitrate and potassium chloride increased nitrate concentration in expanded leaf blades, but their effect on nitrate concentration in petioles depended on concentration of potassium chloride.

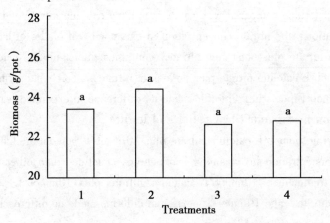

Figure 1 Fresh biomass of lettuce samples

Note: different letters represent significant difference at $P < 0.05$. The same as below.

Table 1 Treatments designed in the experiment

Treatment number	Solution components
1	10mmol/L calcium nitrate
2	10mmol/L calcium nitrate and 10mmol/L potassium chloride
3	5mmol/L calcium nitrate and 10mmol/L potassium chloride
4	10mmol/L potassium chloride

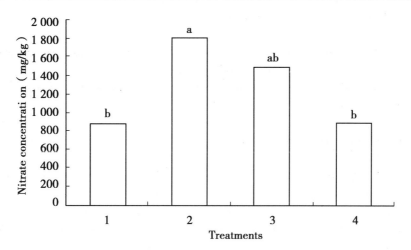

Figure 2 Nitrate concentrations in leaf blades of lettuce of different treatments

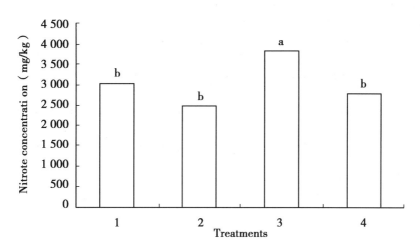

Figure 3 Nitrate cocentrations in petioles of lettuce of different treatments

References

[1] Boroujerdnia, M., Ansari, N. A. & Dehcordie, F. S. Effect of cultivars, harvesting time and level of nitrogen fertilizer on nitrate and nitrite content, yield in romaine lettuce. Asian J. Plant Sci., 2007, 6 (3): 550~553

[2] Casadesus, J., Tapia, L. & Lambers, H. Regulation of K^+ and NO_3^- fluxes in roots of sunflower (*Halianthus annuus*) after changes in light intensity. Physiologia Plantarum, 1995, 93: 279~285

[3] Chen, B. M., Wang, Z. W., Li, S. X., Wang, G. X., Song, H. X. & Wang, X. N. Effect of nitrate supply on plant growth, nitrate accumulation, metabolic nitrate concentration and nitrate reductase activity in three leafy vegetables. Plant Sci., 2004, 167 (3): 635~643

[4] Dong, X. Y. & Li, S. J. Effect of nutrient management on nitrate accumulation of pakchoi under solution culture. Journal of Plant Nutrition and Fertilizer Science, 2003, 9: 447~451

[5] Eichholzer, M. & Gutzwiller, F. Dietary nitrates, nitrites, and N-Nitroso compounds and cancer risk: a review of the epidemiologic evidence. Nutrition Review, 1998, 56 (4): 95~105

[6] Gangolli, S. D., van den Brandt, P., Feron, V. J., Janzowsky, C., Koeman, J. H. & Speijers, G. J. Nitrate, nitrite and N-nitroso compounds, European Journal of Pharmacology, Environmental Toxicology and Pharmacology Section, 1994, 292: 1~38

[7] Gunes, A., Inal, A., & Aktas, M. Reducing nitrate content of NFT grown winter onion plant by partial replacement of NO_3 with amino acid in nutrient solution. Scientia Horticulturae, 1996, 65, 203~208

[8] Ivashikina, N. V. & Feyziev-Ya, M. Modulaiton of nitrate reductase activity by polyamines in leaves of wheat (*Triticum aestivum* L.) seedling. Indian Journal of Experimental Biology, 1998, 131: 25~34

[9] Konstantopoulou, E., Kapotis, G., Salachas, G., Petropoulos, S. A., Karapanos, I. C. & Passam H. C. Nutritional quality of greenhouse lettuce at harvest and after storage in relation to N application and cultivation season. Scientia Horticulturae, 2010, 125 93. e1~93. e5

[10] Liu, W. K, Du, L. F. & Yang, Q. C. Biogas slurry added amino acid decrease nitrate concentrations of lettuce in sand culture. Acta Agriculturae Scandinavica Section B-Soil and Plant Science, 2009, 59: 260-264

[11] Liu, W. K. & Yang, Q. C. Short-term treatment with hydroponic solutions containing osmotic ions and ammonium molybdate decreased nitrate concentration in lettuce. Acta Agriculturae Scandinavica, Section B - Soil & Plant Science, In press, 2011

[12] Mozafar, A. Decreasing the NO_3^- and increasing the Vc contents in spinach by a nitrogen deprivation method. Plant Foods for Human Nutrition, 1996, 49: 155~162

[13] Müller, K. & David, J. J. Influence of differences in nutrition on important quality characteristics of some agricultural crops. Plant and Soil, 1987, 100: 35~45

[14] Poulsen, N., Johansen, AS & S. rensen, J. N. Influence of growth conditions on the value of crisphead lettuce. 4. Quality changes during storage. Plant Foods Hum Nutr., 1995, 47 (2): 157~62

[15] Riens, B. & Heldt, H. W. Decrease of nitrate reductase activity in spinach leaves during a light – dark transition. Plant Physiol., 1992, 98 (2): 573~577

[16] Santamaria, P. Nitrate in vegetables: toxicity, content, intake and EC regulation. Journal of the Science of Food and Agriculture, 2006, 86 (1): 10~17

[17] Wang, Z. H. & Li S. X. Effects of N forms and rates on vegetable growth and nitrate accumulation. Pedosphere, 2003, 13 (4): 309~316

[18] Wolff, I. A. & Wasserman, A. E. Nitrates, nitrites, and nitrosamines. Science, 1972, 177 (4043): 15~19

[19] Ysart, G., Clifford, R. & Harrison, N. Monitoring for nitrate in UK grown lettuce and spinach. Food Addit. Contam., 1999, 16 (7): 301~306

EM 菌对水培生菜生长和品质的影响*

琚志君**,刘厚诚***,陈日远,孙光闻,宋世威

(南方设施园艺研究中心,华南农业大学园艺学院,广州 510642)

摘要:通过在营养液中添加不同 EM 菌浓度(稀释 800、1 000 和 1 200 倍)和不同频率的通气处理(间隔 15、30 和 45min 通气 15min),研究了 EM 菌对水培生菜产量和品质的影响。结果表明:营养液中添加 EM 菌和通气能提高水培生菜的产量和品质,其中 1 000 倍的 EM 菌处理能显著提高植株的株高,鲜重和比叶重等生长指标,以及维生素 C、可溶性糖和可溶性蛋白的含量。15:45 + 1 000 倍菌结合处理的话,则增产和改善品质的效果更好。

关键词:EM 菌;生菜;通气

Effect of Effective Microorgnisms on Growth and Quality of Lettuce in Hydroponics

Ju Zhijun, Liu Houcheng, Chen Riyuan, Sun Guangwen, Song Shiwei

(*Lab of Protected Horticulture in South China*, *College of Horticulture*, *South China Agriculture University*, *Guangzhou* 510642, *P. R. China*)

Abstract: Effects of different concentraion of Effective Microorganisms (EM), which are 800, 1000 and 1200 times in dilution, and different frequency of aeration, which are 15 minute aeration in 15, 30 and 45 minute interval, on growth and quality of lettuce in hydroponics were investigated. The results indicated that the growth and quality of lettuce were improved by EM and aeration. The plant height and fresh weight and specific leaf weight, and the concentration of vitamin C, soluble sugar and soluble protein were observably increased by addition of 1000 times EM. And the growth and quality of lettuce were more improved by the combination of aeration frequency of 15:45 and 1 000 times of EM.

Key words: Effective Microorganisms, lettuce, aeration

　　水培条件下蔬菜生长整齐、生育期短、商品性好,水培技术已经被广泛应用于蔬菜生产特别是绿叶蔬菜的生产。有效微生物群(Effective Microorganisms,EM)自 20 世纪 80 年代初被提出来之后,已经在生态农业中得到广泛应用(Schulz,1992),关于 EM 能促进作物的生长发育,使作物增产增收的报道很多;目前 EM 菌在蔬菜生产上主要用于喷施、浸种和处理堆肥。EM 能有效地改善水体水质,增强水体中溶解氧的含量(黄永春,2009)。本试验设置营养液添加 EM 菌和充气泵供氧两个处理因素,研究其对生菜生长和品质的影响,为

---

* 基金项目:广东高校科技成果转化重大项目(cgzhzd0809)。
** 琚志君 硕士研究生,研究方向:设施蔬菜栽培。
*** 通讯作者:E-mail:liuhch@ scau. edu. cn。

EM 菌在蔬菜水培上的应用提供参考。

1 材料与方法

1.1 供试材料及试验设计

试验于 2010 年 9～10 月在华南农业大学园艺学院试验基地温室内进行。供试的生菜（*Lacuca sativus* L.）品种为意大利精选生菜。EM 菌由中山东益新农业有限公司提供。

试验设计不同 EM 菌浓度和不同的通气量 2 个因素，16 个处理，处理如表 1 所示。每处理设 3 个重复，随机区组排列。每周更换一次营养液。营养液配方采用 1/2Hoagland 配方（pH≈6.2）。

表 1 处理组合

EM 菌稀释倍数	通气 15min			0 通气
	间隔 15min	间隔 30min	间隔 45min	
1/800	C1 + G1	C1 + G2	C1 + G3	C1
1/1 000	C2 + G1	C2 + G2	C2 + G3	C2
1/1 200	C3 + G1	C3 + G2	C3 + G3	C3
0	G1	G2	G3	CK

9 月 8 日以珍珠岩为基质撒播育苗，苗期浇灌 1/2 Hoagland 营养液。幼苗两叶一心时，选取生长一致的幼苗洗净根部基质，在茎基部裹海绵条后，将苗定植于泡沫板孔中。移栽至塑料容器（长×宽×高：61cm×42cm×8cm）（内装 15L 营养液），每盆 11 株。

1.2 取样与测定方法

植株具 18 片叶时采收，每处理取 9 株测定生菜的株高、植株鲜重和干重、地上部鲜重以及比叶重（总叶干重/总叶面积），鲜样测定品质：维生素 C 含量用 2,6-二氯酚靛酚滴定法测定，可溶性蛋白含量采用考马斯亮蓝 G-250 染色法测定，可溶性糖含量采用蒽酮比色法测定（李合生，2000）。

数据采用 SPSS17.0 统计软件和 EXCEL2003 进行数据统计分析。

2 结果与分析

2.1 不同处理对水培生菜生长的影响

不同处理对生菜的生长有明显的影响（表 2）。各处理的植株生长均显著高于对照。加 EM 处理之间的株高、单株鲜重、地上部鲜重和比叶重之间的差异均不显著，而处理 1 200 倍菌的单株干重显著低于另外两个处理；但均呈显著高于对照。与对照相比，800 倍、1 000 倍和 1 200 倍菌的株高分别增加了 10.7%、12.6% 和 11.3%；单株鲜重分别增加了 165.1%、157.1% 和 151.6%；地上部鲜重分别增加了 166.2%、152.2% 和 152.2%；单株干重分别增加了 158.9%、160.7% 和 123.8%；比叶重分别增加了 16.9%、19.0% 和 20.6%。总体上 800 倍菌和 1 000 倍菌的处理生菜的生长较好。

表2 不同处理对水培生菜生长指标的影响

通气频率	EM稀释倍液	株高（cm）	单株鲜重（g）	地上部鲜重（g）	单株干重（g）	比叶重（g/cm²）
15∶15	1/800	22.09 ± 0.27 ef	94.52 ± 3.71 de	83.97 ± 3.25 efg	4.02 ± 0.15 def	33.97 ± 1.23 abc
	1/1 000	23.51 ± 0.23 d	116.31 ± 1.13 ab	103.59 ± 1.17 ab	4.81 ± 0.08 abc	36.80 ± 0.91 ab
	1/1 200	24.71 ± 0.38 c	109.86 ± 2.82 bc	37.71 ± 2.93 bc	4.55 ± 0.13 bcd	37.07 ± 0.74 a
	0	23.34 ± 0.09 d	91.64 ± 6.36 e	83.16 ± 5.53 fg	3.66 ± 0.25 fg	34.80 ± 0.30 abc
15∶30	1/800	25.40 ± 0.35 abc	124.31 ± 0.34 a	111.90 ± 0.41 a	4.99 ± 0.18 ab	34.90 ± 1.21 abc
	1 000	21.92 ± 0.28 f	108.74 ± 4.79 bc	95.67 ± 4.18 bcd	5.16 ± 0.22 a	33.53 ± 0.66 abc
	1/1 200	24.92 ± 0.53 bc	104.71 ± 4.12 bcd	94.11 ± 4.22 bcde	4.50 ± 0.24 bcd	32.63 ± 0.84 c
	0	22.81 ± 0.22 def	95.14 ± 1.82 de	85.66 ± 1.37 def	3.67 ± 0.10 fg	32.77 ± 1.73 bc
15∶45	1/800	22.00 ± 0.37 ef	102.49 ± 0.09 cde	92.10 ± 0.39 cdef	4.37 ± 0.17 bcd	36.73 ± 1.58 abc
	1/1 000	26.10 ± 0.28 a	104.98 ± 5.65 bcd	95.24 ± 5.05 bcd	4.55 ± 0.34 bcd	35.80 ± 2.17 abc
	1/1 200	25.91 ± 0.30 ab	108.11 ± 5.28 bcd	94.34 ± 4.65 bcde	4.30 ± 0.25 cde	33.77 ± 0.58 abc
	0	23.42 ± 0.43 d	81.13 ± 4.09 f	74.21 ± 3.89 g	3.32 ± 0.29 g	32.87 ± 1.36 bc
0	1/800	23.11 ± 0.53 de	100.91 ± 1.04 cde	90.84 ± 0.36 cdef	4.35 ± 0.05 cde	27.47 ± 0.67 d
	1/1 000	23.50 ± 0.36 d	97.88 ± 4.60 cde	86.07 ± 4.09 def	4.38 ± 0.16 bcd	27.97 ± 1.49 d
	1/1 200	23.23 ± 0.22 d	95.77 ± 1.70 de	86.07 ± 1.10 def	3.76 ± 0.04 efg	28.33 ± 1.60 d
	0	20.87 ± 0.58 g	38.07 ± 1.17 g	34.13 ± 1.19 h	1.68 ± 0.09 h	23.50 ± 0.83 e

注：表内数据是三次重复的平均值，同列相同字母表示处理间差异不显著，不同字母表示差异显著（$P<0.05$）。

三个通气处理的株高、地上部鲜重、单株干重以及比叶重之间差异不显著，而处理G3的整株鲜重显著低于其余两个通气处理。三个通气处理的株高、植株鲜重和干重、比叶重和地上部鲜重则显著高于对照。

在加菌和通气结合处理中，C1+G2处理的生长总体优于其他各个处理。其株高、单株鲜重、地上部鲜重、单株干重以及比叶重分别比对照增加了21.7%、226.5%、227.9%、197.0%和48.5%，并且显著高于纯加菌和纯通气的处理。

2.2 不同处理对水培生菜品质的影响

不同处理对水培生菜维生素C（Vc）含量的影响较大（图1）。加EM菌处理均显著高于对照，且1/1 000菌的处理Vc含量最高，其次是1/800菌和1/1 200菌；与对照相比，1/1 000、1/800和1/1 200菌的处理分别升高了35.6%、21.6%和19.0%。纯通气的三个处理对生菜Vc含量影响不显著。加菌和通气结合的9个处理中，只有C1+G3处理和C2+G3处理显著高于对照，且与对照相比，分别升高了32.3%和22.1%，其他各处理间以及与对照之间无显著差异。

不同处理对水培生菜可溶性蛋白含量的影响不大（图2）。加EM菌处理中，1/1 000菌的处理最高，其次是1/1 200菌处理，两者之间差异不显著；与对照相比，两者分别显著增加了84.9%和62.4%；而1/800菌处理只增加了48.4%，与对照相比差异不显著。纯通气的3个处理则与对照差异不显著。

加EM菌和通气结合处理中，C2+G2处理可溶性蛋白含量最高，且显著高于其他8个处理；其次是C3+G2处理和C1+G3处理、C1+G2处理和C2+G3处理，分别比对照显著

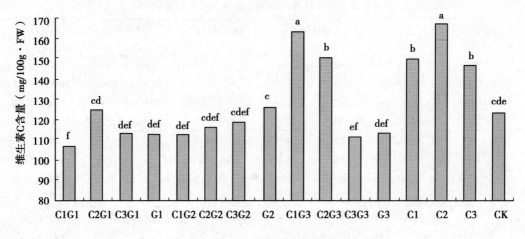

图1 不同处理对水培生菜维生素C含量的影响

增加了127.9%、70.9%、68.8%、65.6%和55.9%。

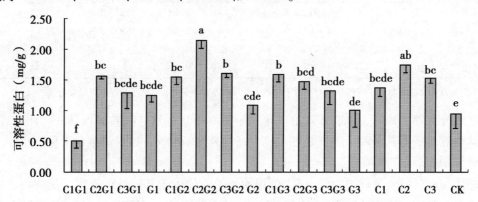

图2 不同处理对水培生菜可溶性蛋白含量的影响

不同处理对水培生菜可溶性糖含量的影响较大（图3）。加EM菌处理中，1/1 000菌处理可溶性糖含量最高，其次是1/800菌处理，两者之间差异显著；与对照相比，显著增加了139.4%和69.7%；而1/1 200菌处理和对照之间差异不显著。纯通气的三个处理中，只有G1处理的可溶性糖含量显著高于对照，增加了72.7%；而G2和G3两个处理与对照差异不显著。在加EM菌和通气结合处理中，C3+G1处理的可溶性糖含量最高，其次是C1+G1处理，两者之间差异不显著，且显著高于其他各处理，分别比对照增加了257.6%和230.3%；其次是C1+G3处理、C2+G1处理和C2+G3处理，3个处理差异显著，与对照相比，分别显著增加了193.9%、145.5%和109.1%。其余各处理与对照相比差异不显著。

3 讨论

EM菌是一种新型的复合微生物制剂，在促进植物生长等方面有明显的效果。喷施1/400EM液芹菜产量提高18.1%（车豪杰等，2007），喷施1/300EM菌液西芹产量提高25%（金柏年、刘红霞，2009）。EM处理能使玉米叶面积和比叶重有所增加，提高植株可

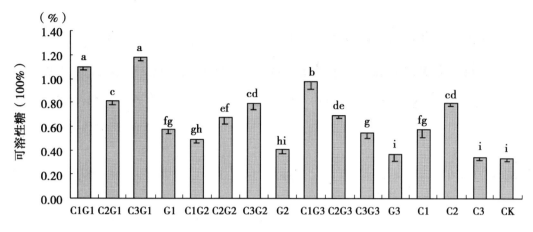

图 3　不同处理对水培生菜可溶性糖含量的影响

溶性蛋白的含量（张　烈、戴俊英，1999）。

本试验发现，添加 EM 菌、通气以及通气和加 EM 菌结合处理均能促进水培生菜植株的生长（表 2），其中 1/800 菌的处理和 1/1 000 菌的处理能够明显的提高生菜的株高、比叶重以及地上部鲜重。加 EM 菌和通气处理的交互作用显著，其中通气频率 15:30 另 +1/800 EM 菌的结合处理效果最好，其次是通气频率 15：45 另 + 1/1 000EM 菌的处理。

EM 菌处理能提高生菜的品质，1/1 000 菌处理的 Vc 含量，可溶性糖和可溶性蛋白的含量最高的，而且 EM 菌处理的生菜品质优于纯通气的处理。加 EM 菌和通气处理的交互作用显著，通气频率 15：45 另 +1/800 菌和 15：45 另 + 1/1 000 菌的处理其品质总体是较好。

从本试验结果看，EM 菌在水培叶菜上能够代替通气，保持叶菜较好的生长和品质，节省成本。通气的叶菜水培营养液中添加 EM 菌也能提高产量和品质。其机理还有待进一步的研究。

参考文献

[1] Schulz D G. Effective micro organisms for organic agri-culture—a case study from Sri Lanka. In：Sangakara UR，Higa T，Kopke U（ed.），Proceedings 9th Intemational Scientific Conference IFOAM：Organic agriculture，a key to a sound development and a sustainable enveounment. Sao Paulo，Brazil. 1992：152～159
[2] 黄永春. 有效微生物菌群对养虾水体细菌生态和水质的影响. 广东海洋大学学报. 2009，29（1）：44～48
[3] 李合生. 植物生理生化实验原理和技术. 北京：高等教育出版社，2000
[4] 车豪杰，华龙洲，陶新颖，袁小强. EM 益生菌在设施蔬菜生产中的应用效果. 山东蔬菜，2007，(4)：32～33
[5] 金柏年，刘红霞. EM 菌素在蔬菜上的应用效果. 新农业，2009，(10)：44
[6] 张　俊，戴俊英等. 有效微生物群对玉米植株生长与代谢的影响. 农业现代化研究. 1999，20（6）：355～358

Crop Management in Greenhouses: Adapting the Growth Conditions to the Plant Needs or Adapting the Plant to the Growth Conditions?[*]

L. F. M. Marcelis[**] S. De Pascale

(*Wageningen UR Greenhouse Horticulture, Department of Agricultural Engineering P. O. Box 16, and Agronomy, University of Naples 6700 AA, Wageningen, Federico II-Via Università, 100 ~ 80055 The Netherlands Portici, Naples Italy*)

Abstract: Strategies for improving greenhouse crop production should target both developing advanced technological systems and designing improved plants. Based on greenhouse experiments, crop models and biotechnological tools, this paper will discuss the physiology of plant-greenhouse interactions. It is discussed how these interactions can be applied to control the production process at Northern and Mediterranean climatic conditions.

Absorption of light by the leaves is important for maximum crop photosynthesis. For this, it is important to have plants that develop as fast as possible a sufficient leaf area index. The question is: what leaf area index is needed for optimal crop performance? Most of the light is absorbed by the upper part of the canopy. Can we improve the light distribution in the canopy and, moreover, does this increases yield or quality? Virtual plant models may help to address this question. In some cases removal of older leaves can improve yield, while in other cases removal of young leaves may accomplish the same objective.

In summer time the light transmission of the greenhouse is often reduced by growers to avoid plant stress. However, in several cases this stress is only an indirect effect of light, because other growth factors (e. g. temperature, humidity) tend to be suboptimal.

In Northern countries CO_2 supply is commonly used. The introduction of semi-closed greenhouses allows to maintain high CO_2 concentrations all year round. 2 In Mediterranean countries, a large yield increase is still feasible by CO_2 supply.

Optimum growth conditions means that there is a good balance among different climate conditions. The source/sink ratio of a crop (ratio between production and demand of assimilates) often reflects whether these conditions are balanced. Variation in the source/sink balance affects formation and abortion of organs, product quality and production fluctuations. Some examples are shown on temperature control based on the source/sink balance of a crop.

Drought and salinity may limit production especially in the Mediterranean. Morphological and metabolic traits, with known genetic bases, can be functionally altered to test current hypotheses on plant-environment interactions and eventually design a greenhouse plant. Reasonably, such a plant should have specific shoot vs. root developmental patterns, efficient water and nutrient uptake systems as well as other specific features that have not been sufficiently explored. Elucidation of the complex plant-greenhouse interactions would establish a physiological basis to improve both product quality and resource use efficiency in greenhouse.

---

[*] Proc. IS on Prot. Cult. Mild Winter Climate. Eds.: Y. Tüzel *et al.* Acta Hort. 807, ISHS 2009.
[**] Leo. marcelis@ wur. nl.

Key words: light, CO_2, temperature, leaf area, source-sink, salt stress, drought

Introduction

Greenhouse production allows growers to improve growth conditions for maximizing crop production, product quality and resource use efficiency. Strategies for these improvements should target both developing advanced technological systems and designing improved plants. The development of advanced systems should aim at controlling growth conditions such that they meet the demand of the plant. At the same time, the design of the plants should aim at a plant that is better suited to cope with the growth conditions in the greenhouse.

Physiological crop models are powerful tools to identify the desired growth conditions, to explore effects of growth conditions related to the introduction of new technologies as well as to identify the target traits of a crop that are particularly important for a specific environment. Biotechnology provides tools to generate plants with improved traits as well as to explore the potentials of crop improvement.

Based on greenhouse experiments, crop models and biotechnological tools, this paper will discuss the physiology of plant-greenhouse interactions. It is discussed how these interactions can be applied to control the production process at Northern and Mediterranean climatic conditions.

Results and discussion

Light

Light forms the basis for growth of plants, as it is the driving force for photosynthesis. Besides, light quality and length of photoperiod may affect developmental processes in the plant. . For most crops a 1% light increment results in 0.5 to 1% increase in harvestable product (Marcelis et al., 2006). This is an average value, which depends on several factors. For instance, the relative effect of light on growth is greater at lower light levels, at higher CO_2 concentrations and at higher temperatures. Consequently, the relative effect is larger in winter than in summertime and the effect is larger in Northern than in Mediterranean regions. The effect of light on growth also depends on the duration and moment that the light level is changed. Besides a positive effect on yield quantity, light usually has a positive effect on quality as well. Light should not be considered as a separate growth factor in greenhouse horticulture, as it forms an integral part of the total farm management. Many growers, for instance, choose a higher temperature, a lower plant density and different cultivar when the light level is increased.

Photosynthesis shows a saturating response to light. At low light levels photosynthesis increases rapidly, but at higher levels effects of light diminish. The level at which photosynthesis saturates is not a constant, but may depend on amongst others CO_2 concentration, nutrient (especially N) concentrations or the season (most likely the main factor is light level during the previous weeks). A common mistake made by many authors is neglecting the difference between response curves at the leaf level and the crop level. Many authors measure photosynthesis of single leaves (or a few cm^2 of a leaf) and apply this curve to predict the response of the whole canopy. This, however, can only be done when a simulation model is used that accounts for light penetration in a canopy. The top leaves may saturate at $500 \sim 1\,000\,\mu mol$ PAR $m^{-2}\,s^{-1}$ (depending on growth conditions), but the leaves below the top leaves are not saturated and can use additional light. For instance when leaf photosynthesis saturates at about $600\,\mu mol$ PAR $m^{-2}\,s^{-1}$, a crop with LAI of 6 only saturates at $1\,100\,\mu mol$ PAR $m^{-2}\,s^{-1}$ (Figure 1). The higher the LAI the higher the light intensity at which saturation occurs (Figure 1).

In Northern countries lamps are used to improve growth and quality under poor light conditions. However, even in these Northern countries in summer often screens or white wash is used to prevent too high radiation levels in the greenhouse. In Mediteranean countries a large fraction of light is prevented to enter greenhouse by shading nets, screens or white wash. Considering the photosynthesis response of canopies far too much shading is applied. In fact large amounts of light which could drive growth, are unused. Nevertheless, if less shading was applied production of high quality produce in the present systems would reduce. In most cases the growth impairment at high radiation levels is not the result of a too high light intensity, but rather because of a too high heat load or too high vapour pressure deficit of the air. In addition most shading measures increase the diffuseness of the light. If we can control temperature and air humidity by other means than shading measures, more light could be allowed to enter the greenhouse, which can increase yield. Options for temperature and humidity control under high light intensity are for instance fogging which cools and increases air humidity, cooling greenhouse by pad and fan systems, mechanical cooling, growing a crop with high leaf area index and sufficient water supply to increase crop transpiration which cools the greenhouse and increases air humidity. Fogging systems and semi-closed greenhouses with mechanical cooling (and storing the heat in aquifer) are recently gaining interest of growers in the Netherlands. Use of screens or cover materials that reflect NIR radiation, can be helpful to prevent too much heat load in the greenhouse (Kempkes et al., 2008). Furthermore, very high light conditions may lead to an unbalanced sink/source ratio (too high source in relation to sink). Hence, measures must be taken to increase sink strength of the plant. For instance, in tomato at low stem density in summer, short leaf syndrome may occur which may be related to an unbalanced source/sink ratio (Nederhoff, 1994). Formation of trusses per stem (the main sink organs) mainly depends on temperature and is hardly affected by stem density (number of stems per m^2). Therefore, the number of sinks per m^2 can easily be increased by increasing number of stems, as is common practice in Northern Europe.

Beside the effects of light intensity the light distribution within the canopy can help to further improve crop production. Top leaves may be close to light saturation while lower leaves receive insufficient light. Converting the incoming direct radiation into diffuse radiation improves the light distribution in the canopy and hence the crop production (e. g. Hemming et al., 2008; Heuvelink and Gonzalez-Real, 2008).

In summary a substantial fraction of the available light is not used in greenhouse production. An improved control of temperature and humidity in greenhouses might allow to make better use of the available light.

CO_2

For several decades CO_2 supply is common practice in winter period in the Netherlands. At day time when windows are closed or not far open, most growers supply CO_2 up to levels of about 800~1 000ppm. In most cases flue gases are used as source of CO_2. Especially when CO_2 is supplied from a heat and power generator, cleaning of the flue gases for NOx and ethylene can be very critical. Recent measurements at commercial farms show that the NOx dosage might sometimes be a risk for optimal crop growth. Besides CO_2 supply from flue gases more and more CO_2 is obtained from industry (e. g. OCAP) or by supplying pure CO_2 from a tank.

The primary effect of CO_2 on plant production is on photosynthesis. At low CO_2 concentrations photosynthesis increases rapidly with increasing CO_2, while it usually saturates at levels of about 800~1 000ppm. Based on measurements of canopy photosynthesis of several fruit vegetable crops, Nederhoff (1994) proposed a generic rule to estimate effects of CO_2 on canopy photosynthesis:

$$X = (1\ 000/C)^2 \times 1.5$$

where X is the percentage increase in photosynthesis when the CO_2 concentration is raised by 100ppm; C is the CO_2 concentration expressed as ppm.

This implies an increase in photosynthesis of 12% when the CO_2 concentration is raised from 350 to 450ppm or 4% when the concentration is raised from 600 to 700ppm. In greenhouses without CO_2 enrichment the CO_2 concentration may drop well below the ambient outside concentration (Nederhoff, 1994). A drop form 350 to 250ppm would lead to 19% decrease in photosynthesis. It should be noted that this a generic rule which on average yields reliable results, but effects of CO_2 concentration also depend on several other growth factors. Therefore, effects in specific circumstances may deviate from the rule. It is well known that effects of CO_2 on photosynthesis interact with light intensity and temperature. The measurements of Nederhoff were performed under moderate Dutch light conditions under conditions of higher light intensities stronger effects of CO_2 are expected. Furthermore, a higher CO_2 concentration may result in a larger fraction of dry matter partitioned into the fruits in cucumber and pepper (Nederhoff, 1994; Dieleman, unpublished data), which may result in an even stronger response as predicted by the generic rule. This positive effect on dry matter partitioning is probably the result of increased source strength on fruit set (Marcelis, 1994; Marcelis et al., 2004). When plants are exposed to high CO_2 concentrations during a prolonged period, the photosynthetic capacity may decrease which is likely to be the result of feedback inhibition (e. g. Sims et al., 1998). Feed-back inhibition occurs when the assimilates produced in photosynthesis are insufficiently used by the sink organs of the plant. Therefore it is important to maintain sufficient sink organs. Despite the numerous studies with plants showing feed back inhibition, so far we did not find indications that feedback-inhibition commonly occurred in plants grown in commercial greenhouses in the Netherlands (Marcelis, 1991; Heuvelink and Buiskool, 1995). Probably because in these growing conditions the source/sink ratio is usually quite low (e. g. Marcelis, 1994).

Recently closed and semi-closed greenhouses have been introduced, which keep the windows closed at most times and cool the greenhouse by use of heat pumps and using cooling capacity from aquifers (Heuvelink et al., 2008). Due to the closure of the windows, high CO_2 concentrations can be maintained throughout the whole year, whereas in conventional greenhouses with opened windows CO_2 concentrations higher than 400~500ppm in summer time are not feasible. Based on calculations by a crop model and measurements at a commercial farm maintaining a CO_2 concentration of about 900~1 000ppm in summer increased annual tomato yield by about 16%~17% compared to a conventional greenhouse (400~500ppm).

In Mediterranean countries with higher light levels, effects of CO_2 on production are expected to be bigger than in Northern countries. In these countries only a limited number of growers use CO_2 enrichment. The need for opening of windows (to cool down the greenhouse or prevent too high air humidity) and limited availability of cheap CO_2 are factors that hamper the introduction of CO_2 enrichment. Stanghellini et al. (2008) concluded that in Mediterranean greenhouses growers should aim at concentrations of 1 000ppm CO_2 in the absence of ventilation, and gradually decrease to maintaining the external value of CO_2 concentration when ventilation rates are exceeding 10 per hour. The trend between these two extremes depend on value of produce and price of CO_2.

In general it can be concluded that still quite some yield increase can be realized by better managing the CO_2 concentration. This is not just true for Mediterranean countries where CO_2 enrichment is not common practice yet. In Northern countries also improvements are foreseen in the summer half year, by keeping the windows more closed and thus keeping the CO_2 inside in combination with some additional CO_2 supply. Closure of windows is only possible when too high greenhouse temperatures are prevented by cooling, either mechanically or by fogging.

Leaf Area

Crops need a sufficiently high leaf area to intercept the light. For optimal light interception an LAI of about 3 to 4 is needed (Heuvelink et al., 2005). The top leaves have the highest rate of photosynthesis, but also because they have the highest photosynthetic capacity. Light response curves of leaf photosynthesis showed that photosynthesis, transpiration respiration decreased from top to bottom in the canopy even when measurements are performed at the same light intensity at the leaf (Figure 2; Dueck et al. 2007; González-Real et al., 2007). These reductions in gas exchange lower in the canopy likely result from adaptation to lower ambient light conditions as well as leaf aging. Dueck et al. (2007) studied the contribution of different leaf layers in a sweet pepper crop, which reaches an LAI of up to 8 in summer when the crop is planted in winter. At a low light intensity of $50 \mu mol \cdot m^{-2} \cdot s^{-1}$ above the canopy, only the top 25% of the leaves ($2m^2 m^{-2}$) contributed positively to canopy photosynthesis, while at a higher light irradiance, $200 \mu mol \cdot m^{-2} \cdot s^{-1}$, the top 50% ($4m^2 m^{-2}$) contributed positively. From the middle of August onwards, the net photosynthesis of the lower half of the crop was negative. Based on these measurements, the contribution of each leaf level to the net crop photosynthesis and transpiration was calculated. On an annual basis, the lower half of the crop made a 0.5% negative contribution to net photosynthesis, while making a

10% positive contribution to crop transpiration. As in the winter half year energy is needed to prevent too high air humidity, reducing transpiration can save energy. Therefore, removal of leaves from the lower levels might increase the efficiency of energy utilization.

Optimization of the light distribution and photosynthesis in the canopy may include adaptations in row structure and leaf pruning. Functional structural plant models that simulate the plant architecture, 3D distribution of light and photosynthesis in the canopy can be powerful tools to explore possibilities (Vos et al., 2007).

Young growing leaves compete for assimilates with other sinks. Removal of young leaves favored partitioning to the fruits in tomato but decreased LAI and total yield (Heuvelink et al., 2005; Table 1). However, if removal of old leaves was delayed such that an LAI of $3m^2 m^{-2}$ was maintained, removal of every second young leaf improved yield by 10% (Table 1). An alternative means to removal of young leaves would be breeding for varieties that form two leaves in between trusses, while tomato cultivars generally have three leaves in between two trusses. Model calculations showed that a genotype with two instead of three leaves between trusses indeed will improve yield. To maximize the benefit of this trait it is important to keep the LAI sufficiently high by delaying removal of old leaves or increasing plant density. Whether breeding can realize the predicted extra yield for a genotype with two leaves between trusses is not clear. Tomato genotypes with only two leaves between trusses do exist, but this plant characteristic seems to be linked to a determinate growth pattern (W. H. Lindhout, pers. comm.), whereas for greenhouse cultivation plants with indeterminate growth pattern are needed.

Source/Sink Balance

Optimum growth conditions means that there is a good balance among different climate conditions. The source/sink ratio of a crop (ratio between production and demand of assimilates) often reflects whether these conditions are balanced. The source strength mainly depends on the amount of light intercepted by the leaves and the CO_2 concentration and to a lesser extent on temperature and air humidity; the sink strength mainly depends on temperature and the number of sink organs (rapidly growing such as fruits) on a plant. The source/sink ratio can vary strongly from day to day. This variation in source/sink balance affects formation and abortion of organs (Marcelis et al., 2004; Carvalho & Heuvelink 2004; Figure 3), which leads to fluctuations in production and quality. If we could stabilize the source/sink ratio we can achieve a more balanced growth, a more regular biomass partitioning, a more regular fruit or flower size, more regular production in time and prevent feed-back inhibition of photosynthesis (e. g. Marcelis 1994; Heuvelink et al., 2004). Elings et al. (2006) developed a temperature control that reduces the day to day fluctuations in source/sink ratio of a cucumber crop. This led to a more stable and greater dry matter partitioning into the fruits and more stable fruit size and age at harvest and increase. On an annual basis this strategy resulted in 5% yield increase or 13% energy use, depending on the optimization criteria. Van Henten et al. (2006) also showed how production fluctuations in sweet pepper can be reduced and 10% of energy saved by optimizing the temperature setpoints.

Crop Physiology in Relation to Molecular Biology

The study of plant response to environmental stressors has made significant progress in the last twenty years (Maggio et al., 2006). During this time, while new experimental tools have been proposed to better understand how plants respond to environmental stimuli, standard research approaches have greatly benefited of new discoveries in the field of molecular biology. Today we have quite a clear picture of many physiological mechanisms that may affect plant production in stressful environments, however we are still far from being able to design effective strategies to substantially improve plant stress tolerance by using all the available tools, including traditional breeding, genetic engineering and specific cultural techniques. A step forward in this process is the identification of *specific* plant/environment interactions for a given cultural system, since these may involve different approaches to improve plant adaptation to environmental constraints. Ultimately, a thorough comprehension of these interactions will support the *design* of plants able to efficiently use the available resources in each specific environment.

Abiotic Stress. Drought, salinity and extreme temperatures are common stresses in agriculture productions and they may limit optimal yield in both open field and controlled environment (Flowers, 2004). In addition, multiple stresses may co-exist in most agricultural contexts. Quite often drought and/or salinization are associated to exposure to high temperatures (Maggio et al., 2002). Salinity is particularly critical for many horticultural crops. In the field, salinization often occurs as a transitory event that may be controlled by both suitable irrigation volumes and the seasonal rainfall that leaches out the excess of salt from the root zone. In greenhouse, soil salinization may reach critical levels as well. In this case the control of soil salinization is strictly dependent on an efficient irrigation management. Salinization may result from a cumulative effect of nutrients added in the irrigation water and additional ions such as Na^+ and Cl^-, both of which may overall generate a hyperosmotic environment.

To cope with a continuous exposure to biotic and abiotic stresses, which generally interfere with the normal growth and development, plants have evolved a complex response system that is largely mediated by phytohormones (Fujita et al., 2006). The most accredited sequence of events, includes the perception of the signal (stressor), the transduction of the signal and finally the activation of downstream components, i.e. exnovo synthesis and/or activation of molecules that would facilitate adaptation under stress conditions (Zhu, 2001). Common response mechanisms are usually initiated by different stressors and intermediate signalling components, such as Ca^{2+} and reactive oxygen species (ROS) which are involved in the signal transduction cascade activated by both abiotic and biotic inducers (Torres et al., 2005). Advancement in molecular techniques have allowed to dissect the transcriptional profile of abiotic and biotic stress responses (Bohnert et al., 2006) and have confirmed that these regulate a recurrent set of common genes. Recent efforts using the model plant *Arabidopsis* have revealed that genes are expressed in response to a various range of stresses (Ma et al., 2007). Consequently, the identification of *upstream* stress components, i.e. intermediate molecules common to different response pathways, has become the research focus of many molecular engineers interested in the generation of multi-stress tolerant plants.

The identification of molecules responsible for sensing and transducing the stress signals (Guo et al., 2002) has therefore become the new approach after the modest results attained through the over-expression of single downstream components (Maggio et al., 2002). In horticultural productions, tomato has been the model species for molecular engineering of abiotic stress tolerance traits. First attempts to improve the tolerance to drought, salinity and extreme temperatures included the generation of plants that 1) overproduced compatible solutes such as glycinebetaine (Park et al., 2004); 2) were capable of controlling cytoplasmic Na^+ accumulation (Zhang and Blumwald, 2001); 3) manifested an improved chilling and heat tolerance (Hsieh et al., 2002; Mishra et al., 2002). In most cases these plants showed a reduced growth in absence of stress, revealing that the control of cell enlargement and division is part of the complex stress adaptation process and is likely mediated by these metabolites (Ruggiero et al., 2004).

Cross Talk Biotic-Abiotic. Plants are quite often required to respond to overlapping environmental stimuli of biotic and abiotic origin and, consequently, they have developed mechanisms to integrate their signal transduction pathways leading to adaptation. These responses, in most cases, do not function independently, yet a concerted activation of different pathways controls the response specificity to biotic and abiotic stress (Ludwig et al., 2005). We have recently confirmed the existence of a cross-talk between stress adaptive mechanisms by analyzing the response of tomato plants over-expressing the prosystemin cDNA to salt stress. Chewing insects and mechanical wounding cause the release of a highly mobile peptide called systemin (Schilmiller and Howe, 2005). In tomato, this 18-amino acid molecule is synthesized as a 200-aa precursor protein named prosystemin (McGurl and Ryan, 1992). This molecule is typically involved in mechanical stress responses such as those induced by some insects. Evidence gained through mutational analysis suggests that long-distance defence signalling mediated by systemin involves jasmonic acid, a plant hormone that also controls growth regulation (Schilmiller and Howe, 2005). Jasmonic acid and its methyl-ester have both a role in stress responses, including water-, osmotic-, and wound-stress (Reindbothe et al., 1992). The interaction between jasmonic acid and other stress hormones, including ABA, has also been demonstrated (Staswick et al., 1992). Preliminary results indicated that a constitutive overproduction of systemin had some positive effects on tomato tolerance to salinity. This response was associated to a constitutively reduced stomatal conductance of systemin overxpressing plants, which was most likely responsible for a better control of the plant water homeostais in saline environment. This conclusion was consistent with the lower level of stress metabolites, such as ABA and proline that we found in systemin overxpressing plants respect to their relative control (unpublished results).

Looking Forward. The existence of multiple stress responses that ultimately lead to plant stress adaptation indicates that there are margins to dissect the contribution of these physiological mechanisms in each specific environment. In this respect, the greenhouse environment would be particularly suitable to promote strategies of "precision" horticulture (Maggio et al., 2008). To pursuit this approach it should be first defined which kind of tolerance we need in each specific agricultural system and which function should be improved. This may involve identifying the most efficient combination of genes, generating better alleles of the most promising genes for a specific cultivation process and assessing cultivation protocols that would potentiate constitutive physiological responses. In addition, the isolation of new tolerance determinants by using novel screening techniques should also be considered. This may surprisingly lead to isolate unconventional stress tolerance traits that may be important in a specific agricultural context. Strategies to improve water use efficiency and salinity tolerance should look for example at morphological and physiological traits, such as leaf characteristics (hairiness, waxiness, leaf angle), root architecture, root hydraulic conductivity and other characteristics that may have a particular/specific value under certain cultivation systems (Maggio et al., 2008). The control of water fluxes through the stomata and membrane aquaporins is also important and may play a critical role especially when other environmental parameters can be modulated, as it may occur in greenhouse cultivation.

Conclusions

There is still quite some room for increasing yield and quality, while improving sustainability of greenhouse production systems. To do so we need to control growth conditions in the greenhouse such that they meet the demand of the crop. At the same time the crop management should aim at a plant that is better suited to cope with the growth conditions in the greenhouse.

Light is the most important growth factor in greenhouse production. A substantial fraction of the available light is not used in greenhouse production. Better control of temperature and humidity in greenhouses might allow to make better use of the available light. Quite some yield increase can also be realized by a better management of the CO_2 concentration in both Mediterranean and Northern countries. Optimization of leaf area includes maximization of light interception for photosynthesis and minimization of assimilate use for leaf formation. In addition it considers effects of leaf area on transpiration which is needed for cooling in hot summer conditions, but leads to energy consumption under cool winter conditions. Growth conditions need to be balanced such that the source/sink ratio is balanced. This can be used to reduce yield fluctuations or to increase yield.

In this contest, biotechnology should be seen as a powerful tool to identify both physiological traits and metabolic components that may be 'potentiated' to improve greenhouse plant resource use efficiency.

Tables

Table 1. Simulated cumulative dry weight of fruits (DW_{fruit}) and plant (DW_{total}), fraction partitioned to the fruits (F_{fruits}), average LAI for a tomato crop grown from Dec. till Nov. Young leaves were removed at appearance (from Heuvelink et al., 2005).

Number of young leaves removed	DW_{fruit} (kg m^{-2})	DW_{total} (kg m^{-2})	F_{fruits}	LAI (m^2 m^{-2})
Control: no removal	2.92	4.25	0.69	2.41
1 out of 6	3.01	4.24	0.71	2.38
1 out of 3	3.11	4.22	0.74	2.33
1 out of 2	3.22	4.18	0.77	2.25

Figures

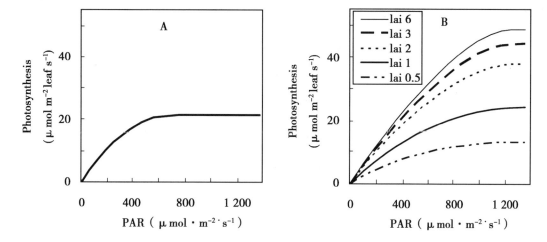

Figure 1 Light response curves for gross photosynthesis of a leaf (A) and of a crop (B) with different leaf area indices (LAI). Leaf photosynthesis was calculated by a biochemical model. Based on this model crop photosynthesis was calculated by the crop model INTKAM (Marcelis *et al.*, 2000) Calculations at 21℃ and 400ppm CO_2

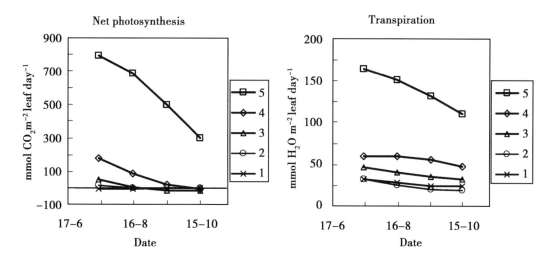

Figure 2 Net leaf photosynthesis and transpiration of leaves at different heights in a sweet pepper canopy with LAI =6. Layer 1 is bottom and Layer 5 is top. From Dueck *et al.* (2007)

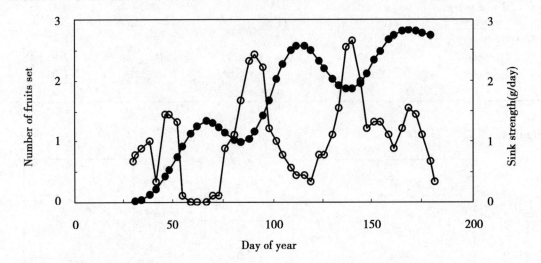

Figure 3 Fruit set (o; number of young fruits, less than 10 days from anthesis) of sweet pepper plants and plant sink strength (●) during a growing season. From Heuvelink *et al.* (2004)

Literature Cited

[1] Bohnert, H. J., Gong, Q., Li, P. and Ma, S. 2006. Unravelling abiotic stress tolerance mechanisms-getting genomics going. Curr. Opin Plant Biol. 9: 180~188

[2] Carvalho, S. M. P. and Heuvelink, E. 2004. Modelling external quality of cut chrysanthemum: achievements and limitations. Acta Hort. 654: 287~294

[3] Dueck, T. A., Grashoff, C., Broekhuijsen, G. and Marcelis, L. F. M. 2007. Efficiency of light energy used by leaves situated in different levels of a sweet pepper canopy. Acta Hort. 711: 201~205

[4] Elings, A., de Zwart, H. F., Janse, J., Marcelis, L. F. M. and Buwalda, F. 2006. Multiple day Temperature Settings on the Basis of the Assimilate Balance: a Simulation Study. Acta Hort. 718: 219~226

[5] Flowers, T. J. 2004. Improving crop salt tolerance. J. Exp. Bot. 55: 307~319

[6] Fujita, M., Fujita, Y., Noutoshi, Y., Takahashi, F., Narusaka, Y., Yamaguchi-Shinozaki, K. and Shinozaki, K. 2006. Crosstalk between abiotic and biotic stress responses: a current view from the points of convergence in the stress signaling networks. Curr. Opin Plant Biol. 9: 436~442

[7] Gonzalez-Real, M. M., Baille A. and Gutiérrez Colomer, R. P. 2007. Leaf photosynthetic properties and radiation profiles in a rose canopy (*Rosa hybrida* L.) with bent shoots. Scientia Hortic. 114: 177~187

[8] Guo, Y., Xiong, L., Song, C. P., Gong, D., Halfter, U. and Zhu, J. K. 2002. A calcium sensor and its interacting protein kinase are global regulators of abscisic acid signaling in *Arabidopsis*. Dev. Cell 3: 233~244

[9] Hemming, S., Dueck, T. A., Janse, J. and Van Noort, F. 2008. The effect of diffuse light on crops. Acta Hort. 801: 1 293~1 300

[10] Heuvelink, E. and Buiskool, R. P. M. 1995. Influence of sink-source interaction on drymatter production in tomato. Ann. Bot. 75: 381~389

[11] Heuvelink, E. and Gonzalez-Real, M. M. 2008. Innovation in plant-greenhouse interactions and crop management. Acta Hort. 801: 63~74

[12] Heuvelink, E., Bakker, M., Marcelis, L. F. M. and Raaphorst, M. 2008. Climate and yield in a closed greenhouse. Acta Hort. 801: 1 083~1 092

[13] Heuvelink, E., Bakker, M. J., Elings, A., Kaarsemaker, R. and Marcelis, L. F. M. 2005. Effect of leaf area on tomato

yield. Acta Hort. 691: 43~50

[14] Heuvelink, E., Marcelis L. F. M. and Korner, O. 2004. How to reduce yield fluctuations in sweet pepper? Acta Hort. 633: 349~355

[15] Hsieh, T. H., Lee, J. T., Yang, P. T., Chiu, L. H., Charng, Y. Y., Wang, Y. C. and Chan, M. T. 2002. Heterology expression of the Arabidopsis C-Repeat/Dehydration Response Element Binding Factor1 gene confers elevated tolerance to chilling and oxidative stresses in transgenic tomato. Plant Physiol. 129: 1 086~1 094

[16] Kempkes, F., Stanghellini, C. and Hemming, S. 2008 Cover materials excluding Near Infrared radiation: what is the best strategy in mild climates? Acta Hort. 807: 67~72

[17] Ludwig, A. A., Saitoh, H., Felix, G., Freymark, G., Miersch, O., Wasternack, C., Boller, T., Jones, J. D. G. and Romeis, T. 2005. Ethylene-mediated cross-talk between calcium-dependent protein kinase and MAPK signaling controls stress responses in plants. Proc. Nat. Acad. Sci. U. S. A. 102 (30): 10 736~10 741

[18] Ma, S. and Bohnert, H. 2007. Integration of Arabidopsis thaliana stress-related transcript profiles, promoter structures, and cell-specific expression. Genome Biol. 8: R49

[19] Maggio, A., De Pascale, S. and Barbieri, G., 2008. Designing a Greenhouse Plant: Novel Approaches to Improve Resource Use Efficiency in Controlled Environments. Acta Hort. 801: 1 235~1 242

[20] Maggio, A., Matsumoto, T., Hasegawa, P. M., Pardo, J. M. and Bressan, R. A. 2002. The long and winding road to halotolerance genes. In: Salinity: environment-plantsmolecules. Eds. A. Lauchli and U. Luttge. Kluwer Acad. Pub., Netherlands, pp. 505~533

[21] Maggio, A., Zhu, J. K., Hasegawa, P. M. and Bressan, R. A. 2006. Osmogenetics: Aristotle to Arabidopsis. Plant Cell 18: 1 542~1 557

[22] Marcelis, L. F. M. 1991. Effects of sink demand on photosynthesis in cucumber. J. Exp. Bot. 42: 1 387~1 392

[23] Marcelis, L. F. M. 1994. A simulation model for dry matter partitioning in cucumber. Ann. Bot. 74: 43~52

[24] Marcelis, L. F. M., Broekhuijsen, A. G. M., Nijs, E. M. F. M. and Raaphorst M. G. M. 2006. Quantification of the growth response of light quantity of greenhouse grown crops. Acta Hort. 711: 97~103

[25] Marcelis, L. F. M., Heuvelink, E., Baan Hofman-Eijer, L. R., Den Bakker, J. and Xue, L. B. 2004. Flower and fruit abortion in sweet pepper in relation to source and sink strength. J. Exp. Bot. 55: 2 261~2 268

[26] McGurl, B. and Ryan, C. A. 1992. The organization of the prosystemin gene. Plant Mol. Biol. 20: 405~409

[27] Mishra, S. K., Tripp, J., Winkelhaus, S., Tschiersch, B., Theres, K., Nover, L. and Scharf, D. K. 2002. In the complex family of heat stress transcription factors, HsfA1 has a unique role as master regulator of thermotolerance in tomato. Genes Dev. 16: 1 555~1 567

[28] Nederhoff, E. M. 1994. Effects of CO_2 concentration on photosynthesis, transpiration and production of greenhouse fruit vegetable crops. Thesis Wageningen University. 213pp

[29] Park, E. J., Jeknić, Z., Sakamoto, A., De Noma, J., Yuwansiri, R., Murata, N. and Chen, T. H. H. 2004. Genetic engineering of glycinebetaine synthesis in tomato protects seeds, plants, and flowers from chilling damage. Plant J. 40: 474~87

[30] Reinbothe, S., Reinbothe, C., Lehmann, J. and Parthier, B. 1992. Differential accumulation of methyl jasmonate-induced messenger-rnas in response to abscisicacid and desiccation in barley (Hordeum-vulgare) Physiol. Plant. 86: 49~56

[31] Ruggiero, B., Koiwa, H., Manabe, Y., Quist, T. M., Inan, G., Saccardo, F., Joly, R. J., Hasegawa, P. M., Bressan, R. A. and Maggio, A. 2004. Uncoupling the effects of ABA on plant growth and water relations: analysis of sto1/nced3, ABA deficient salt stress tolerant mutant in *Arabidopsis* thaliana. Plant Physiol. 136: 3 134~3 147

[32] Schilmiller, A. L. and Howe, G. A. 2005. Systemic signalling in the wound response. Curr. Opin Plant Biol. 8: 369~377

[33] Sims, D. A., Luo, Y. and Seemann, J. R. 1998. Comparison of photosynthetic acclimation to elevated CO_2 and limited nitrogen supply in soybean. Plant Cell Environ. 21: 945~2 952

[34] Stanghellini, C., Kempkes, F. and Incrocci, L. 2008. Carbon dioxide fertilization in mediterranean greenhouses: when and how is it economical? Acta Hort. 807: 135~142

[35] Staswick, P. E., Su, W. P. and Howell, S. H. 1992. Methyl jasmonate inhibition of rootgrowth and induction of a leaf protein are decreased in an *Arabidops* thaliana mutant. Proc. Nat. Acad. Sci. U. S. A. 89 (15): 6 837~6 840

[36] Torres, M. A. and Dangl, J. L. 2005. Functions of the respiratory burst oxidase in biotic interactions, abiotic stress and development. Curr. Opin Plant Biol. 8: 397~403

[37] Van Henten, E. J., Buwalda, F., de Zwart, H. F., de Gelder, A., Hemming, J. and Bontsema, J. 2006. Toward an optimal control strategy for weet pepper cultivation. 2. Optimization of the yield pattern and energy saving. Acta Hort. 718: 391~398

[38] Vos, J., Marcelis, L. F. M., de Visser, P. H. B., Struik, P. C. and Evers, J. B. 2007. Functional Structural Plant Modelling in Crop Production. Springer, Dordrecht. 268pp

[39] Zhang, H. X. and Blumwald, E. 2001. Transgenic salt-tolerant tomato plants accumulate salt in foliage but not in fruit. Nature Biotech. 19: 765~768

[40] Zhu, J. H. 2001. Cell signalling under salt, water and cold stress. Curr. Opin Plant Biol. 4: 401~406.

温室番茄营养诊断研究初报*

郭建华,武新岩,王钟扬,王 秀,马 伟,张瑞瑞,徐 刚

(国家农业信息化工程技术研究中心,北京 100097)

摘要:温室番茄营养诊断是提高番茄产量,减少肥料用量以及提高番茄品质的重要手段之一,本研究利用叶绿素仪 SPAD-502 测定番茄不同生育期不同叶位 SPAD 读数值,结果表明番茄无论新叶还是成叶全氮含量、叶绿素含量都与 SPAD 测定值有很好的相关性。随着施肥量的增加,番茄植株体内的叶绿素含量、全氮含量都在增加,番茄 SPAD 读数在逐渐的增大,番茄的 Vc 含量呈上升趋势,在 600kg/hm² 时,果实 Vc 含量达到最大值,当施氮量为 900kg/hm² 时,番茄体内的固形物达到了最大值。SPAD 能够对番茄的营养状况作出诊断。

关键词:番茄;SPAD;叶绿素;全氮

The Application Research of Nondestructive Test Technology On Tomato Nutrition Diagnosis

Guo Jianhua, Wu Xinyan, Wang Zhangyang, Wang Xiu, Ma Wei, Zhang Ruirui, Xu Gang

(*National Engineering Research Center for Information Technology in Agriculture, Beijing 100097 P. R. China*)

Abstract: The aim of this study was to develop a non-destructive, fast, simple nutritional diagnosis methods of tomato nitrogen status in the greenhouse. The experiments were conducted during 2009 ~ 2010 at the Xiaotangshan Precision Agriculture Experimental Base, Changping District, Beijing. SPAD-502 were used to determine new leaves and mature leaves of tomato under different growth stages. SPAD readings have a good correlation with nitrogen and chlorophyll contents. The research results showed that SPAD readings and nitrogen, chlorophyll contents have positive correlation. With quantity fertilizer increase, chlorophyll contents and nitrogen of tomato are increasing, SPAD readings in gradual enlargement. When the nitrogen is 600 kg/hm², the VC content of tomato is max; when the nitrogen is 900 kg/hm², the soluble solids is max. SPAD could be inter-grated to diagnose tomato nitrogen nutritional status.

Key words: Tomato; SPAD; Chlorophyll; Nitrogen

番茄(*Lycopersicon esculetum*)是我国设施栽培中面积最大的蔬菜之一。研究表明,作物生长与其氮素吸收和氮浓度关系密切[1]。由于其具有很高营养价值和药用价值,风味独特,深受广大消费者青睐。近年来,叶绿素仪在作物叶片养分的间接速测上应用广泛,取得了较好的效果,利用叶绿素仪测定的 SPAD 值可以间接反映作物叶片的叶绿素含量及含氮量[2~6]。对番茄最经济施肥量的研究发现,合理施肥在一定程度上可以提高番茄的产量,并且能够改善番茄的品质和风味[7]。氮素营养对番茄生长发育及产量有很大的影响,氮肥的合理施用比传统施用有更大的经济可行性和生态可行性。

---

* 基金项目:863(SQ2010AA1000764006)。

合理氮素营养调控取决于对作物氮素营养水平的精确判定[8]。由于植株全氮的化学分析方法普遍基于通过破坏土壤和作物获取样本，历经采样、烘干、研磨、称重、化验分析等多道程序，花费时间长，测试结果不具有实时性，不能满足日趋发展的农业信息化的要求[9]。近年来，一种基于测定叶绿素相对含量的手持式叶绿素仪（SPAD-502）较为广泛地应用于植物的氮素检测，叶绿素检测仪SPAD-502是通过不同叶绿素含量的叶片对两种不同波长光的吸收不同来确定其叶绿素含量，与传统的叶绿素测定方法：乙醇—丙酮浸提法相比，它能够快速、简便、较精确、非破坏性地监测植物氮素营养水平并能及时提供追肥所需的信息。本研究利用叶绿素仪对番茄不同氮素水平下各生育期成叶和新叶叶片的SPAD读数进行测定，分析氮素不同用量、叶绿素、全氮与SPAD值的关系，旨在寻找一种快速、简单、无损的氮素营养状况诊断方法。

1 材料与方法

1.1 试验地基本情况

试验于2009年春在北京市昌平区小汤山国家精准农业示范基地的日光温室内进行。番茄移栽前，大棚按照S型进行取样，取样深度0~20cm，土壤类型为石灰性褐土，试验土壤的基础地力：有机质32.2g/kg，全氮1.63g/kg，有效磷179mg/kg，速效钾272mg/kg，pH值为6.87，硝态氮98.8mg/kg。供试番茄品种为京丹，2009年3月10号定植，株距40cm，行距40cm，每小区共28株。田间浇水以及病虫害管理与其他大棚相同。

1.2 试验设计

试验设5个氮素处理，CK 0kg/hm²、MF1 300kg/hm²、MF2 600kg/hm²、MF3 900kg/hm²、MF4 1 200kg/hm²；每处理重复3次，随机排列，每个小区面积为5.2m²，共15个小区。供试氮肥为硫酸铵（N 21%），磷肥为过磷酸钙（P_2O_5 17%），钾肥为硫酸钾（K_2O 50%），磷钾肥在播种前作为底肥施入。试验氮肥施用方法：1/3作为基肥施入，2/3作为追肥冲施。

1.3 样品取样和测定

取样时分为下部叶片（成叶）和上部叶片（新叶），使用SPAD-502型叶绿素计测定单叶片5次，取平均值，该叶片准备用做叶绿素含量测定。

土壤有机质采用重铬酸钾容量法—外加热法进行测定，叶绿素采用丙醇—乙酮浸提法，全氮的测定采用凯式定氮法测定。

2 结果与讨论

2.1 施肥对番茄叶片SPAD测定值的影响

番茄新叶的SPAD读数在43.3~59.1之间，成叶的SPAD读数在45.20~68.18之间。无论新叶还是成叶，在番茄整个生育期间，SPAD值随施肥量的变化而变化（图1）。同一施肥量的新叶还是老叶，差异显著，苗期SPAD值比较低，到结果期SPAD值达到最大值，以后逐渐下降，这种变化与番茄营养生长的变化一致。不同施肥处理，随着施肥量的增加，番茄叶片SPAD读数值也在增大，整个生育期间N1 200kg/hm²处理的SPAD测定值最大，N 0kg/hm²处理的SPAD测定值最小。N 0 kg/hm²处理，新叶叶片SPAD读数值在苗期比成叶叶片SPAD读

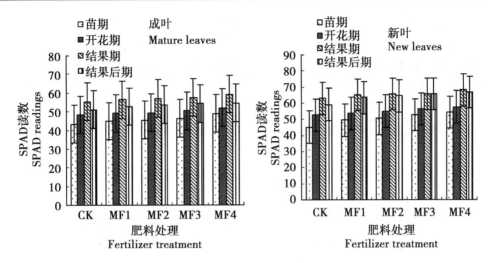

图1 同施肥处理在不同生育期对番茄新叶、成叶 SPAD 值影响
Figure 1 In different growth stages, the effect of different fertilizer treatment on new leaves, mature leaves of tomato

数值要小1.9，在结果后期，新叶叶片的测定值比成叶叶片的 SPAD 测定值下降快，其读数值相差8.9。不同施氮处理对苗期叶片 SPAD 值影响相对较小，到开花期这种影响逐渐加大；由于施肥量的增加促进了叶片含氮量、叶绿素含量的增加，因此 SPAD 读数值也随之增加。番茄的成叶和新叶在不同施氮量的处理下，两个不同叶位的变化趋势基本一致。

2.2 番茄叶片叶绿素含量与 SPAD 测定值的关系

绿色植物光能利用率高低取决其叶片叶绿素含量的多少，而作为植物生理指标之一的叶片叶绿素含量能确定植物在整个生育期间的营养状况，因此，研究绿色植物叶片叶绿素含量至关重要[10]。

表1 番茄叶片叶绿素含量及 SPAD 测定值的测定结果
Table1 SPAD readings and chlorophyll contents in leaves of tomato

叶位 Leaves position	SPAD 读数 SPAD readings	鲜叶叶绿素含量（mg/g）chlorophyll contents		
		chl-a	chl-b	chl
新叶	53.68	1.60	0.61	2.21
新叶	55.25	1.69	0.64	2.34
新叶	56.30	1.78	0.68	2.46
新叶	57.10	1.81	0.65	2.46
新叶	58.03	1.82	0.71	2.53
成叶	62.85	1.23	0.48	1.71
成叶	64.40	1.26	0.49	1.75
成叶	65.63	1.40	0.60	2.00
成叶	66.70	1.41	0.54	1.95
成叶	67.18	1.55	0.63	2.18

从表 1 可以看出，随 SPAD 值增加，叶绿素含量（Chl）亦呈增加趋势。由雷泽湘[11]的研究表明，其散点图图形符合对数函数和线性函数，两函数相比较，线性函数计算简单，相关系数略高于乘幂函数和指数函数。根据表 1 按线性函数分别对叶绿素 a、叶绿素 b 和叶绿素含量进行统计分析，相关趋势图见图 2。

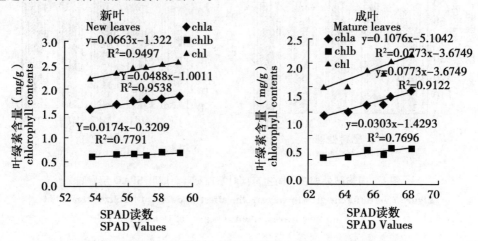

图 2 番茄新叶、成叶 SPAD 测定值与叶绿素含量的相关图
Figure 2 Correlogram between SPAD readings and chlorophyll contents in new leaves and mature leaves of tomato

从图 2 可以看出，成叶和新叶叶片的叶绿素 a 都与 SPAD 测定值有很好的相关性，相关系数分别是 0.9122、0.9588；叶绿素 b 与 SPAD 测定值的相关性较差，相关系数分别为 0.7696、0.7794；叶绿素与 SPAD 测定值的相关性介于叶绿素 a 与叶绿素 b 之间，分别是 0.8869、0.9497。表明无论新叶还是成叶叶片的叶绿素 a 与 SPAD 测定值的相关性大于叶绿素 b 与 SPAD 测定值的相关性，且叶片内的叶绿素 a 的含量高于叶绿素 b 的含量。这是由于 SPAD-502 型叶绿素计测定原理是以 660nm 左右固定波长测定叶绿素的含量，而叶绿素 a 的吸收峰为 660nm，与之非常接近；而叶绿素 b 的吸收峰在 643nm，与 SPAD-502 型叶绿素计的波长相差较远，所以叶绿素 b 与 SPAD 测定值的相关性要差一些。

2.3 番茄不同生育期叶片全氮含量与 SPAD 测定值的关系

表 2 番茄不同生育期成叶、新叶叶片全氮含量与 SPAD 读数的相关性
Table 2 In different growth stage, the correlation between nitrogen and SPAD value of tomato

生育期 Growth stage	叶位 Leaves position	方程式 Equation	R^2	N
苗期 Seeding	新叶	$y = 0.0248x^2 - 0.047x + 4.7348$	0.9273	15
	成叶	$y = -0.0293x^2 + 0.3129x + 2.8665$	0.9882	15
开花期 Flowering	新叶	$y = 0.0411x^2 + 1.0161x + 51.53$	0.9991	15
	成叶	$y = 0.1714x^2 - 0.1886x + 48.49$	0.9810	15
结果期 Fruiting	新叶	$y = -0.0054x^2 + 1.1496x + 62.115$	0.8709	15
	成叶	$y = 0.05x^2 + 0.59x + 54.73$	0.9691	15
结果后期 After fruiting	新叶	$y = -0.4304x^2 + 4.3196x + 55.67$	0.9667	15
	成叶	$y = -0.1964x^2 + 2.0186x + 49.31$	0.9981	15

番茄叶片全氮含量的测定结果表明（表2），番茄叶片无论成叶还是新叶都与全氮具有很好的相关性。苗期新叶、开花期新叶、开花期成叶和结果期成叶全氮含量与SPAD测定值呈正相关；苗期成叶、结果期新叶、结果后期新叶、结果后期成叶全氮含量与SPAD测定值呈负相关。在开花期时，新叶全氮含量与SPAD测定值的相关性达到了极显著，其函数表达式为 $y = 0.0411x^2 + 1.0161x + 51.53$，相关系数 $R^2 = 0.9991$。结果期新叶全氮含量与SPAD测定值的相关性显著，但相对较差，其函数表达式为 $y = -0.0054x^2 + 1.1496x + 62.115$，相关系数 $R^2 = 0.8709$。其他时期的无轮新叶还是成叶都达到了极显著相关。所以用SPAD测定时无论选新叶还是成叶都能很好地进行番茄氮素诊断。

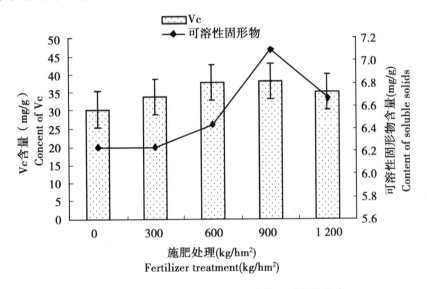

图3 不同施肥处理对番茄Vc、可溶性固形物的影响

Figure 3　The influence on Vc, soluble solids of tomato in different fertilizer treatment

2.4　不同施肥处理对番茄果实品质的影响

果实维生素C含量和可溶性固形物含量的测定结果（图3）表明，不同施肥处理果实维生素C含量差异不显著，随着施氮量的增加，番茄果实体内的维生素C含量也在逐渐增加，当施氮量达到600kg/hm² 时，果实的维生素C含量达到最大值，随着施氮量的继续增加，果实体内的维生素C含量逐渐降低。不同施肥处理果实硝态氮含量差异显著，随着施肥量的增加，果实体内的可溶性固形物逐渐上升，当施氮量为900kg/hm² 时，番茄体内的固形物达到了最大值，1 200kg/hm² 时开始下降。

3　结论

用无损测试技术对土壤或作物的氮素状况的精确评价是作物营养诊断技术的发展趋势[2]。应用SPAD仪，在番茄生长的各个生育期，对其氮营养状况进行诊断，结果表明它们之间具有很好的相关性，用SPAD能够估测番茄氮素丰缺。

番茄成叶叶片、新叶叶片全氮含量、叶绿素含量与SPAD测定值有很好的相关性，

SPAD 能够动态的监测番茄的生长状况，随着施肥量的增加，番茄体内的全氮含量、叶绿素含量都在增加，番茄 SPAD 读数在逐渐的增大。与传统的作物氮素营养状况诊断方法——番茄叶片全氮含量相比，叶绿素仪读数与番茄无论新叶还是成叶全氮含量之间为极显著的函数关系，表明叶绿素仪测试与植株全氮分析两种方法是相类似的氮营养诊断方法，叶绿素仪测试方法可以反映番茄的氮营养状况。与李志宏[10]在冬小麦上叶绿素仪测定值与作物全氮有很好的相关性这一研究结果比较接近。

番茄的品种繁多，不同品种间的 SPAD 值、果实品质往往相差很大，要根据品种的不同建立对应的诊断指标。另外，由于叶色是许多因素综合影响的结果，并不是单纯因氮含量不同而存在差异，当植物缺磷、缺钾或缺乏微量元素时，叶色也会发生变化，影响测定结果。今后尚需做大量的实验来完善利用叶绿素计进行蔬菜合理施肥这项技术。

参考文献

[1] Cui R Y, Lee B W. Spikelet number estimation model using nitrogen nutrition status and biomass at panicle initiation and heading stage of rice. Korean J Crop Sic, 2002, 47: 390~394

[2] 唐延林, 王人潮, 张金恒等. 高光谱与叶绿素计快速测定大麦氮素营养状况研究. 麦类作物学报, 2003, 23 (1): 63~66

[3] 杨暹, 陈晓燕, 冯红贤. 氮营养对菜心炭疽病抗性生理的影响. 华南农业大学学报, 2004, 25 (2): 26~30

[4] 王强, 姜丽娜, 符建荣. 氮素形态、用量及施用时期对小青菜产量和硝酸盐含量的影响. 植物营养与肥料学报, 2008, 14 (1): 126~131

[5] 唐延林, 王人潮, 张金恒等. 高光谱与叶绿素计快速测定大麦氮素营养状况研究. 麦类作物学报, 2003, 23 (1): 63~66

[6] 姜丽芬, 石福臣, 王化田等. 叶绿素计 SPAD2502 在林业上应用. 生态学杂志, 2005, 24 (12): 1 543~1 548

[7] 张艳玲, 宋述尧. 氮素营养对番茄生长发育及产量的影响. 北方园艺, 2008 (2): 25~26

[8] Balasubramanian V, Morales A C, Cruz R T. On-farm adaptation of knowledge-intensive nitrogen management technologies for rice systems. Nutr Cycling Agroecosyst, 1999, 53 (1): 59~69

[9] 郭建华, 赵春江, 王秀等. 作物氮素营养诊断方法的研究现状及进展. 中国土壤与肥料, 2008 (4): 10~14

[10] 李志宏, 刘宏斌, 张福锁. 应用叶绿素仪诊断冬小麦氮营养状况的研究. 植物营养与肥料学报, 2003, 9 (4): 401~405

[11] 雷泽湘, 艾天成, 李方敏等. 草莓叶片叶绿素含量、含氮量与 SPAD 值间的关系. 湖北农学院学报, 2001, 21 (2): 138~140

Influences of Air Inflated Film Covering and CO_2 Enrichment on Greenhouse Microclimate, Growth and Yield of Tomato (*Lycopersicon esculentum* Mill)

Min Wei[1,2], Yuki Tanoue[1], Toru Maruo[1]**, Masaaki Hohjo[1] and Yutaka Shinohara[1]

(1. Graduate School of Horticulture, Chiba University, 648 Matsudo, Chiba 271~8 510, Japan
2. College of Horticultural Science and Engineering/State Key Laboratory of Crop Biology, Shandong Agricultural University, Taian, Shandong 271018, P.R.China)

Introduction

Accompanying the world crisis of energy resources and increase in emission of greenhouse gases, energy-saving, ecological and sustainable greenhouse cultivation is becoming important issue. Many efforts have been made to reduce the energy cost, including the improvement of greenhouse structures, materials, heating methods and temperature control strategies, etc. Air inflated film covering has been used for greenhouses and extended into some countries. However, there have been few detailed studies on the environmental characteristics, energy saving potential of air inflated film greenhouse, as well as its influences on growth, yield and quality of vegetable crops, especially in east Asian area. The environmental factors in greenhouses and effects of air inflated film covering and CO_2 enrichment on tomato growth and production were investigated in our experiments.

Materials and Methods

· Experiment period: October 2006 to June 2007, in the experimental field of the Faculty of Horticulture of Chiba University (Matsudo city, Chiba, Japan)

Plant material: Tomato 'Reiyoh', Sakata Seed Corporation

· Plant density: 2400 plant/ 1 000m^2

· Growing conditions: Three greenhouses (16m × 4.5m × 2.4m) were used. Side curtains were set. Temperature set points for ventilation and heating were 25℃ and 16℃ respectively.

Seedlings were transplanted to plastic bag (Containing 20L "Tsuchitaro" mixed with slow-release coated fertilizers 8g/L and dolomitic limestone 10g/L).

· Treatments: Single film (0.15mm, MKV Platech Co.), CO_2 1 000μl/L
Air inflated (0.15mm, MKV Platech Co.), CO_2 1 000μl/L
Air inflated (0.15mm, MKV Platech Co.), CO_2 3 000μl/L

· CO_2 enrichment: Around 4hrs in the morning, 2hrs in the afternoon.

· Irrigation: Drip irrigated with water in accordance with accumulated solar radiation

($1MJ/m^2$), 50 ml/plant at early growth stage, and 100 ml/plant at late stage.

Results and Discussion

- Compared with Control, the cumulative solar radiation in air inflated film greenhouses decreased by 17.2%~29.9%, the average relative humidity was 6.1%~7.8% higher.
- To maintain same temperature, 38.4%~41.8% oil saving could be achieved in air inflated film greenhouses (11, Jan. ~19, Apr.), Showing better heat insulating property (Figure 1).
- The growth of tomato in air inflated film greenhouse could be promoted. However, the marketable yield decreased by 33.1% owing to the reduced fruit number caused by weak radiation, higher humidity and serious diseases.
- CO_2 enrichment at 3 000μl/L could compensate the yield loss to a certain degree, despite the delayed harvesting date and declined fruit soluble solid content (Table. 1).

Conclusion

Air inflated plastic film greenhouses are of better heat insulating property and obvious energy-saving effectiveness. However, measures must be taken in practice to prevent growth and yield loss caused by weak radiation, higher air humidity and consequent diseases.

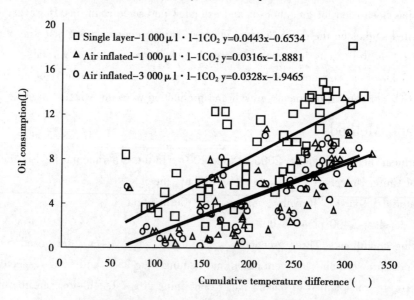

Figure 1 Relationships between daily cumulative temperature differences and oil consumption in single-layered and air inflated film greenhouses (19th, Feb-19th, Apr)

Daily cumulative temperature difference = $\sum$ (Txin-Txout), where Tin and Tout were represented as hourly average air temperature in greenhouse and open field, respectively.

Table 1 Growth, yield and fruit quality of tomato in single film and air inflated film greenhouses

Treatments	Plant height (cm)	Days to harvesting	Fruit number per plant	Marketable yield (g/plant)	Fruit quality (Brix%)
Single film, 1 000 μl/L CO_2	156.9b	84.5ab	19.5a	4 637.0a	5.4b
Air inflated, 1 000 μl/L CO_2	162.6a	82.4b	13.0b	3 101.0b	6.4a
Air inflated, 3 000 μl/L CO_2	159.7ab	86.9a	19.5a	4 295.3a	5.3b

影响日光温室 CO_2 浓度变化的因素与增施效果研究[*]

马 俊[1][**]，贺超兴[1][***]，闫 妍[1]，张志斌[1]，尹宏峰[2]

(1. 中国农业科学院蔬菜花卉研究所，北京 100081；
2. 中国人民解放军总参通信部阳坊生活供应服务中心，北京 102205)

摘要：研究了不同日光温室中 CO_2 的变化规律和增施效果，结果表明天气状况，蔬菜个体大小、生育期和栽培方式均可对日光温室中 CO_2 浓度的日变化规律产生影响。表现为阴天植物光合强度较弱，CO_2 下降缓慢；一天中温室中 CO_2 浓度变化均表现为昼间下降夜间升高的变化趋势，在下午 13：30 左右降至最低值；番茄苗期温室在一天内的大部分时间都处于 CO_2 亏缺状态，而温室番茄在开花期和结果期其 CO_2 浓度均表现为上午较大幅度下降和晚上缓慢上升的变化趋势。晴天上午施用 CO_2 气肥可明显增加温室 CO_2 浓度，从而使植物光合作用处于最佳 CO_2 浓度，相应地增施 CO_2 增加了温室黄瓜植株的叶绿素含量，根系活力，并对促进果实产量增加和果实品质改善都有明显作用。

关键词：日光温室；CO_2 变化规律；影响因子；CO_2 亏缺；增施 CO_2；增施效果

The Influence Factors to CO_2 Concentration and Increased CO_2 Application Study in Chinese Solar Greenhouse

Ma Jun[1], He Chaoxing[1], Yan Yan[1], Zhang Zhibin[1], Yin Hongfeng[2]

(*Institute of Vegetables and Flowers, Chinese Academy of Agriculture sciences, Beijing 100081, P. R. China*)

Abstract: The CO_2 concentration changing regularity and increased application effect were studied. The results showed that the weather, the varieties of vegetable, the development stage of vegetable and cultivated substrate has different effects to CO_2 concentration in Chinese solar greenhouse. The plants may have weak photosynthesis in a cloudy day. In general close solar greenhouse, the CO_2 concentrations will reach the lowest value at 13：30pm which present the trend as decrease in daytime and rise at night. In most time of a day, the tomatoes seedling are deficient in CO_2. Application CO_2 in solar greenhouse will increase CO_2 concentration which make the plants in better CO_2 concentration during the photosynthesis time, which will increase the chlorophyll content, root vigor, yield and quality of cucumber.

Keys word: Chinese solar greenhouse; Carbon dioxide changing regularity; Influence factors; CO_2 deficiency; Increased CO_2 application; Effects of application

CO_2 是光合作用的主要原料，是影响日光温室蔬菜产量的重要因素，由于温室的相对密

---

[*] 基金项目：国家科技支撑计划项目 (2011BAD12B03) (2009BADA4B04)。
[**] 作者简介：马俊 (1986—)，女，硕士研究生，主要从事设施蔬菜栽培生理研究。E-mail: yongyuanxinxiangeini@126.com。
[***] 通讯作者：贺超兴，博士，研究员，主要从事设施园艺有机栽培和蔬菜栽培生理等研究。
E-mail: hechaoxing@126.com　Tel: 010-82109588。

闭性，一方面造成温室 CO_2 亏缺，另一方面也为 CO_2 施肥提供了可能[1]。国外从 20 世纪 20 年代就开始了有关 CO_2 施肥的研究，并在生产中取得了显著的效果，其中日本、荷兰等国的发展较快[2]。国外对温室作物全天增施 CO_2 的研究表明延长增施 CO_2 时间可明显提高黄瓜、番茄产量。通过对 CO_2 施肥的机理、效果、技术、温室 CO_2 浓度分布的时空规律的研究 70 年代以来，国外在设施栽培的 CO_2 施肥方面达到研究和应用高潮，挪威有 75%、荷兰有 65% 的温室施用 CO_2，其他如丹麦、日本、英国、美国等在温室中施用 CO_2 也相当普遍。我国从 20 世纪 80 年代开始进行 CO_2 施肥的试验，在北方日光温室对蔬菜包括黄瓜、番茄等进行了推广应用，研究明确了日光温室内 CO_2 的变化规律，取得了很好的经济效益[3]。在日光温室中施用 CO_2 的方法上也开发出了诸如钢瓶法，酸反应法及燃烧过滤施用法、增施有机肥法和生物生态法等多种对日光温室蔬菜增施 CO_2 的技术[4,5]。但是由于成本较高或施用不方便，因此在我国大部分日光温室栽培中，对 CO_2 施肥重要性的认识还不足，很多人认为增施有机肥就能提高温室 CO_2 浓度满足蔬菜生长需要。为了更好地明确日光温室的 CO_2 变化规律和不同因子的影响，本研究分析了不同日光温室的 CO_2 浓度变化规律，并且对黄瓜施用 CO_2 的效果进行了分析研究，以期明确日光温室施用 CO_2 的必要性，对温室风口管理和增施提供科学依据。

1 材料与方法

试验于 2010 年秋冬季在北京中国农业科学院蔬菜花卉研究所试验农场的日光温室进行。供试日光温室 10 个，分别种植不同的作物，采用加温或非加温方法进行生产，在研究温室 CO_2 浓度变化规律的基础上，对日光温室黄瓜（Cucumissativus L.）座果期增施 CO_2 处理，比较其在与不施用 CO_2 相比研究了增施 CO_2 对黄瓜生理指标、产量及品质的影响。黄瓜品种为中国农业科学院蔬菜花卉研究所的中农 27，于 8 月 27 日播种，9 月 22 日定植，11 月 15 日始收。将日光温室用玻璃隔分为东中西 3 段，黄瓜在东段与中段种植，CO_2 处理（东段）和对照（中格）分别种植 6 畦，双行种植，每行 14 株，行株距为 60cm × 30cm。土壤理化性质状况为，pH 值 6.29；容重：0.56g/cm^2；土壤孔隙度：76.8%；有机质含量：1.51%；速效氮：0.33 mg/g；速效磷：0.34mg/g；速效钾：0.37 mg/g。施用 CO_2 时间为果实膨大期的 11 月 9 日，在晴天上午 9：00 左右施用，每周 3~4 次。

增施 CO_2 试验采用 CO_2 钢瓶进行；用美国产 ZT-7000 型手持便携式 CO_2 测定仪对温室 CO_2 进行监测，日光温室 CO_2 浓度动态变化和温湿度环境数据观测用旗硕"农用通"农业环境远程无线监控管理平台测定；植株光合速率采用 LI-6400 光合测定仪测定。

试验调查研究在温室中按时间进行，测定 CO_2 浓度的部位为植物冠层，即种植带中部，距温室墙体 4~5m，同时测定温度和湿度。不同日光温室的 CO_2 调查试验选用 ZT-7000 型手持便携式 CO_2 测定仪在晴天（阴天）早上揭被后 8：30 到放被 16：00 之间每隔 1h 测定 1 次，以比较 CO_2 随时间变化测定环境 CO_2 浓度。增施 CO_2 通过钢瓶和有孔塑料管在晴天上午 9：00 左右将气体直接施放于植物顶部，并通过测定温室 CO_2 浓度来确定开关释放时间。

测定处理前后叶片的叶绿素含量（丙酮法），根系活力（TTC 法）等指标，用美国 LI-6400 型光合仪测量净光合速率、胞间 CO_2 浓度、气孔导度等。测定的品质指标为：可溶性蛋白（考马斯亮蓝 G-250 染色法）；Vc（钼蓝比色法）；可溶性固形物（手持测糖仪）；硝酸盐（水杨酸法）；可溶性糖含量（蒽酮乙酸乙酯比色法）。

2 结果与分析

2.1 影响日光温室 CO_2 浓度变化的主要因子

2.1.1 光照强度对日光温室 CO_2 浓度变化的影响

通过对黄瓜栽培的日光温室 CO_2 浓度日变化分析可以看出，黄瓜温室其 CO_2 浓度明显在晚上呈升高趋势，究其原因与植物和土壤的呼吸作用有关，而白天由于光照引起的光合作用影响，温室 CO_2 迅速下降，晴天光合起点早于阴天，CO_2 下降速度也快，表明强光下植物的光合能力很强，另一方面，在阴天由于光合作用减弱，CO_2 浓度下降较晚，且降幅不大致使 CO_2 浓度较高。

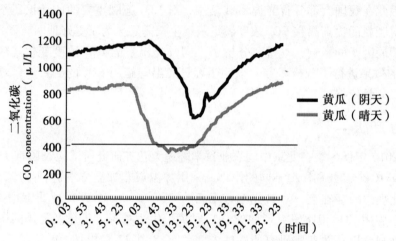

图 1 日光温室结果期黄瓜晴阴天 CO_2 浓度日变化曲线

从图 1 可以看出，日光温室 CO_2 浓度日变化曲线呈不规则"U"形，在下午 13：30 左右达最小值，有时呈不规则"W"形。晴天中温室中 CO_2 浓度变化幅度明显高于对照（阴天），并且晴天 CO_2 浓度最高值明显低于对照（阴天）。

2.1.2 日光温室不同蔬菜采收期的 CO_2 浓度日变化

番茄、甜椒和黄瓜等蔬菜在座果期的 CO_2 日变化规律如图（图2）所示，从中可见不同蔬菜的光合特性有所差别，其中黄瓜、甜椒的初始光合速率较快，而 CO_2 吸收以甜椒最多，表明甜椒的 CO_2 补偿点较低。光合同化强度以番茄最快，与其叶片多，叶面积较大和 CO_2 光合速率较高有关，甜椒的 CO_2 利用率最高。从图中还可看出揭帘被后，温室中 CO_2 浓度迅速下降，10 点半后的温室 CO_2 明显低于室外对照且相对走平线，在中午很长时间内处于 CO_2 亏缺状态，说明温室的 CO_2 饥饿引起了光合停滞，因此在日光温室适当增施 CO_2 对于提高设施蔬菜的产量品质具有很好应用前景。

对叶菜类蔬菜的日 CO_2 浓度变化曲线结果见表（表1），由于叶菜类蔬菜为降低温度，温室风口大开，造成室内外 CO_2 趋同，但成熟期的羽衣甘蓝光合吸收要强于大白菜。

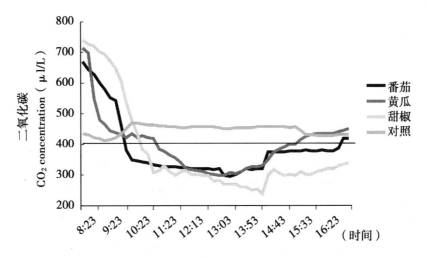

图2 果菜类不同作物日光温室 CO_2 浓度日变化曲线

表1 日光温室不同叶菜的 CO_2 浓度日变化（μl/L）

时间	蔬菜	
	白菜	羽衣甘蓝
08：30	439	422
09：00	399	415
09：30	436	397
10：00	449	393
10：30	430	392
11：00	422	383
11：30	422	391
12：00	421	409
13：00	418	417
14：00	396	407
15：00	417	406

注：风口处于开启状态。

2.1.3 日光温室番茄不同生育期的 CO_2 浓度日变化

番茄在苗期、开花期、结果期的 CO_2 浓度日变化如图（图3）所示，从图中可以看出，番茄在开花结果期时。揭开帘被后 CO_2 浓度迅速降低，并且对温室内 CO_2 消耗量较大，使得温室中 CO_2 浓度较苗期有很大幅度的变化。苗期的 CO_2 浓度始终在 400μl/L 左右，并且在很长的一段时间处于 CO_2 亏缺状态。结果期番茄在 9：20 时，温室中的 CO_2 浓度迅速下降，并且在 10：30 达到最小值，在 10 点半后的温室 CO_2 明显低于室外对照且相对走平线。相比而言结果期的 CO_2 浓度迅速下降速度缓慢一些，在 13：00 达到最小值。现象说明，温室的 CO_2 饥饿引起了光合停滞，在结果期更为严重。所以在开花结果期适当增施 CO_2 对于提高设施蔬菜的产量品质是很有必要的。

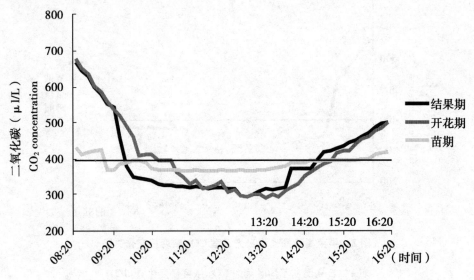

图3 日光温室番茄不同生育期 CO_2 浓度日变化曲线

2.1.4 日光温室黄瓜不同基质的 CO_2 浓度日变化

黄瓜在不同基质中的日光温室 CO_2 浓度日变化情况如图（图4），可以看出在揭开帘被之后，有机土栽培基质中日光温室的 CO_2 明显高于普通土日光温室 CO_2 浓度，在有机土基质种植的黄瓜所处 CO_2 浓度下降缓慢，到14：00达最小值。而揭开帘被后，普通土基质的温室 CO_2 浓度迅速下降到11：00达最小值，并在10：00到16：30期间处于 CO_2 亏缺状态。说明有机土中富含微生物群落，在不断的进行呼吸作用，所以有机土对温室 CO_2 有一种 CO_2 补充的作用，有机土较普通土更能有效提高温室中 CO_2 浓度。

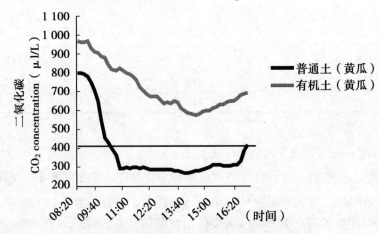

图4 不同基质的日光温室 CO_2 浓度日变化曲线

2.2 日光温室增施 CO_2 对黄瓜生长和叶片生理指标的影响

2.2.1 施用 CO_2 对温室 CO_2 浓度的影响

在黄瓜的开花结果期选择晴天上午对栽种黄瓜的温室进行施肥，施肥时间点选在温室 CO_2 浓度日变化曲线中突然下降的位置（上午9：00左右）。用 CO_2 钢瓶施肥，施用3min，温室中 CO_2 浓度从1 000μl/L上升到1 300μl/L，过一段时间 CO_2 下降后再施用1分钟，使温室 CO_2 浓度在14：00之前保持在800μl/L以上（图5）。可见增施 CO_2 对温室的 CO_2 浓度有明显提高作用，从而提高了光合效率和植物对 CO_2 的光合吸收量。

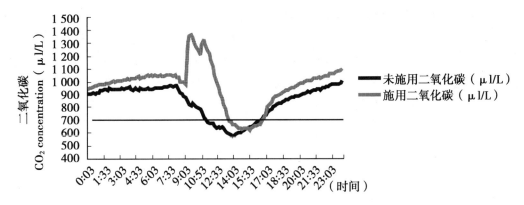

图5 施用 CO_2 与未施用日光温室 CO_2 浓度日变化曲线

2.2.2 增施 CO_2 对日光温室黄瓜叶片叶绿素和根系活力的影响

对增施 CO_2 处理和普通对照的黄瓜叶片叶绿素及根系活力进行的测定结果见表2，从中可见增施 CO_2 提高了黄瓜叶片的叶绿素含量包括Chla，Chlb，Chla+Chlb，降低了Chla/Chlb值。并且植株根活也高于对照。

表2 中农27黄瓜增施 CO_2 后对叶绿素和根系活力的影响

	叶绿素 Chla (mg/g)	叶绿素 Chlb (mg/g)	Chla/Chlb	叶绿素 Chl_{a+b} (mg/g)	根活 [mg/(g·h)]
增施 CO_2 处理	20.49 ±0.021	8.31 ±0.025	2.47 ±0.013	0.50 ±0.045	0.52 ±0.041
对照	18.08 ±0.045	7.10 ±0.033	2.55 ±0.022	0.25 ±0.021	0.48 ±0.011

2.2.3 增施 CO_2 对日光温室黄瓜光合作用的影响

增施 CO_2 处理可以提高黄瓜的净光合速率（表3），而且对叶片气孔导度有明显增加作用，使胞间 CO_2 浓度明显低于对照，而蒸腾速率会明显低于对照。

表3 CO_2 处理对黄瓜光合作用的影响

	CO_2 处理	对照
净光合速率 Photo/ [μmol/ ($m^2 \cdot s$)]	23.7a	20.67b
气孔导度 Cond [mol H_2O/ ($m^2 \cdot s$)]	2.1a	1.92b
胞间 CO_2 浓度 Ci (μmol CO_2/mol)	246.67a	256b
蒸腾速率 Tm [mmol H_2O/ ($m \cdot s$)]	22.37a	25.2b

2.3 温室增施 CO_2 对黄瓜风味品质的影响

经补充 CO_2 处理后的黄瓜品质与普通黄瓜的对比如图（图6）所示，CO_2 处理黄瓜 Vc 含量显著高于对照，粗蛋白和可溶性固形物的含量均大于对照，其中，但是总糖含量处理和对照没有差别。与施用 CO_2 对光合作用的影响相联系起来，光合作用的提高有利于营养物质的积累，所以处理后黄瓜的可溶性固形物高于对照。说明施用 CO_2 对改善黄瓜品质有一定的作用，在日光温室适当增施 CO_2 对于提高设施蔬菜的产量品质具有很好应用前景。

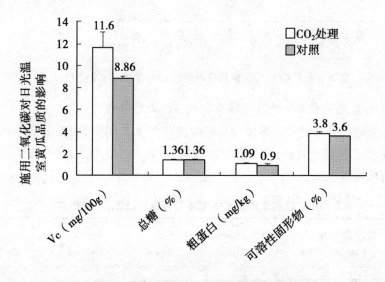

图6 增施 CO_2 对日光温室黄瓜品质的影响

3 讨论

白天光照引起的光合作用可使温室 CO_2 迅速下降，晴天光合起点早于阴天，CO_2 下降速度也快，表明强光下植物的光合能力很强，另一方面，在阴天由于光合作用减弱，CO_2 浓度下降较晚，且 CO_2 浓度较高。阴天中温室中 CO_2 浓度明显高于晴天，而且高于对照。

果菜类蔬菜需要大量 CO_2 以备光合作用需要，相比较，叶菜类蔬菜对温室 CO_2 浓度的影响较小，这是因为叶菜类比果菜类植株高大，根茎叶对环境 CO_2 都有影响，而叶菜类相对体积较小，所以对环境 CO_2 浓度影响较小。幼苗期群体光合较弱、土壤呼吸旺盛，温室 CO_2 浓度较高；结果期群体光合旺盛、土壤呼吸衰竭，CO_2 亏缺严重[5]。所以温室中番茄在开花结果期时,对温室内 CO_2 消耗量较大,在揭开帘被后 CO_2 浓度迅速降低,而苗期对温室 CO_2 浓度也有影响,但是较开花结果期很弱。

有机土中富含微生物群落,在不断的进行呼吸作用,所以有机土对温室 CO_2 有一种 CO_2 补充的作用,在揭开帘被之前与之后,温室中 CO_2 都明显高于普通土温室 CO_2 浓度。有机土对温室 CO_2 有一种 CO_2 补充的作用,大气 CO_2 体积分数升高对植物光合色素质量分数的影响目前尚无定论。以往很多研究表明,经高体积分数 CO_2 处理后[5],植物叶片中的叶绿素质量分数下降,Chla 和 Chlb 值下降或不变[6]。本实验表明 CO_2 施肥能增加叶绿素总含量,降低叶绿素 a 和叶绿素 b 值,更好地利用散射光[6]。植物进行光合作用的能量来源主要是光合色素捕获的光能,所以叶绿素的高低与光合功能关系密切。而且可以进一步提高光能利用率以提高光合作用。在一定范围内,CO_2 的浓度越高,植物的光合作用也越强,这样便导致地下根部要吸收营养以供地上部分生长,所以 CO_2 处理的黄瓜植株根系活力较强。初瓜期至盛瓜初期是营养生长与生殖生长并进阶段,两者的积累量都较大,光合能力增强后,经过 CO_2 处理的植株在结果期自然会较对照单果重大[7]。光合作用的提高有利于营养物质的积累,所以处理后黄瓜的品质比对照的稍好些。

日光温室中 CO_2 亏缺是不可避免的,即使使用的有机土,也只能起到微小的作用,传统的栽种模式是不能起到很好的作用的,放风等操作虽然能降低温室湿度,但同时也放走了光合作用的原料——CO_2。所以迫切的需要 CO_2 施肥等一系列技术来增加温室中 CO_2 浓度,以提高作物产量及品质。

参考文献

[1] 贺超兴,张志斌,王怀松,王耀林. 蔬菜设施栽培 CO_2 施肥技术,中国蔬菜. 2000(4):51~52
[2] 高素华,郭建平. 施用 CO_2 的现状及前景. 气象科技. 1995(1):59~64
[3] 魏珉,邢禹贤,王秀峰,马红. 日光温室 CO_2 浓度变化规律研究. 应用生态学报. 2003,14(3):354~358
[4] 贺超兴,张志斌,王怀松,王耀林. 蔬菜设施栽培 CO_2 施肥技术,中国蔬菜,2000(5):51~53
[5] 魏珉,王秀峰,邢禹贤,张衍鹏,王纪银. 几种主要 CO_2 施肥肥源性能的比较与评价. 农业工程学报. 2001,17(3):10~14
[6] 高阳俊,张乃明,张玉娟. 设施栽培增施 CO_2 对西芹影响研究. 农业现代化研究. 2004,25(4):317~320
[7] 邱莉萍,刘军,王益权,等. 土壤酶活性与土壤肥力的关系研究. 植物营养与肥料学报. 2004,10(3):277~280
[8] 王修兰,徐师华,梁红. CO_2 浓度增加对 C3、C4 作物生育和产量影响的实验研究. 中国农业科学. 1998,31(1):0
[9] 卢育华,申玉梅. 冬季日光温室黄瓜光合作用. 山东农业大学学报:自然科学版. 1996,27(1):39~43
[10] 陈双臣,邹志荣,贺超兴,张志斌. 温室有机土栽培 CO_2 浓度变化规律及增施 CO_2 对番茄生长发育的影响. 西北植物学报,2004,24(9):1 624~1 629

Exogenous Polyamines Enhance Cucumber (*Cucumis sativus* L.) Resistance to NO_3^- Stress, and Affect the Nitrogen Metabolism and Polyamines Contents in Leaves of Cucumber Seedlings

Wang xiuhong[1], Yang fengjuan[1,2], Wei Min[1,2], Shi QingHua[1,2], Wang xiufeng[1,2]*

(1. College of Horticulture Science and Engineering, Shandong Agricultural University, Tai'an, 271018, P. R. China;
2. State key laboratory of Crop Biology, Tai'an 271018, Shandong, P. R. China)

Purpose

Nitrogen fertilizer plays an important role in crop yield and quality. To satisfy the nitrogen demand, farmers use very large amounts of nitrogen fertilizers to obtain maximum yields. These high fertilizer inputs lead to marked deterioration in soil and cause secondary salinization in Chinese greenhouse. The excessively accumulated salinity in soil of the greenhouse were mainly composed of anion NO_3^-, accounting for 67% to 76% of total anions and limited the growth of plants significantly. Polyamines are low-molecular-weight aliphatic amines that are not only involved in the regulation of plant developmental and physiological processes, but also play important roles in modulating the defense response of plants to diverse environmental stresses, especially salt stress. But most of salt stresses regarded NaCl as means of research, while the effects of polyamines on plant growth under excess NO_3^- was seldomly investigated.

Materials and Methods

The cucumber (*Cucumis. sativus* L. cv. Xintaimici) is tolerant to salinity stress. The germinated seeds were sowed in plastic plugs (50holes) filled with nursery substrate (peat: vermiculite: perlite = 2 : 1 : 1). When there was one completely expanded true leaf, the plants were transplanted to hydroponic boxes (40cm × 30cm × 12cm, 6 plants per box) containing a complete cucumber nutrient solution [Ca(NO_3)$_2$·4H_2O 3.5mmol/L, KNO_3 7mmol/L, KH_2PO_4 1mmol/L, $MgSO_4$·7H_2O 2mmol/L, H_3BO_3 46.3 μmol/L, $MnSO_4$·H_2O 10 μmol/L, $ZnSO_4$·7H_2O 1.0 μmol/L, $(NH_4)_6Mo_7O_2$·4H_2O 0.3 μmol/L, $CuSO_4$·5H_2O 0.75 μmol/L, EDTA-FeNa 100 μmol/L] with continuous aeration by a electric pump.

When seedlings had three fully expanded true leaves, they were subjected to five different treatments. The treatments are as follows: (1) CK: complete nutrient solution (14mmol/L NO_3^-), (2) N: 140mmol/L NO_3^- treatment, (3) Spm: sprayed with 1mmol/L Spm under 140mmol/L NO_3^- treatment, (4) Spd: sprayed with 1mmol/L Spd under 140mmol/L NO_3^- treatment, (5) Put:

sprayed with 1mmol/L Put under 140mmol/L NO_3^- treatment. KNO_3 and $Ca(NO_3)_2 \cdot 4H_2O$ provided the same mol of NO_3^- in all treatments. The polyamines spray once a day at 4:00 pm.

After 7 days of treatment, nitrogen metabolism parameters and polyamines content were assayed. Free polyamines (Spm, Spd and Put) were estimated according to high performance liquid chromatography (HPLC) technique.

Results and Discussion

In this study, cucumber seedlings were cultivated in nutrient solution with foliar spraying different kinds of polyamines (PAs) (Spermine, Spermidine and Putrescine). The effects of exogenous PAs on growth of cucumber seedlings, the content of $NO_3^- - N$, $NH_4^+ - N$, proline, soluble protein and polyamines and the activities of nitrogen metabolism in cucumber leaves under 140mmol/L NO_3^- stress were investigated. The results showed that the accumulation of $NO_3^- - N$, $NH_4^+ - N$, proline and soluble protein in leaves were increased, the increment of plant height and leaf area were decreased, the avtivities of nitrogen reductase (NR), glutamate synthase (GOGAT) and glutamate dehydrogenase (GDH) were significantly decreased for 7 days under 140mmol/L NO_3^- treatment. Foliar spraying with 1mmol/L Spd, Spm or Put under 140mmol/L NO_3^- treatment, the content of Put and Spd in leaves increased significantly, while the accumulatiom of Spm only increased under Spm treatment. The activities of NR, GS, GOGAT and GDH are siginificantly increased, and the content of proline and soluble protein in leaves significantly enhanced, while the accumulation of $NO_3^- - N$ and $NH_4^+ - N$ are significantly decreased. These findings suggest 1mmol/L exogenous Spm, Spd and Put could enhance the capacity of nitrogen metabolism, promote the growth and increase resistance to high concentration of NO_3^- stress. But the Spd effect is better than Spm and Put.

Table 1 Effects of exogenous polyamines on cucumber seedlings growth potential under NO_3^- stress
Data are means ± SD of three replicates. Mean values followed by different letters (a ~ d) are significantly different ($P < 0.05$). The same as below

Treatment	Plant height (cm)			Leaf area (cm²)		
	0d	7d	Increment	0d	7d	Increment
CK	9.50 ± 1.00a	46.12 ± 1.23a	36.62 ± 0.96a	208 ± 11.77a	709 ± 15.44a	501 ± 3.93a
N	8.44 ± 0.89a	20.40 ± 0.67d	11.96 ± 0.98d	200 ± 10.52a	418 ± 21.61b	219 ± 11.83c
Spm	8.36 ± 0.68a	22.92 ± 1.16bc	14.56 ± 1.09bc	199 ± 15.06a	456 ± 30.53b	257 ± 15.56b
Spd	8.94 ± 0.35a	24.02 ± 1.15b	15.08 ± 0.92b	200 ± 23.02a	456 ± 30.59b	256 ± 16.65b
Put	8.50 ± 0.81a	22.1 ± 1.00c	13.6 ± 0.68c	189 ± 10.49a	427 ± 16.76b	238 ± 23.81bc

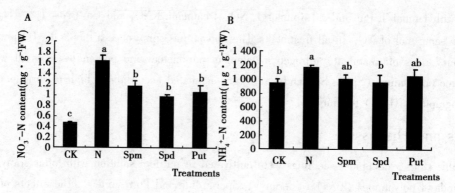

Figure 1 Effects of exogenous polyamines on content of $NO_3^- - N$ (A) and $NH_4^+ - N$ (B) in leaves of cucumber seedlings under NO_3^- stress

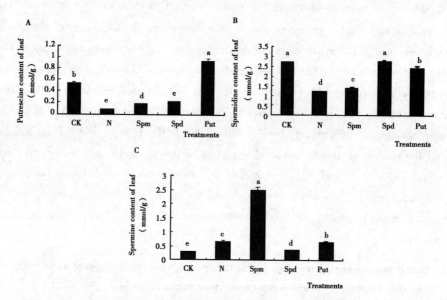

Figure 2 Effects of exogenous polyamines on Put (A), Spd (B) and Spm (C) content in leaves of cucumber seedlings under NO_3^- stress

微生物有机肥在辣椒育苗中的应用效果研究

杨伟国*, 孙光闻**, 刘厚诚, 宋世威

(华南农业大学园艺学院, 广州 510642)

摘要: 以辣椒为试验材料,采用珍珠岩、椰壳粉(1∶1)为育苗基质。研究微生物有机肥不同施用量对辣椒育苗的效果,并将微生物有机肥与复合肥(N、P、K 15-15-15)的育苗效果进行了比较。结果表明,与对照相比,T_6(复合肥)处理辣椒幼苗的各项生态指标表现最好。微生物有机肥的5个处理中,T_4处理(50g/L基质)的施用量效果最佳。

关键词: 微生物有机肥; 辣椒; 育苗

Effects of Microbial Organic Fertilizer on Pepper Seedlings

Yang Weiguo, Sun Guangwen*, Liu Houcheng, Song Shiwei

(*College of Horticulture, South China Agricultural University, Guangzhou 510642, P. R. China*)

Abstract: The purpose of this experiment is to study the effects of microbial organic fertilizer on pepper seedlings. The substrate was perlite and coconut coir (1∶1 in volume). The microbial organic fertilizer was added with six levels: CK (without microbial organic fertilizer added), T_1-T_5 (with microbial organic fertilizer 5, 10, 20, 50, 100g/L added, respectively) and T_6 2.5g/L compound fertilizer (N、P、K15-15-15) added. The results showed that T_6 treatment 2.5g/L (N、P、K15-15-15) compound fertilizer was superior to all other treatments for increasing the growth of the pepper seedlings. Among five treatments application with microbial organic fertilizer, T_4 treatment (50g/L) is the best treatment, as compared with the control treatment.

Key words: Microbial organic fertilizer; Compound fertilizer; Pepper seedling

  生物有机肥是传统有机肥的继承和发展,是在科学地研究了有机肥的作用与机理后,把有益的微生物与农作物生长、开花、结实所需的必要元素科学组合而形成的一种新型复合微生物肥[1]。施用微生物有机肥可以促进蔬菜增产、改善品质[2,3],增加土壤有机质含量,降低温室土壤盐分积累[4]。辣椒育苗时间相对较长,利用基质进行育苗时有必要添加一些肥料,以促进幼苗生长。本试验在辣椒育苗基质中添加不同量的微生物有机肥,并对微生物有机肥和传统的复合肥育苗效果进行比较,旨在探讨微生物有机肥在辣椒育苗上的应用效果,为微生物有机肥在蔬菜育苗上的应用提供参考。

---

* 第一作者,杨伟国,男,云南人。华南农业大学园艺学院设施系大四学生。
** 通讯作者,孙光闻,女,吉林人,华南农业大学园艺学院,副教授。E-mail: sungw1968@scau.edu.cn。

1 材料与方法

1.1 试验材料

试验以"牛角椒"辣椒品种为试材,供试肥料复合肥为 N、P、K 复合肥(15-15-15),微生物有机肥为广东金饭碗有机农业发展有限公司生产的"金饭碗微生物有机肥"。金饭碗微生物有机肥为固体肥,为粉末状,总养分(氮、磷、钾)≥4%,有效活菌:≥0.20 亿个/g,适用于多种作物。

1.2 试验设计

本试验在华南农业大学蔬菜实验基地塑料大棚内进行,采用复合基质(珍珠岩:椰壳粉 =1:1)和容积为 0.7L 的黑色塑料育苗袋的无土育苗方式。随机区组设计,实验设 6 个处理和 1 个不施肥对照,每个处理 10 株,6 个处理分别为生物有机肥的不同施用量和施用三元复合肥 2.5g/L,将肥料与基持充分混匀后装袋备用。具体处理情况见表1。

表1 各处理的肥料施用量

处理	微生物有机肥（g/L）	复合肥（g/L）
CK	—	—
T_1	5.0	—
T_2	10.0	—
T_3	20.0	—
T_4	50.0	—
T_5	100.0	—
T_6	—	2.5

2010 年 4 月 22 日播种。具体方法是先将辣椒种子温汤浸种:常温水浸泡 15min 后换 55~60℃温水浸泡 15min,并不断搅拌,让水自然冷却再浸泡数小时;再置到人工气候室(宁波江南仪器厂,RXZ 智能型)催芽:催芽温度为 29℃,催芽 5d,90% 以上辣椒种子冒出白芽。播种前一天浇湿基质,每袋基质播一粒,尽量挑选出芽程度相近的种子播种,并覆盖相应基质 0.5~1.0cm 厚。播种后浇湿基质,控制浇水量,浇湿基质即可,避免浇水太多冲走养分。5 月 23 日结束育苗,对幼苗进行各项指标的测定。

1.3 测定项目与方法

株高测定从袋内基质表面开始,至植物的生长点,用直尺测量;茎粗在袋内基质表面上部用游标卡尺进行测量。壮苗指数 = [茎粗(mm)/株高(cm)] × 单株干重(g)。叶片数从子叶到顶端展开的叶片的数目;叶绿素含量采用 SPAD-502 叶绿素仪进行测定。地上部分、根部鲜重采用精度为 0.1g 的电子天平称取,地上部分、根部干重采用精度为 0.0001g 的电子天平称取。干质量用烘干法,105℃杀青 15min,75℃烘干至恒重。试验数据采用 SPSS 和 EXCEL 软件进行分。

2 结果与分析

2.1 不同处理对辣椒叶片数和叶绿素含量的影响

表2 不同处理对辣椒叶片数和叶绿素含量的影响

处理	叶片数	叶绿素 SPAD 值
CK	8.0 ± 0.0c	30.8 ± 0.3b
T_1	6.3 ± 0.2d	27.0 ± 0.1d
T_2	6.7 ± 0.2d	25.9 ± 0.2d
T_3	9.3 ± 0.2b	30.6 ± 0.0b
T_4	10.0 ± 0.0b	30.7 ± 0.2b
T_5	6.3 ± 0.2d	28.3 ± 0.3c
T_6	11.0 ± 0.0a	34.2 ± 0.3a

由表2可知，各处理中，复合肥处理 T_6 的辣椒叶片数最多，和其他处理差异均显著。$T_1 \sim T_5$ 各生物有机肥处理中，T_4 处理的叶片数最多，T_3 次之，二者与 T_1、T_2、T_5、CK 处理的差异均达到显著水平。各处理中，复合肥处理 T_6 的辣椒叶片叶绿素含量最高，比CK增加11.3%，与其他各处理差异均达到显著水平。$T_1 \sim T_5$ 各生物有机肥处理中，处理 T_4 的叶绿素含量最高，略低于对照，但与对照相比没有达到显著差异。

2.2 不同处理对辣椒茎粗、株高及壮苗指数的影响

表3 不同处理对辣椒茎粗、株高及壮苗指数的影响

处理	茎粗（mm）	株高（cm）	壮苗指数
CK	2.77 ± 0.11b	11.2 ± 0.3c	0.049b
T_1	2.18 ± 0.04c	10.4 ± 0.2c	0.025c
T_2	2.22 ± 0.02c	10.9 ± 0.2c	0.027c
T_3	2.82 ± 0.01b	13.2 ± 0.1b	0.055b
T_4	2.86 ± 0.04b	12.6 ± 0.2b	0.058b
T_5	2.15 ± 0.03c	8.7 ± 0.2d	0.028c
T_6	3.68 ± 0.09a	15.0 ± 0.3a	0.122a

从表3的数据可知，T_6 处理的辣椒茎粗、株高在各处理中最大，比CK分别增加32.8%、33.5%，与其他处理差异均达到显著水平。$T_1 \sim T_5$ 各处理中，T_3、T_4 的茎粗略高于对照，但没有达到差异显著水平。处理 T_3、T_4 处理的株高显著高于对照。其余各处理的株高和茎粗均低于对照。各处理中，壮苗指数最大的是 T_6 处理，与其他处理差异均达到显著水平。$T_1 \sim T_5$ 处理中，壮苗指数最大的 T_4 处理，次之的为 T_3 处理，二者与其余差异均达

到显著水平,但与 CK 差异不显著。
2.3 不同处理对辣椒干重及鲜重的影响

表 4 不同处理对辣椒干重及鲜重的影响

处理	地上部鲜重(g/株)	根鲜重(g/株)	地上部干重(g/株)	根干重(g/株)
CK	1.9±0.1b	0.3±0.0c	0.1590cd	0.0292bc
T_1	1.0±0.0c	0.3±0.0c	0.0952e	0.0252cd
T_2	1.1±0.0c	0.3±0.0bc	0.1041de	0.0310bc
T_3	2.2±0.0b	0.5±0.1b	0.2078bc	0.0440b
T_4	2.6±0.1b	0.4±0.0bc	0.2221b	0.0310bc
T_5	1.2±0.0c	0.2±0.0c	0.0985e	0.0141d
T_6	5.0±0.3a	1.2±0.1a	0.4162a	0.0848a

由表 4 数据可知,各处理中,T_6 处理的地上部分鲜重、根部鲜重的值最大,分别比 CK 增加 160.3%、300.0%,与其他处理差异均达到显著水平。处理 $T_1 \sim T_5$ 中,处理 T_4 的地上部分鲜重高于其余处理,与 T_1、T_2、T_5 处理差异显著,与 T_3、CK 差异不显著;处理 T_3 的根部鲜重,高于其余处理,与 T_1、T_5 及 CK 处理差异显著,与 T_2、T_4 差异不显著。处理 T_6 的地上部分干重、根部干重均为各处理中的最大,比 CK 分别增加 161.7%、190.3%,与其他各处理差异均达到显著水平。处理 $T_1 \sim T_5$ 中,T_4 处理的地上部分干重最大,与对照比差异显著,其余各处理与对照比差异不显著或显著低于对照。T_3 处理的根部干重最大,但与对照比差异不显著,其余各处理与对照比差异不显著或显著低于对照。

3 结论

本试验以微生物有机肥为主要研究对象,并选用空白、复合肥作对照,研究微生物有机肥不同施用量对辣椒育苗的效果,并对微生物有机肥和复合肥的育苗效果进行比较。与不施肥的基质对照相比,在试验的 6 个处理中,以处理 T_6 的叶片数、叶绿素 SPAD 值、茎粗、株高、地上部分鲜重与干重、根部鲜重与干重、壮苗指数最高,与其他处理存在着显著差异。$T_1 \sim T_5$ 处理中,处理 T_4 的叶片数、叶绿素 SPAD 值、茎粗、地上部分鲜重与干重、壮苗指数最好,处理 T_3 的株高、根部鲜重与干重最好,除 T_4、T_3 处理的地上部分鲜重和干重与根部鲜重和干重指标中只与处理 T_1、T_5 差异显著外,其余指标中 T_4、T_3 与 T_1、T_2、T_5 差异均显著。由此可见,施用三元复合肥更有利于辣椒植株生长,培育壮苗;在施用微生物有机肥的处理中,处理 T_3、T_4 的效果最好,可以得出比较适合的微生物施用量范围是 20~50g/L。但综合各项指标,施肥量以 50g/L 为最佳。施肥量过高或过低,效果均不理想。本试验的微生物有机肥在辣椒上的育苗效果不如三元复合肥效果好,可能因为微生物有机肥具有肥效缓释的特点,肥效不如化肥迅速。但总体上施入合适量的生物有机肥育苗效果优于不施肥的对照。

参考文献

[1] 唐小付,龙明华,于文进.生物有机肥在蔬菜生产上的应用现状及展望.广西热带农业,2010,128(3):16~19
[2] 陈克农,张鹏.微生物有机肥在小白菜上应用研究.北方园艺,2001,(3):9~10
[3] 龙明华,于文进,唐小付等.复合微生物肥料在无公害蔬菜栽培上的效应初报.中国蔬菜,2002,(5):4~6
[4] 曹林奎,陆贻通,林玮.生物有机肥料对温室蔬菜硝酸盐和土壤盐分累积的影响.农村生态环境,2001,17(3):45~47

新型超声波雾培装置及其系统设计简报*

程瑞锋[1]，杨其长[1]**，魏灵玲[1]，闻婧[2]

(1. 中国农业科学院环境与可持续发展研究所，北京 100081；
2. 江苏省农业科学院蔬菜研究所，南京 210014)

摘要：雾培作为所有栽培技术中根系的水气矛盾解决的最好的一种栽培模式，是一种重要的无土栽培模式。本文在阐述雾培优势的基础上，简要介绍了一种新型超声波雾培装置的及其系统设计，分系统阐述了新型雾培栽培装置的组成，并对雾培的应用前景进行了分析。

关键词：雾培；超声波；栽培系统设计

A Preliminary Report on the New Ultrasonic Aeroponics Device and its System Design

Cheng Ruifeng[1], Yang Qichang[1]**, Wei Lingling[1], Wen Jing[2]

(1. Institute of Envirnment and Sustainable Develofment in Agricultare, CAAS, Beijing 100081;
2. Institute of Vegetable Crops, Jiangsu Acadcmy of Agricnlture Scieces Nanjing 210014 P. R. China)

Abstract: The earoponics is an important soilless cultivation mode. It is the best cultivation pattern to solve the contradiction between nutrition and oxygen than others. The article introduce a new ultrasonic aeroponics. The composition and characteristic of the aeroponics was stated, and the application of aeroponics was also outlined in the future.

Key words: Aeroponics; Ultrasonic Device; System Design

1 雾培及其优势

植物雾培（即气培，Aeroponics）是把植物的根系全部或部分直接裸露在空气中，定期向根系喷洒营养液（雾），从而满足植物生长的培养方式[1]。植物根系可以从营养液（雾）中摄取需要的水分和养分，从空气中直接获取充足的氧气，是解决植物根系水气矛盾的最佳方式。只要营养液配方正确，环境条件适宜，雾培法可获得高产和优质的产品。雾培法的各个环节较易精准控制，便于实现农业工厂化生产。另外，雾培的优点还体现在以下方面：雾培不用基质，且可采用各种轻质扳材组成栽培槽，节省投资；利于高密度栽培和立体空间栽培；雾培作物产量高，产品质量好；营养液用量少，节水节肥；根系脱离土壤和液层，减免病害发生；雾培系统可作为新的研究工具，十分有利于研究作物根系发育、营养物质吸收运输以及根系环境对植株生长发育的影响[2]。

---

* 基金项目：中央级公益性科研院所基本科研业务费专项（BSRF201109）；科技部国际合作重点项目（2010DFB30550）；农业部948项目（2011-Z7）；北京市科委重点项目（D111100000811001）。
** 通讯作者 Author for correspondence：E-mail：yangq@cjac.org.cn。

目前，在设施园艺产业发达的国家，用雾培生产蔬菜、花卉以及快繁育苗的技术已比较完善，并利用雾培模式开展了多种类的植物根系及其环境的研究[3,4,5]；国内雾培模式应用较晚，只是在部分蔬菜、马铃薯育苗与种薯培育方面有所报道[6,7]。

2 新型超声波雾化栽培装置及其系统设计

目前植物雾培主要是采用压力式雾化和机械高速旋转离心雾化进行根系培养，存在雾化雾滴大、雾化不均匀等问题。利用超声波雾化原理进行水的雾化已在生产和生活中广泛应用，如环境加湿器等，但应用于营养液雾化，进行植物雾化栽培还没得到实质性发展。

为此，中国农业科学院环发所探索了一种新型植物雾培栽培系统（图1）。新型超声波雾化栽培装置核心组成部分为：超声波雾化器、雾化器固定片、超声波雾化栽培槽、种植板、营养液循环系统。其中，超声波雾化器长125mm，宽88mm，高56mm，表面均匀分布10个直径为20mm圆形雾化片，可将营养液进行均匀雾化。雾化器固定片为宽90mm，厚5mm的PVC板，可将超声波雾化器固定其上。超声波雾化栽培槽，由多个宽500mm，高100mm，长500mm的栽培槽组合而成，底面从边缘到中间有5度的下倾角，防止种植槽积水；底部有宽90mm高60mm的矩形凹槽，用于放置超声波雾化器和固定片。种植板，宽500mm，长500mm，嵌合在栽培槽上，板面均匀分布有直径为20mm种植孔，种植板底面中间有长125mm宽88mm高50mm的矩形凸起，用于防止雾化器喷溅水滴。营养液循环系统包括水箱、水泵、供液管、回液管，实现营养液的循环使用。本超声波雾培装置比水耕栽培装置节省用水30%，提高植物根系环境温度5℃，同时可显著提高植物根系活力。该栽培装置可大可小，操作简单方便，利于推广应用，是适合叶菜生产的一种新型超声波雾化栽培装置。

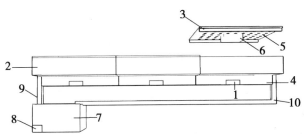

1. 超声波雾化器　2. 雾化器固定片　3. 超声波雾化栽培槽　4. 种植板
5. 圆形雾化片　6. 矩形凹槽　7. 种植孔　8. 矩形凸起　9. 水箱
10. 水泵　11. 供液管　12. 回液管

图1　超声波雾培装置结构示意图

与新型超声波雾培装置相关的其他配套系统构成还包括：

2.1 营养液池

超声波雾培与其他雾培相似，对于规模较大的雾培可用水泥砖砌成较大体积的营养液池，而规模较小的可用大塑料桶来代替。液池的体积取决于栽培用量，以及水泵的供液能力，要保证水泵有一定的供液时间而不至于很快就将池中营养液抽干，至少能满足植物一二天的耗水需要。

2.2 管道

供回液管道主要由 UPVC、PE 等塑料管构成。各级供回液管道的大小应根据选用的喷雾装置，以及喷头所要求工作压力的大小而定（图2、图3）。

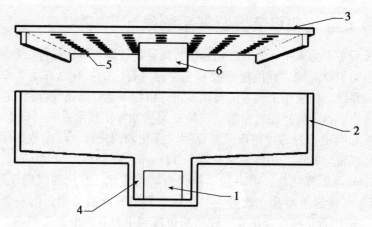

图 2 超声波雾培栽培装置结构示意图

图 3 超声波雾培生菜及其根系部分

2.3 种植槽

雾培的种植槽可以用硬质塑料、泡沫塑料板、木板或水泥混凝土制成，形状可多种样式。种植槽要求能够盛装营养液，并能够将喷雾后多余的营养液回流到营养液池中。种植槽的形状和大小要考虑到植株的根系伸入到槽内之后，有充分的空间将营养液均匀喷到各株根系上。因此，种植槽不能太狭小而使雾状营养液扩散不开，但也不能太宽大，否则不能将营养液雾化到所有根系上[8,9]。

3 雾培系统中的栽培装置

在雾培生产系统中，栽培装置是其主体。目前用于雾培生产的常用栽培装置主要包括："A"型栽培装置、箱式栽培装置、网式栽培装置、床式栽培装置、柱式立体栽培装置等。超声波雾培可借鉴上述装置类型，均可利用。

3.1 "A"型栽培装置

"A"型装置采用双面立体栽培模式，大大提高了植株种植密度，而且装置下部有足够的空间放置供液管及喷头，甚至营养液罐或营养液池也可置于其下，大大提高了空间利用率。这些优势使"A"型栽培装置在雾培系统中得以广泛应用。根据栽培植物的自身特点，如株高、重量、株幅等，相应调整栽培面板和支撑框架的用材。此种栽培装置主要用于叶菜类作物及花卉生产。

3.2 箱式栽培装置

箱式栽培装置给每个植株都提供了一个相对独立的生长环境，加强了生产者对植株生长的控制能力，而且便于移动。

采用箱式培养，营养液供给时间可定为每分钟喷雾30s。随着植株株高增加，收拢茎蔓于栽培箱中，茎蔓会萌发新的气生根，如此经多次收拢茎蔓并着生新的根系后，可去除旧的根系部分。这就省却了重新定植幼苗及其生长过程，适合果菜类等高大植物雾培应用，比较容易实现了长季节可持续栽培。

3.3 网式栽培装置

用塑料网作栽培装置，具有轻捷、节省投资的优势。利于生产姜及百合科的蔬菜，它们的根系生长对外界环境要求不严，且植株较轻。喷头上面是用于根系附着、支撑植株的塑料网，网线与铁架相连，网上有2.5cm厚的多孔渗水物质以固定根系。

3.4 床式栽培装置

床式是用于农业生产中较为普及与最易建造与操作的一种雾培栽培系统。其建造方法为：首先，在平地上建造宽1.2m、高20~25cm、长15~20m的砖砌水泥苗床，床底需铺建成稍带坡度且不渗漏的水泥底面，以利于营养液的回流与循环利用，在苗床稍低洼的一端留置养液回流孔以用于养液的回流与收集；其次，在床底铺设弥雾管道并按每50cm安装一个朝上喷射的雾化喷头（或超声波雾化器），以确保栽培时能让苗床整个内空间均匀雾化，不能安装留有死角雾化不匀的折射式喷头。最后，建好苗床与雾化供液管道后，选择强度较高的泡沫板作为定植用板，用海绵条作为包缚固定植株所用的填塞材料，种植时把植株用海绵包缚植入孔中即可，也可直接把植株卡植入有海绵条填充的条槽中。这种栽培系统大多用于叶菜类或植株较为矮小或者可以引缚固蔓的植株。

3.5 柱式立体栽培装置

柱式立体栽培装置是一种以简易围桶作为根域空间，以大空间雾化作为供液方式的栽培系统，大多用于大植株的栽培，如番茄树、巨型南瓜、西瓜树、黄瓜树、蔬菜树等，具有培育与创造最佳的根域空间发挥最大生长潜能的作用。在建造该栽培系统时，需按目标植株的大小来确定根域空间大小，再去选择设定围桶的大小尺寸，如以巨型的番茄树为例，由于根系的大小与植株的大小全息对应相关，因此需设定一个至少$1m^3$以上的根域空间，才能培养出单株结千果或万果的巨型番茄树。

4 应用前景

气雾培是一种新型的栽培模式，它是基于农业工程技术、设施园艺、计算机控制技术基础上的一种全新栽培模式，据专家分析预测，它将是未来高效农业模式中最为先进，对植物生长潜能开发最为有效的栽培模式，相信不久的将来也会在我国得以普及与运用。

气雾培具有其他任何一种栽培所不能比拟的优势，特别是在植物生长潜能发挥上，可以让植物处于最佳的根域状态下进行生长发育，对于加快植物生长速度，提高产量来说具有独特的技术优势,现已被生产与科研广泛地运用。特别是设施园艺发达国家如日本,已把气雾培技术作为植物工厂建设中的一项重要项目。在提高设施利用率方面,气雾培也具有独特的优势,利于实施立体式的空间栽培以提高单位面积产量,它在未来设施农业中将占有重要地位。在科研方面,如对植物根系的研究,雾培具有直观便于观察与控制根系的特点。美国亚利桑那州大学的根系研究所就是采用大型的气雾室作为研究植物根系的实验室,甚至科研人员可在气雾室中穿梭走动。

气雾培雾化的供液方式更适于失重状态下栽培,气雾培的根域空间环境以其最适的水分、矿质营养及空气环境,为根系的生长发育、生理代谢、功能基因的表达创造了最佳的环境,使植物的生长速度得以最大化地发挥。气雾培可以用于各种农业生产中植物的高产优质栽培,可以用于科研生产上的植物潜能研究开发,也是太空宇宙农业的最佳生产方式,在宇宙太空站、宇宙飞船上生命维持系统中的开发应用研究具有技术优势[10~11]。

参考文献

[1] 高建民,任宁,顾峰等. 低频超声雾化喷头优化设计及试验. 江苏大学学报(自然科学版). 2009. 30 (1): 1~4
[2] 施智雄,董加强,陈晓萍. 超声波雾培装置的设计与使用. 西南园艺. 2006. 34 (1): 16~17
[3] R itter E, Angulo B, Riga P, *et al*. Comparison of hydropon ic and aeroponic cultivati on system s or t he producti on the pota- to minitubers. Potato Research, 2001 (44): 127~135
[4] R. Kamies, M. S. Rafudeen, J. M. Farrant. The use of aeroponics to investigate desiccation tolerance in the roots of the resurrec- tion plant, Xerophyta viscose. SAAB Annual Meeting Abstracts. 2009. 2. 66
[5] Heidi A. Kratsch, William R. Graves, Richard J. Gladon. Aeroponic system for control of root-zone atmosphere. Environmental and Experimental Botany. 2006. 55: 70~76
[6] 李功义,梁杰. 汽雾法生产马铃薯微型薯技术的研究. 吉林农业科学. 2007, 32 (3): 21~22
[7] 杨元军,孙慧生,王培伦等 雾培与基质栽培马铃薯脱毒小薯继代产量比较. 山东农业科学. 2002, (1): 29~30
[8] 郭世荣. 无土栽培学. 北京: 中国农业出版社, 2003
[9] 刘士哲. 现代实用无土栽培技术. 北京: 中国农业出版社, 2001
[10] 徐伟忠,王利炳,詹喜法等. 一种新型栽培模式-气雾培的研究. 广东农业科学. 2006. (7): 30~34
[11] S. Leoni, B. Pisanu, R. Grudina, A new system of tomato greenhouse cultivation: high density aeroponic system, Acta Hor- ticulturae 361, ISHS 1994

设施内蔬菜水旱轮作治理连作障碍新模式*

江解增[1]**,缪旻珉[1],曾晓萍[2],曹光亮[2]

(1. 扬州大学园艺与植物保护学院,江苏扬州 225009;2. 江苏省园艺技术推广站,江苏南京 210036)

摘要:连作障碍已成为设施栽培蔬菜可持续发展的瓶颈问题,水旱轮作是治理连作障碍的有效、生态模式。本文归纳了在常规设施蔬菜基地利用水生蔬菜开展水旱轮作的主要模式及注意事项,并提出了进一步研究的设想。

关键词:水生蔬菜;水旱轮作;连作障碍

New Models for Control of Continuous Cropping Obstacles under Protected Condition by Aquatic-Upland Vegetable Rotations

Jiang Jiezeng, Miao Minmin, Zeng Xiaoping, Cao Guangliang

(1. *Horticulture and Plant Protection College*, *Yangzhou University*, *Yangzhou* 225009, *Jiangsu*, *P. R. China*;
2. *Horticulture Extension Station of Jiangsu Province*, *Nanjing* 210036, *Jiangsu*, *P. R. China*)

Abstract: Continuous cropping obstacles have been a problem for the sustainable production of vegetables under protected condition. Aquatic-upland vegetable rotation is an effective and ecological model for control of continuous cropping obstacles. Main models and their attentions were summarized on aquatic-upland rotation using aquatic vegetables in conventional greenhouse. Further research suggestions were also discussed.

Key words: Aquatic vegetable; Aquatic-upland vegetable rotation; Continuous cropping obstacles

"推进蔬菜、水果、茶叶、花卉等园艺产品集约化、设施化生产",这是中央文件中第一次作出阐述,为设施园艺发展指明了方向。全国各地设施园艺面积呈快速增长态势,设施种类也向钢架大棚、连栋大棚和日光温室、智能温室等固定、永久类型发展。设施功能也由过去以防寒保温为主,进一步拓展到遮阳降温、阻隔防虫、避雨控湿防病等。但蔬菜生产的规模化、专业化和工厂化的发展也导致了一些地区连作障碍的日益严重,并成为一些地区制约蔬菜生产可持续发展的瓶颈问题(喻景权和杜尧舜,2000),尤其是长江流域的大棚设施基地,近年大力推广夏季避雨控湿防病栽培速生叶菜新技术,基本上周年覆盖塑料薄膜,虽然有效地改善了夏季叶菜的供应,但也显著加重了土壤盐渍化的发生。因此,如何有效地综合防治土壤盐渍化、酸化、营养失衡以及病虫害等连作障碍,改善生产条件,保障设施蔬菜生产的可持续发展,已愈显迫切。喻景权和杜尧舜(2000)、吴凤芝 等(2000)、郑军辉 等(2004)、吴玉光 等(2005)、赵尊练 等(2007)等先后发表文章,综述了国内外对设施蔬

---

* 基金项目:江苏省农业三项工程项目[sx(2010)283]。

** 作者简介:江解增,男,教授,博士生导师,江苏省水生蔬菜育种攻关协作组首席专家,江苏省农业科技入户省级技术执行专家,专业方向:水生蔬菜育种与栽培,E-mail:jzjiang@yzu.edu.cn。

菜连作障碍的成因及其防治措施的研究进展。认为通过轮作、间作套种、化学防治、生物防治、选育抗性品种、嫁接、土壤消毒和施肥管理等措施可以部分或全部解决蔬菜作物的连作障碍问题。

水旱轮作是我国劳动人民在长期的实践中总结形成的一种生态、高效的栽培制度。吴玉光 等（2005）认为蔬菜与水田作物轮作是改善连作障碍最经济有效的途径，很多证据显示菜田经长年累月连作造成土壤酸化、盐化、养分失衡、土传病虫害严重等，一旦与水田轮作，前述情况将获得改善和消除。蒋义彩 等（2006）、杨化恩 等（2009）、沈忠才 等（2009）、张产端 等（2010）等在日光温室或塑料大棚中开展水旱轮作，上半年设施种植番茄、西瓜、甜瓜等旱生蔬菜后，下半年则揭除薄膜后露地种植水稻。能有效地克服连作障碍、提高产量和品质，又为实现设施蔬菜可持续、稳定发展找到了有效的途径。何圣米 等（2005）利用生育期较长的荸荠、慈姑等水生蔬菜替代水稻开展设施蔬菜基地的水旱轮作，经济效益显著提高。但上述水旱轮作都是采用旱生蔬菜设施栽培、水生作物露地栽培的模式。其实，莲藕（李士丁，1996）、茭白（李挺和陈可可，2001）、慈姑（钱忠贵，2010）、菱（严志萱 等，2008）、水蕹菜（黄显进和刘义冠，1995）和豆瓣菜（侯俊林 等，2000）等水生蔬菜都已先后试验成功其设施栽培技术，水芹则已育成适于保护地内湿润栽培的专用湿栽水芹新品种（江解增，2010a；2010b），使得水生蔬菜的采收上市期明显延长、或适宜种植区域显著扩大。

因此，在常规设施蔬菜基地开展设施水生蔬菜与设施旱生蔬菜的水旱轮作，既可以延长水生蔬菜的采收上市期、丰富菜篮子的品种类型，又可以进一步提高产量和效益，同时还起到治理连作障碍的作用，可望实现社会、经济、生态效益三丰收。

笔者根据现有水生蔬菜设施栽培技术，结合自身的研究实践，对水生蔬菜在常规的设施蔬菜基地进行水旱轮作的注意点、茬口模式及其栽培要点进行初步归纳，供相关生产单位参考。

1 设施蔬菜基地要求

1.1 具有较好的灌排设施

水生蔬菜生长阶段需要田间保持一定的水位，基地灌排能力、尤其是灌溉能力的设计应高于常规旱生蔬菜一倍以上。

1.2 具有较好的贮水能力

常规旱生蔬菜基地由于底层土壤水分含量较低，刚开始很难贮水，应持续不断灌水、使底层土壤充分湿润，这本身也是水作茬口洗盐压盐的过程。当然，考虑到灌溉成本，实施水旱轮作的设施蔬菜基地应尽可能避免难以贮水的高坡地。

1.3 适当加高田埂

不同水生蔬菜作物生长过程中对田间水位的不同要求，同时考虑到田块间的水分渗透，实施水旱轮作的单元面积不宜过小，一般以灌溉渠所围的田块作一个单元为宜，各单元周围的田埂应适当加高达30cm以上。日光温室、尤其是土墙建筑的简易类型，应在内墙边缘构筑不少于1m宽的围埂，以免灌水后水分渗透、侵蚀墙体。

1.4 提前覆膜升温

由于水的热容量较大,早春或冬季低温季节灌水栽培条件下,设施内土壤温度提升较慢,因此,早春种植水生蔬菜应提前15~20d覆盖薄膜并灌水,使水温及土温逐步提升、满足水生蔬菜生长需要。但是,设施内水温及土温一旦上升,与常规旱生蔬菜设施相比,设施内储热较多、夜晚降温较慢,对寒流的抵御能力也较强;12月中旬在扬州大学蔬菜试验田中,比较湿栽水芹大棚和黄瓜种植大棚中间位置50cm高度的温度,清晨最低温度时的温差达2~5℃,湿栽水芹大棚内0℃以下时间要比黄瓜大棚少2h以上。另外,水生蔬菜对水分要求较高,生长过程中一般不必经常揭棚通风降湿,既减少了设施内水分和热量的散失、又降低了管理成本。

2 水生蔬菜种类及其品种的选择原则

2.1 尽量不打破原有高效种植模式

经过多年的探索和实践,各地都有比较成熟的高效种植模式。因此,在形成新的高效模式之前,尽可能保留原有种植模式。可以在原有主茬之间1~2个月的空茬期间,插空抢栽速生的水生叶菜,如在大棚春提前或日光温室越冬茬口与秋延后茬口之间的盛夏高温季节,灌水种植水蕹菜;而在大棚秋延后茬口与翌年春提前茬口之间的寒冷季节,种植豆瓣菜或湿栽水芹。

2.2 种苗方便易得

目前市场容量较大的莲藕、茭白等水生蔬菜多为无性繁殖类型,如果当地没有种植基地,需从外地引种,则第一次种植的引种成本相对较高,而且以后必须在当地有计划轮流种植以保持种苗供应、或须建立专门的留种区域,因此,如果仅是通过水旱轮作治理连作障碍,可以考虑先种植一些种子播种的种类,如水蕹菜、豆瓣菜和芡实、菱等;或者种苗用量较低的如慈姑、湿栽水芹、荸荠等。

2.3 选择适宜浅水栽培的种类及其品种

莲藕应选择叶柄较短的浅水品种类型,并尽可能选择早熟品种,如珍珠藕、鄂莲一号等;茭白、荸荠、慈姑等水生蔬菜对水位要求不高,生长过程中最高水位仅需10cm左右;水蕹菜、豆瓣菜、水芹等水生蔬菜生长过程中只需浅薄水层,即使土壤充分湿润也能较好生长;只有菱和芡实生长过程中对水位的要求稍高,需达20~30cm。

3 适宜的水生蔬菜种类及其栽培要点

3.1 水蕹菜

3.1.1 品种选择

现有蕹菜品种均可以水栽,南方也有专门水栽的水蕹菜品种。主要有阔叶、柳叶和青梗、白梗等4个品种类型,各地应根据市场消费习惯选择种植。

3.1.2 建议茬口

"大棚春提前或日光温室越冬茬蔬菜(旱)—水蕹菜(水)—大棚秋延后或日光温室越冬蔬菜(旱)",在初夏6月份前茬结束后,灌水种植水蕹菜,在秋季后茬种植之前结束,整地后再种植下茬。水蕹菜在盛夏高温季节实施露地栽培。也可以根据需要分别向前或向后延长水蕹菜种植时间,乃至在设施内自早春到晚秋始终种植水蕹菜。

3.1.3 栽培要点

提前20d左右另田播种育苗,然后根据设施内空茬时间的长短,以行距15~30cm、穴距15~20cm、每穴2~3株移栽,灌水并保持3~5cm浅水位。当群体植株高度达30~40cm时,适当降低水位,近地面留根茬3cm左右收割,扎把销售。残留的根茬应露出水面,以免水淹腐烂。适当施用追肥、促进下茬继续生长。水栽蕹菜与旱栽蕹菜相比,产品生长速度快、产量高、口感脆嫩。

3.2 豆瓣菜(西洋菜)

3.2.1 品种选择

现有豆瓣菜品种主要依据小叶大小,分为大叶品种和小叶品种,一般选择大叶品种。

3.2.2 建议茬口

"大棚秋延后蔬菜(旱)—大棚豆瓣菜(水)—大棚春提前蔬菜(旱)",在秋延后蔬菜结束后,移栽豆瓣菜,在翌年早春茬种植前结束,种植下茬。豆瓣菜为冷凉性速生叶菜,适宜生长温度为15~20℃。可以根据需要延长种植时间,或直接作为大棚秋延后、温室越冬茬栽培。

3.2.3 栽培要点

提前15~20d另田土壤湿润条件下播种育苗,以10cm左右的间距移栽设施内,灌水并保持3cm左右浅薄水层。当群体高度达30~40cm时,近地面留茬3~5cm割收,清理枯黄叶后扎把销售;也可逐株采摘嫩梢,或者每次齐泥割去全田的3/4,把老根踩入泥中,以留下的1/4作种苗再行栽插。追施肥料、促进下茬继续生长;豆瓣菜在5℃以下时生长缓慢,可加盖小拱棚保温促长。

3.3 慈姑

3.3.1 品种选择

根据球茎表皮颜色,慈姑有紫皮、黄皮和白皮等3种类型,考虑到低温高湿条件容易诱发黑粉病,建议选择抗性较强的紫皮慈姑类型。

3.3.2 建议茬口

"早春大棚慈姑(水)—夏季速生叶菜(水或旱)[或秋延后大棚蔬菜(旱)]",早春3月上旬大棚内移栽,6月下旬开始采收,至秋延后茬种植前结束,也可于7月中旬结束后种植越夏茬速生叶菜如水蕹菜、小白菜等。

3.3.3 栽培要点

早春茬于2月上旬开春后在大棚内套小棚双层覆盖育苗,3月上旬移栽,4月中旬揭除棚膜,6月下旬开始采收,如下茬种植水稻或越夏速生叶菜,则须在7月上中旬结束;也可在9月份露地早熟慈姑上市之前结束采收,种植秋延后蔬菜。

3.4 菱

3.4.1 品种选择

菱的品种类型较多,各地应根据当时市场需求选择,一般以本地品种为主。

3.4.2 建议茬口

"早春大棚菱(水)—秋延后大棚蔬菜(旱)",开春前播种菱,初夏开始采收,露地种植菱大量上市前结束采收，接茬秋延后大棚蔬菜；也可进一步延长采收时间，接茬大棚越冬蔬菜；还可在秋季气温下降时再次覆膜、延迟采收至深秋。"春提前大棚蔬菜（旱）—露地菱（水）—越冬茬大棚蔬菜（旱）"，春提前大棚蔬菜结束后，从早春大棚菱田中疏苗移栽，至深秋季节结束，然后接茬大棚越冬蔬菜。

3.4.3 栽培要点

早春播种前 1 个月左右，将菱种置于 5~10℃ 的低温环境中清水浸种并经常换水，以促进破眠；20d 后改在大棚等温暖环境催芽，当有一半菱种露白时，在大棚内按 1m 行距、每 m 7~10 颗菱种的密度播种，田间保持 5~10cm 浅水层，每 3~5d 换水 1/2 以保持水质清洁防止水绵发生；随温度上升，逐步揭棚通风直至揭除薄膜，揭膜后将较密处的菱株疏出一部分，移栽到大棚间的空白处或其他轮作的大棚中，高温季节通过加深水位及换水降温，控制水温不超过 35℃；花后 10~15d 开始采收。

3.5 莲藕

3.5.1 品种选择

宜选择早熟品种类型，如珍珠藕、鄂莲一号等；或直接用当地主栽品种。

3.5.2 建议茬口

"早春大棚莲藕（水）—越夏速生叶菜（水或旱）[或秋延后大棚蔬菜（旱）]"：1 月底至 2 月初种植莲藕，6 月中下旬开始采收，莲藕产区在 7 月中旬露地早熟莲藕上市前结束，种植越夏茬速生叶菜如水蕹菜、小白菜等；也可陆续采收至 9 月份，种植秋延后大棚蔬菜。

3.5.3 栽培要点

大棚内以 1m 行距开深 20cm、宽 30cm 的定植沟，沟内施基肥与土壤混匀使沟底离畦面 10cm 左右、灌入 3~5cm 浅水，以 1m 株距排放种藕，土表面覆盖 0.004~0.006mm 的超微地膜，莲藕放叶后逐步增加水位至畦面之上 10~15cm。4 月上中旬当棚内最高温度超过 35℃ 时逐步揭棚通风，直至揭除棚膜。5 月中下旬开始根据莲藕长势，降低田间水位控制地上部生长、促进地下部膨大。6 月中下旬开始陆续采收上市。

3.6 茭白

3.6.1 品种选择

宜选择低温孕茭类型的双季茭品种，如苏州类型品种小蜡台和葑红早等。

3.6.2 建议茬口

"大棚夏茭（水）—越夏速生叶菜（水或旱）—秋延后大棚蔬菜（旱）"，早春大棚覆盖种植夏茭，6 月底夏茭结束后种植越夏速生叶菜如水蕹菜、小白菜等，秋季种植秋延后蔬菜。"春提前大棚蔬菜（旱）—大棚秋茭（水）—越冬茬大棚蔬菜（旱）"：大棚春提前蔬菜结束后，从夏茭田中选种移栽到大棚内，秋季日均气温降至 15℃ 以下时扣膜保温促长，延长采收期、提高产量，深秋采收结束后接茬越冬蔬菜。

3.6.3 栽培要点

夏茭早熟栽培在上年秋茭采收后自然越冬的夏茭田，或在其中将老墩的一半挖出、按 70cm×50cm 株行距移栽大棚内，在早春气温回升前 20~30 天扣棚覆膜增温促长，田间保持 3~5cm 浅水位；随温度上升，在苗高达 35~40cm 时根据田间茎蘖密度适当疏苗，使每亩总茎蘖数在 2.2 万~2.5 万个之间；并注意通风降温直至日均气温达 15℃ 时揭除棚膜；揭棚前应注意避开棚内较长时间的20~25℃孕茭适温，以防早期小苗孕茭；4月中下旬开始采收，至6月底、7月初高温季节来临时结束。秋茭延迟栽培则在夏茭采收结束后，在其中按茭白选种要求选出种苗，以每亩1 800~2 300穴的较高密度移栽到大棚内，重施分蘖肥促进分蘖萌发，秋季日均气温下降至15℃时覆膜保温，进一步延长茭白生长期及孕茭时期，提高产量。

3.7 湿栽水芹

3.7.1 品种选择

宜选择适于湿润栽培的品种，如湿栽水芹新品系"G0601"和"D07"、江阴湿栽水芹、沙洲竹梗芹等。

3.7.2 建议茬口

"秋延后大棚蔬菜（旱）—大棚湿栽水芹（水）—春提前大棚蔬菜（旱）"，在秋延后大棚蔬菜结束后，育苗移栽湿栽水芹，在翌年开春后结束，接茬种植春提前大棚蔬菜。也可在温室或大棚内在延长湿栽水芹种植季节，自秋季9月下旬直至翌年春季4~5月，前后茬均种植越夏速生叶菜。

3.7.3 栽培要点

9月份气温降至25℃以下时按畦宽1~1.5m、畦沟宽30~40cm、深30cm开沟筑畦，在畦面上以10~15cm间距开2~3cm深的播种沟，将水芹种茎首尾相接排入沟底，上覆细土；播种后在畦沟中灌水，使畦面充分湿润、畦沟保持有水；根据其他设施田块的让茬时间，按10cm间距移栽。当群体植株高度达30~40cm时，近地面留根茬2cm左右割收，清理枯黄叶后扎把销售；适当施用追肥、促进下茬继续生长。

3.8 芡实

3.8.1 品种选择

宜选择果实表面无刺的苏芡类型中的早熟品种，如紫花苏芡。

3.8.2 建议茬口

"早春大棚芡实育苗—越夏速生叶菜"，在芡实产区进行春提前播种育苗，芡苗移栽后种植越夏速生叶菜。"春提前大棚蔬菜—露地栽培芡实"，即以芡实替代水稻，能延长水淹时间并提高经济效益。

3.8.3 栽培要点

开春后芡实催芽，在营养钵中播种，大棚内做平畦，将营养钵在畦面上紧挨摆放，灌水使水层淹没营养钵，出苗后逐步加深水位3~5cm；1个月后当幼苗1~2片真叶时进行分苗，使营养钵间距为20~30cm，水位加深至10~15cm；当外界气温稳定在15℃以上时移栽大田；移栽时水位10cm左右，以后逐步加深，至开花结果期达30cm左右，采收期适当降低水位至20cm左右。春提前大棚蔬菜后露地栽培芡实可适当推迟1~2个月播种育苗（苏州市蔬菜研究所，2005）。

4 设想与展望

水生蔬菜作为我国特色蔬菜，与主要旱生蔬菜已基本育成设施专用品种、并实现周年生产供应相比，除了水芹育成适于常规旱生蔬菜设施内土壤湿润条件下种植的湿栽水芹新品系

"G0601"和"D07"外,莲藕、茭白等主要水生蔬菜至今尚未育成适于设施栽培的专用品种,尤其是适于常规旱生蔬菜基地种植的浅水型、节水型的设施专用品种;水生蔬菜设施早熟栽培技术也只有少量试验报道、且主要针对水田种植。笔者在生产调研中还曾发现:露地种植的莲藕早熟品种在设施栽培条件下并不表现早熟、南方的早熟菱品种引种到北方后并未表现其早熟性能。因此,各地在开展设施内水旱蔬菜轮作治理连作障碍的同时,还需进一步开展在早熟品种的引种筛选,同时在种植时期、种植密度、肥水调控、病虫草害防治等环节进一步开展试验研究,并比较不同水生蔬菜种类及各种水旱轮作模式对连作障碍治理效果的差异,从而形成能较好治理设施蔬菜连作障碍的水生蔬菜设施栽培新技术,为综合防治设施蔬菜连作障碍探索新的模式。

参考文献

[1] 何圣米,杨悦俭,李必元,徐明飞,魏国庆.设施蔬菜—水生蔬菜水旱轮作模式的应用.浙江农业科学,2005,(1)10~12
[2] 侯俊林,祁连,彭秀枝.保护地西洋菜栽培技术.内蒙古农业科技,2000,(3):32
[3] 黄显进,刘义冠.大棚蕹菜反季节水培生产试验.广西农业科学,1995,(2):67~68
[4] 江解增.湿栽水芹G06-01及其保护地栽培(上).农家致富,2010a,(1):31~32
[5] 江解增.湿栽水芹G06-01及其保护地栽培(下).农家致富,2010b,(2):32~33
[6] 蒋义彩,郑晓微,王秀珍,唐长青."水稻—大棚番茄"水旱轮作高效栽培技术,2006,(3):13~14
[7] 李士丁.莲藕拱棚早熟高产技术.长江蔬菜,1996,(3):31~32
[8] 李挺,陈可可.茭白塑料棚覆盖栽培技术.中国蔬菜,2001,(2):42~43
[9] 钱忠贵.大棚慈姑早熟高效配套栽培技术.长江蔬菜,2010,(3):41
[10] 沈忠才,黄建华,姜月霞.大棚西瓜、水稻高效轮作栽培技术.上海农业科技,2009,(2):101~102
[11] 苏州市蔬菜研究所.苏州水生蔬菜实用大全.南京:江苏省科学技术出版社,2005
[12] 吴凤芝,赵凤艳,刘元英.设施蔬菜连作障碍原因综合分析与防治措施.东北农业大学学报,2000,31(3):241~247
[13] 吴玉光,张东兴,霍高智.蔬菜连作障碍防治的研究动态.中国蔬菜,2005(增刊):82~86
[14] 严志萱,杨新琴,俞金龙,方卫星.大棚田菱长季节栽培技术.中国蔬菜,2008,(8):52~53
[15] 杨化恩,王献杰,曹荣利.日光温室番茄与水稻轮作栽培技术.中国蔬菜,2009,(9):46~48
[16] 喻景权,杜尧舜.蔬菜设施栽培可持续发展中的连作障碍问题.沈阳农业大学学报,2000,31(1):124~126
[17] 张产端,仇从宇,史永梅.无公害大棚甜瓜与水稻连作高效栽培技术.陕西农业科学,2010,(4):215~217
[18] 赵尊练,杨广君,巩振辉,郭建伟.克服蔬菜作物连作障碍问题之研究进展.中国农学通报,2007,23(12):278~282
[19] 郑军辉,叶素芬,喻景权.蔬菜作物连作障碍产生原因及生物防治.中国蔬菜,2004,(3):56~58

LED 光源光质比对甘薯组培苗生长及电能消耗的影响*

杨雅婷[1]**, 杨其长[2]***, 肖平[2]

(1. 农业部南京农业机械化研究所农业资源与设施工程技术中心,南京 210014;2. 中国农业科学院农业环境与可持续发展研究所/农业部农业环境与气候变化重点开放实验室,北京 100081)

摘要:以红色 LED(660 ± 20nm)和蓝色 LED(450 ± 20nm)组合制成的 LED 灯管作为组培人工光源,研究光照度为 35μmol·m^{-2}·s^{-1}时,红蓝光质比(R/B)分别为 4、6、8、10 的光环境培养条件下甘薯组培苗的生长情况,以相同光照度的荧光灯为对照,培育甘薯组培苗 28d。结果表明:660nm 红光和 450nm 蓝光组合可有效抑制植物徒长、降低地上部分含水率和提高根冠比;荧光灯下生长的植株地上鲜重、叶片含水率和株高均最大,但是植株徒长,干物质积累不良。高的 R/B 处理能提高植株高度和根冠比,增大地下鲜重,降低地上含水率,有利于干物质积累;R/B 为 8 时甘薯组培苗的地下鲜重和根冠比均最大。另外,LED 光源的电能消耗与 R/B 值呈线性关系,并比荧光灯节能 27.6% ~48.0%。

关键词:LED;组培;R/B 处理;甘薯;能耗

Effects of LED Light Quality to Growth of Sweet Potato Plantlets in Vitro and Energy Consumptions of Lighting

Yang Yating[1]**, Yang Qichang[2]***, Xiao Ping[2]

(*Institute of Environment and Sustainable Development in Agriculture,*
Key Open Laboratory for Agro-environment and Climate Change, The Ministry
of Agriculture of China, CAAS, Beijing 100081, P. R. China)

Abstract: A LED (Light Emitting Diode) lighting tube for tissue culture was developed by using red LEDs (660 ± 20nm) and blue LEDs (450 ± 20nm). The ratio of red light intensity to blue light intensity (R/B ratio) was set at 4, 6, 8 and 10 respectively with a total PPF (Photosynthetic photon flux) of 35 μmol/(m^2·s). The sweet potato plantlets in vitro were grown for 28 days under the four different R/B ratios and the normal fluorescent lamp were used as a control. Effects of the R/B ratio on the growth of sweet potato plantlets in vitro and the energy consumption with different R/B ratios were investigated. The results showed that the combination of red light and blue light could inhibit the gain growth, reduce shoot water content and improve the root-top ratio. The shoot fresh weight, shoot water content and height of plantlet were highest under fluorescent lamp conditions, but with vain growth and poor dry matter accumulation. The stem length, root-top ratio, fresh weight of shoot and shoot water content were improved by the high R/B treatment. The root fresh weight and the root-top ratio were highest in the R/B treatment at 8. The energy

---

* 基金项目:中央级公益性科研院所基本科研业务费项目。
** 作者简介:杨雅婷(1984—),女,甘肃玉门人,硕士生,专业方向农业生物环境与能源工程。
*** 通讯作者:杨其长,研究员、博导,从事设施园艺环境工程研究,E-mail:yangq@cjac.org.cn。

consumption of LED tubes had a linear relationship with the R/B ratios and was reduced by 27.6% ~ 48.0% compared to the fluorescent lamp.

Key words: LED; Tissue culture; R/B treatment; Sweet potato; Energy consumption

光是植物光合作用和获取能量的主要环境因子，是设施内环境的主导因子，决定着作物的生长发育与经济产量。植物对 400 ~ 510nm 蓝紫光段、610 ~ 720nm 红橙光段和 720 ~ 780nm 远红光段反应最为敏感，其中可吸收的波长主要集中在蓝紫光段（波峰为 450nm）和红橙光段（波峰为 660nm），但不同植物的适宜光质都不尽相同。

LED（Light-Emitting Diode）即发光二极管，是一种节能环保、使用寿命长、体积小的新型光源，它发出的半波宽窄（±20nm），能够实现精确的光质配比，将 LED 作为人工光源或者补光光源应用于设施园艺领域已经成为国内外的研究热点。目前，国内外研究已证明 LED 可成功用于香蕉、地黄、草莓、白鹤芋、菊花、百合、葡萄等植物组织培养，以及藻类生产、植物工厂育苗和蔬菜生产，但研究主要集中在单色光对植物品质的影响。研究表明，植物在纯红光照射下，植株干物质积累多，节间较长而茎较细，且叶片细小，总糖含量高；在纯蓝光照射下，植株干物重小，节间较短而茎较粗，伸长生长受到一定抑制。红蓝光质比（R/B）对植物生长的影响也有研究，但各试验中 R/B 取值较为单一或者不精确。同时节能也是使用 LED 代替目前组培常用光源荧光灯的重要原因之一，但综观已有研究，均只利用了 LED 光谱可调的特性，研究了 LED 对植物生理品质的影响，忽略了对 LED 在组培生产中的电能消耗分析。

本研究拟采用 LED 组培专用灯管研究 R/B 对甘薯组培苗生长的影响，并对培育过程中的电能消耗进行分析。通过研究不同 R/B 对甘薯组培苗生长的影响和 LED 组培光源的能耗情况，希望能对今后 LED 在植物组培中的应用和经济可行性分析提供参考。

1 材料与方法

1.1 试验设置

甘薯组培苗由中国农业科学院农业环境与可持续发展研究所组培实验室提供；LED 光源为中国农业科学院设施农业环境工程研究中心与深圳四海电气技术有限公司联合开发的同型号 LED 组培专用灯管，红光（波峰为 660 ± 20nm）、蓝光（450 ± 20nm）交替均匀布置，每层布置 7 根；对照组光源采用松下 36W 三基色荧光灯（YZ36RL），每层布置 3 根。根据前人研究和前期试验结果，将光源的总光照度均设置为 35μmol/（m² · s）。

试验于 2009 年 5 ~ 7 月在组培室进行。选取三叶一心的甘薯组培苗接种在 MS 培养基上（添加 20g/L 蔗糖和 5.3g/L 琼脂），调节 pH 值为 5.8 ~ 5.9，无激素添加，每瓶 3 株。控制每瓶中 3 株组培苗总质量为 0.401 ± 0.016g。组培容器为普通锥形玻璃瓶，容积为 100ml，培养室中温度为 23 ± 0.2℃，相对湿度 80 ± 5%。黑暗预培养 24h 后，将组培苗放在设置不同 R/B 的 LED 灯管和荧光灯管下进行光照处理。试验共设 5 个处理（表 1），每个处理 6 瓶。光周期为 10h/d。采用德国 Avantes 公司生产的光纤光谱仪测定光照度，通过增加红光同时减少蓝光照度的方式完成 R/B 的设定。

光照度分布均匀性是温室太阳直射光环境的评价指标之一，参考连栋温室采光性能测试方法，光照度均匀性 λ（$\leqslant 1$）用以下公式计算得出：

$$\lambda = 1 - \frac{s}{E} \tag{1}$$

$$s = \sqrt{\frac{\sum_{i=1}^{n}(E_i - E)^2}{n-1}} \tag{2}$$

其中：s 为各个光照度测点的标准差，E_i 为第 i 点（$i \leqslant n$，$n = 52$）处的光照度，E 为平均光照度。

表 1 光质处理设置及 LED 物理参数

处理	R/B	峰值波长（nm）	总光照度 [μmol/(m²·s)]	光照度均匀性	灯下 5cm 平均温度（℃）
CK	0.62~1.1		35	0.89	26.5
LED4	4	660/450	35	0.56	23.8
LED6	6	660/450	35	0.55	23.6
LED8	8	660/450	35	0.67	22.9
LED10	10	660/450	35	0.68	23.1

1.2 项目观测

培养 30d 后，取出甘薯组培苗，用游标卡尺（±0.02mm）测定株高、茎粗和根长等形态指标，株高是指从地面至植株顶端的最大距离，根长是指从地面至根末端的最大距离；采用德国 Sartorius 天平（型号 BP221S，±0.1mg）分别称地上、地下鲜重后放入烘箱中 75℃ 烘干 72h 至恒重，称干重。同时取鲜样品，采用比色法测量叶绿素和类胡萝卜素含量，每瓶选取植株中间层完全展开的叶片 0.2g，将叶片剪碎并置于 1:1 的乙醇—丙酮混合液中，封口，在常温黑暗中浸提 48h，直至叶片完全变白后用 UNIC-7 200 分光光度计测量其吸光度值，通过计算得到叶绿素和类胡萝卜素含量。同一指标 3 个重复，试验重复 2 次。试验数据采用 3 个重复的平均值 ± 标准差，用 SAS v8.2 数据处理软件进行统计分析，用 Duncan 检验进行显著性差异分析。

2 结果与分析

2.1 LED 光源物理性能分析

由表 1 可见，与对照（CK）相比，LED 光源的光照度均匀性较低，保持在 0.55~0.68，说明 LED 灯管的构造和 LED 芯片的封装角度还需要改进，如加装反光罩、提高芯片封装角度等。灯管下垂直温度采用在灯管垂直下方不同高度上布置热电偶进行连续 24h 监测的方法来测量。发现在灯下 5cm 处，荧光灯照射区域空气的温度比 LED 处理区高 3℃ 以上。荧光灯散发的这部分热量会提高空气温度，一方面，导致组培瓶内明显的结露现象，当组培苗叶片紧贴瓶壁时，水珠会造成叶片的玻璃化，叶片发脆，出苗时会自动落下，严重影响组培苗品质；另一方面，这部分热量必须由空调制冷带走，相应会增加制冷费用。

2.2 不同 R/B 的 LED 光源对甘薯组培苗生长的影响

由表2可见,与对照(CK)相比,LED光源处理的甘薯组培苗的植株高度、地上部分质量差值均明显降低和减少,茎略粗、根略长但差异不明显,说明试验所用LED光源有抑制植株徒长的作用,荧光灯有促使甘薯组培苗徒长的作用,其中增重是指植株鲜重与原重的差值。

表2 不同处理甘薯组培苗的生长情况

处理	株高(mm)	茎粗(mm)	根长(mm)	增重(g)
CK	82.2±9.7a	2.25±0.23a	159.3±28.5a	1.50±0.15a
LED4	54.6±7.4c	2.38±0.44a	168.4±35.5a	1.18±0.20b
LED6	57.7±8.9c	2.50±0.43a	170.3±30.5a	1.35±0.17ab
LED8	62.7±8.6bc	2.31±0.24a	174.6±18.5a	1.40±0.15ab
LED10	66.0±5.9b	2.21±0.33a	180.2±9.4a	1.50±0.38a

注:增重是指培养30d后植株鲜重与原重的差值;同一列内不同字母代表差异显著,采用Duncan检验,$P<0.05$。

进一步分析LED和荧光灯的光谱发现(图1),LED光谱中400~510nm的蓝光和610~720nm的红光的光照度之和占总照度的98.5%以上,而荧光灯中仅为52.8%,其余为510~610nm的绿光。据研究,绿光有使幼苗徒长的作用,本试验中荧光灯处理(CK)中甘薯组培苗确实表现出徒长现象,而LED光源处理则对此有一定的抑制作用。

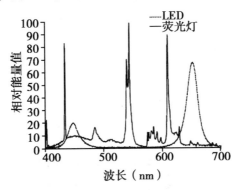

图1 LED和荧光灯光谱图

由表2还可看出,各LED光源处理中,随着R/B的增大甘薯组培苗的各项指标表现出一定的增加趋势,植株高度、地上部分质量增加差异较显著,而茎粗、根长的差异不明显。据研究,纯蓝光有抑制植物地上部分伸长的作用,植株干物重小、节间较短,根部生长差,不利于葡萄试管苗的生根;纯红光有促进地上生长、减小茎粗的作用,植物干鲜重、干物质比重较大,易诱导植物外植体发生不定根,有利于生根。本试验与上述研究结果相一致,高的R/B有提高植株高度和地上部分增重、增大根长的作用,低的R/B则有增大茎粗的作用。

2.3 不同 R/B 的 LED 光源对甘薯组培苗品质的影响

由表3可见,与对照(CK)相比,LED光源处理的甘薯组培苗的地上含水率、根冠比均明显降低和提高,地上鲜重略轻、地上干重和地下鲜重略高,但差异不明显。结合表2数据,可见试验所用LED光源可能有降低地上部分含水率、抑制地上部分徒长进而提高根冠比的作用,荧光灯可能有增大地上部分含水率、促使甘薯组培苗徒长、降低根冠比的作用。

表3 不同处理甘薯组培苗的品质指标

处理	地上鲜重(g)	地上干重(g)	地上含水率(%)	地下鲜重(g)	根冠比(FW)
CK	1.64 ± 0.14a	0.10 ± 0.02ab	93.72 ± 0.51c	0.30 ± 0.03ab	0.19 ± 0.02c
LED4	1.32 ± 0.12b	0.11 ± 0.02ab	92.02 ± 0.74abc	0.28 ± 0.04c	0.21 ± 0.03bc
LED6	1.35 ± 0.14b	0.10 ± 0.02b	92.83 ± 1.29ab	0.32 ± 0.08ab	0.23 ± 0.04bc
LED8	1.40 ± 0.07ab	0.12 ± 0.02ab	91.52 ± 0.74bc	0.39 ± 0.02a	0.28 ± 0.03a
LED10	1.45 ± 0.21ab	0.13 ± 0.01a	90.77 ± 1.12a	0.38 ± 0.03a	0.26 ± 0.03ab

注:同一列内不同字母代表差异显著,采用Duncan检验,$P<0.05$。

由表3还可以看出,各LED光源中,随着R/B的增大,地上干鲜重、地下鲜重、根冠比以及地上含水率都表现出一定增加或者减小的趋势,地下鲜重、根冠比增加和地上含水率降低差异较显著,而地上干鲜重变化不显著,说明高的R/B能促进组培苗干物质的积累和根部生长,也与已有研究相一致。根冠比是植物地下部分与地上部分鲜重的比值,为了提高组培苗在驯化期的抗逆能力,应该提高根冠比。随着R/B的提高,根冠比先上升后略下降,LED8的根冠比最大,说明对根冠比而言,最适宜的R/B是8,低R/B和绿光都可能抑制组培苗生根。

综合根长、根部干鲜重数据,处理LED8的R/B能更好地促进根部生长,而荧光灯对照处理的根长、地下鲜重和根冠比都较小,根部发育不好,可能是因为较多的绿光造成地上部分徒长,向下输送的光合产物减少,影响了根部发育。

2.4 不同R/B的LED光源对电能消耗的影响

试验采用的LED灯管由224颗功率为0.04W的红色LED灯珠和44颗功率为0.06W的蓝色LED灯珠组成,灯管内置直流电源效率为85%,其额定功率为13.6W。测量4个LED处理的耗电量分别为0.47kW·h/d、0.54kW·h/d、0.68kW·h/d、0.75kW·h/d,每一层的耗电量与R/B呈线性增大关系,线性相关系数为0.98。同时测量CK耗电量为1.03 kW·h/d。说明与对照(CK)相比,LED光源能明显节约电能,节电量为27.6%~48.0%,以处理LED8为例,比荧光灯节省电能34.0%。试验周期为28d,则培养每株甘薯苗耗电量见表4。

表4 LED灯管和荧光灯耗电量比较

	CK	LED4	LED6	LED8	LED10
耗电量(kW·h/株)	0.098	0.045	0.051	0.065	0.071

3 结论与讨论

(1)LED光源有抑制植株徒长、降低地上含水率、提高根冠比的作用,荧光灯有促使甘薯组培苗徒长、增大地上含水率并降低根冠比的作用。本试验所用 LED 的光谱中，400～510nm 的蓝光和 610～720nm 的红光的光照度之和占总照度的 98.5% 以上，而荧光灯中仅为 52.8%，其余为 510～610nm 的绿光，使甘薯苗表现出徒长，这与绿光易使幼苗徒长的研究结果相符。LED 光源处理均对徒长有一定的抑制作用，植物能合理分配向下输送的光合产物，进而提高根冠比。

（2）LED 红蓝光对甘薯组培苗的生理状态有一定的调节作用，随着 R/B 值的提高，植株高度、增重、根长、地上干鲜重、地下鲜重表现出一定增大的趋势，而茎粗、地上含水率均表现出一定减小的趋势，但是根长、地上干鲜重、茎粗的差异不明显。这与已有研究认为红光可以促进地上生长、提高干物质比重、有助于生根的研究结果相符合，也与蓝光抑制地上部分伸长、减小干物重的研究结果相符合。对 LED 光源配置高的 R/B，有利于提高植株高度、增重、地下鲜重，降低地上含水率。

（3）LED 光源 R/B 值为 8 时，根部发育最好，植物能合理向下输送光合产物，地下鲜重和根冠比最大，地上部分和地下部分的相关性得到较好的协调。

LED 是新兴的节能光源，其光质可调、低耗电量、低放热量的突出优点和代替荧光灯的发展趋势已经得到共识。但是不同植物的适宜 R/B 值不尽相同，今后的试验中应选用不同作物，进一步探索红蓝光对植株生长发育的作用规律。试验中对耗电的研究还是初步的、粗浅的，同时由于 LED 放热量小而节省的空调耗电量也没有考虑，有待于进一步探索。

参考文献

[1] 杨其长，张成波. 植物工厂概论. 北京：中国农业科学技术出版社，2005
[2] 杨其长. LED 在农业与生物产业的应用与前景展望. 中国农业科技导报，2008，10（6）：42～47
[3] Lund J B, Blom T J, Aaslyng J M. End-of-day Lighting with Different Red/Far-red Ratios Using Light-emitting Diodes Affects Plant Growth of Chrysanthemum x morifolium Ramat. 'Coral Charm'. HortScience, 2007, 42（11）: 1 609～1 611
[4] 魏灵玲. LED 光源在密闭式植物苗工厂的应用研究. 北京：中国农业大学，2007：62～65
[5] 鲍顺淑. 密闭式植物工厂中药用铁皮石斛组培生产的适宜光照环境. 北京：中国农业大学，2007：49～58
[6] 闻婧. LED 红蓝光波峰及 R/B 对密闭植物工厂作物的影响. 北京：中国农业科学，2009：20～40
[7] 邱秀茹，焦学磊，崔瑾等. 新型光源 LED 辐射的不同光质配比光对菊花组培苗生长的影响. 植物生理学通讯，2008，44（4）：661～664
[8] Hunter D C, Burritt D J. Light quality influences adventitious shoot production from cotyledon explants of lettuce（*Lactuca sativa* L.）. In Vitro Cellular & Developmental Biology-Plant, 2004, 40（2）: 215～220
[9] Moon H K, Park S Y, Kim Y W, et al. Growth of Tsuru-rindo（*Tripterospermum japonicum*）cultured *in vitro* under various-sources of light-emitting diode（LED）irradiation. Journal of Plant Biology, 2006, 49（2）: 174～179
[10] Heo J W, Shin K S, Kim S K, et al. Light quality affects *in Vitro* growth of grape 'Teleki 5BB'. Journal of Plant Biology, 2006, 49（4）: 276～280

甜樱桃促成设施栽培调查报告*

孙玉刚**, 魏国芹, 李芳东, 秦志华, 安　淼

（山东省果树研究所，泰安　271000）

摘要：甜樱桃是当前种植效益最好的果树之一，因果实外观艳丽，营养丰富，深受消费者喜爱，各适宜产区积极规划发展。但甜樱桃花期容易遭受低温冻害，成熟期遇雨易裂果，严重制约了栽培发展。近十年来山东、辽宁等地开展了甜樱桃设施栽培，不仅使果品提早成熟 1~2 个月，同时还解决了冻害和裂果问题，种植效益高，销售价格高达 60~400 元/kg，市场空间巨大，各主产区积极推广。但设施栽培投资高、风险大、管理技术尚不成熟，各地均出现扣棚失败的情况，挫伤了种植者的积极性。针对生产实际，笔者在多年研究的基础上，于 2010~2011 年对我国甜樱桃主要设施栽培区进行了系统调查，总结了适宜各地发展的设施类型，提出了适合设施栽培的良种良砧，集成了授粉、设施内环境因子调控、促进花芽形成、预防落花落果以及揭棚后树体营养管理等关键技术措施，以期提升我国甜樱桃设施栽培科技水平，为设施栽培区的高效生产提供借鉴。

关键词：甜樱桃；设施栽培；品种；环境因子

Investigation Report on Facilities Cultivationin Sweet Cherry

Sun Yugang*, Wei Guoqin, Li Fangdong, Qin Zhihua, An Miao

(*Shandong institute of pomology*, *Taian*, 271000, *P. R. China*)

Abstract: Sweet cherry is currently one of the bestprofitable tree species which is deeply loved by consumer because of its gorgeousappearance and abundant nutrition. Nowadayssomesuitable division actively plan for development. But cold damage caused by low temperature during florescence and fruit cracking arise from rainfall before harvest are both limit the developmentof sweet cherryproduction, and that's why for nearly a decade some areas such as Shandong and Liaoning launched facility cultivation. Facility cultivationnot only advance fruit mature period 1 to 2 months earlier whose price could reach 60 to 400 Yuan per kilogram but also protect sweet cherry from cold damage and fruit cracking effectively. Therefore, facility cultivationwas spreadin major producing areas as its high planting benefit and enormous market space. However, unsuccessfulfacility culture practiceswere turned out almostin all promoting areas because of its high investment, high risk and immature management technology, which dampened the enthusiasm of growers. Based on many years researchonthis problem, a systematic surveywas madein main sweet cherry facilities cultivation areasin China during last two years which summarized themain facility types and accurate covering time. Rightly varieties and stock suitable to facilities cultivationand pollinationtechnology were also put forward in the conclusion as well as environmental factors control technology in facility andbud formation promoting technology. Flower and fruit drop preventingtechnology together with tree body nutrition management after unveiling plastic

---

* 基金项目：公益性行业（农业）科研专项（200903019）；2010 年泰安市科研专项（20093010）。
** 第一作者简介：孙玉刚（1964—　），男，山东省诸城市人，研究员，主要从事果树资源与栽培研究。Tel：0538 - 8261223。

were also suggested in this article which could upgrading sweet cherry facilities cultivation technology leveleffectively so as to provide reference for getting high productionin facility cultivation areas.

Key words: Sweet cherry; Facility cultivation; Cultivar; Environment factors

  甜樱桃，原产于欧洲东南部黑海沿岸和亚洲西部，喜温暖，不耐寒、不抗旱、不耐涝，不抗风、喜光性强。其成熟期早，果实色泽艳丽、营养丰富，深受消费者喜爱，被誉为"果中珍品"。甜樱桃设施栽培是指利用日光温室、塑料大棚或其他设施，通过调控环境因子，为甜樱桃生长发育提供适应的条件，实现果品成熟期的人工调节和品质改善。甜樱桃设施栽培研究始于20世纪70年代的瑞士、意大利、德国和日本等，目前，日本设施栽培面积约占总面积的25%。我国甜樱桃栽培最早是1871年由美国引入烟台，1991年烟台甜樱桃设施栽培取得初步成功，辽宁、北京、河北等地不同程度发展。目前，山东和辽宁已成为甜樱桃设施栽培的主产区，主要为促成栽培，约占总面积的2.5%左右。设施栽培不仅扩大了种植范围，避免了露地栽培中经常遇到的花期低温、阴雨、大风等不良天气造成授粉不良及遇雨裂果等问题，而且由于成熟期提前，上市早，果品价格较露地栽培高几倍甚至十几倍，经济效益极大提高，具有广阔的发展前景。

  我国甜樱桃设施栽培面积计约3 333hm^2，其中：山东约1 667hm^2，分布在烟台（333hm^2），青岛（266.7hm^2），潍坊（933hm^2）；辽宁约1 000hm^2，主要分布在大连瓦房店（667hm^2）、普兰店；近几年山东各樱桃产区也在积极开展设施栽培，呈大发展趋势。山东省果树研究所樱桃课题组在多年科研基础上，于2010～2011年对山东省和辽宁省的甜樱桃主要设施栽培区进行了考察，对甜樱桃设施促成栽培成功经验进行了系统总结，以期提升甜樱桃设施栽培科技水平，为新产区提供借鉴。

1 调查对象及内容

  在辽宁和山东选择19个有代表性的栽培设施进行了调查和统计分析，样本选自瓦房店大冯村、旅顺土城子、大连营城子、烟台垆上、莱山南堼、栖霞下孙家、平度铁岭庄、临朐月庄和泰安孙楼村及岱岳朱家庄，调查内容包括设施类型、建棚时期、结构参数、扣棚、升温时间、升温设备、保温材料、设施内环境因子调控技术、上市时间和产量。

2 调查结果及分析

2.1 设施类型及结构

  调查结果见表1。大连地区的主要设施类型为日光温室（图1），山东地区为塑料大棚（图2），两者为我国主要促成栽培设施类型。

  设施结构参数（长度、跨度、脊高、肩高）尚无统一数据，大多根据地形、果园面积等来确定设施的长度和跨度，根据树体高度和设施的跨度来确定脊高和肩高。日光温室长度一般为70～120m，跨度为7～15m，脊高4～5.8m，肩高3～4m，具体参数视甜樱桃生长情况，灵活掌握，如：瓦房店大冯村某日光温室长度112m，跨度15m，脊高5.8m，肩高4m，面积1 680m^2；大连营城子镇某日光温室长度70m，跨度10m，脊高4m，肩高3m，面积700m^2。大连地区日光温室的走向均为东西方向，保温材料多采用保温被，一般不需要加温

图1 日光温室（辽宁）

图2 塑料大棚（山东）

设备。

塑料大棚是山东省主要设施类型，其结构比较灵活，形式多样，有单个农户构建的单栋、双连栋、三连栋塑料大棚，也有2家合作建造一栋大棚，更有多家农户组建多连栋，如：平度铁岭庄的10连栋塑料大棚是由多家农户共同构建的，每家一栋或仅半栋，总面积为15 300m²，单栋长度为90m，跨度为8.5m，脊高6m，肩高4m。每栋大棚之间用塑料膜隔开，方便管理，同时由于多层塑料防护作用，保温效果较好，除两边的大棚需要加温外，其他中间的大棚均不需要加温设备。塑料大棚的走向多为南北方向，覆盖材料多为草苫，一般需要加温设备，烟台地区多为燃煤炉子加空中烟筒，临朐多为烧柴地炉加地龙式瓷炉管，少数采用暖气片加热。骨架材料多为钢管和竹竿，也有采用水泥柱作为支撑材料的。

塑料大棚的长度因园片长度而异，调查结果中最小的为43m，最长的达120m，单栋跨度为5~17m不等。大棚的高度一般较日光温室高，其脊高受树体高度、修剪方式和跨度不同而异，多为6~9m，例如临朐地区，甜樱桃主要采用考特砧木，树体生长健壮、高大，脊高普遍较高，有的高达8m。莱山地区使用大青叶砧木，并采用矮化栽培方式，树体较小，因而扣棚较矮，脊高仅4.7~5.6m。一般的，脊高为肩高加跨度的20%。

表1 甜樱桃设施栽培调查表（2010~2011年）

编号	地点	设施类型	建棚时间	结构参数（m）				品种/砧木	扣棚时间	升温时间	上市时间	产量（kg）
				长度	跨度	脊高	肩高					
1	瓦房店大冯村	日光温室	2009	112	15	5.8	4	美早、佳红、拉宾斯、萨米豆/马哈利	11月上旬	12月下旬	4月初	2 000
2	旅顺土城子1	日光温室	1997	120	8	4.8	3.2	红灯、美早、佳红/大青叶、马哈利	11月中旬	1月上旬	4月中旬	1 500
3	旅顺土城子2	日光温室	1994	100	7	4.8	3.2	红灯、美早、佳红、红密/大青叶	11月上旬	1月上旬	4月中旬	1 500

(续表)

编号	地点	设施类型	建棚时间	结构参数（m）				品种/砧木	扣棚时间	升温时间	上市时间	产量(kg)
				长度	跨度	脊高	肩高					
4	大连营城子镇	日光温室	1999	70	10	4	3	红灯、佳红、雷尼、萨米豆、晚红珠、美早/草樱、马哈利	10月下旬	12月下旬	4月底5月初	1 000
5	烟台炉上1	单栋大棚	2001	54	17	6.8	2.6	红灯、先锋/大青叶	12月下旬	1月中旬	6月上旬	500
6	烟台炉上2	单栋大棚	2004	43	12	5.8	2	红灯、巨红、先锋/大青叶	12月下旬	1月中旬	5月中旬	500
7	莱山南堍1	3连栋大棚	2005	56	6	5.6	3.5	红灯、意大利早红、先锋、美早、雷尼/大青叶	12月20日	12月下旬	4月中旬	400
8	莱山南堍2	3连栋大棚	2007	65	5.5	4.7	2.5	红灯、先锋、美早、意大利早红、芝罘红、斯得拉/大青叶	12月4日	1月上旬	4月下旬	200
9	栖霞下孙家	3连栋大棚	2009	43	5.0	6.7	3	红灯、斯得拉、早生凡、先锋、雷尼、美早、岱红、意大利早红、黄玉	12月18日	1月下旬	4月底5月初	1 100
10	栖霞下孙家	单栋大棚	2006	160	12	6	3	红灯、美早、拉宾斯/大青叶	12月中旬	1月中旬	4月下旬	1 000~1 500
11	平度铁岭庄	10连栋大棚	2001	90	8.5	6	3	红灯、雷尼、先锋、红艳/考特	11月中旬	1月上旬	4月中旬	1 200
12	临朐月庄1	单栋大棚	2007	80	10	8	4	红灯、先锋、红蜜、红艳、美早/考特	12月初	12月上旬	5月初	1 200
13	临朐月庄2	单栋大棚	2008	115	14.5	6.8	4.8	红灯、先锋、美早、红蜜、雷尼/考特	12月底	1月下旬	4月上旬	1 000
14	临朐月庄3	2连栋大棚	2007	70	15	7.5	4	红灯、先锋、斯得拉、美早、佐藤锦、雷尼/考特	12月底	1月下旬	4月上旬	1 000
15	临朐月庄4	3连栋大棚	2007	48	12.5	6	4	红灯、先锋、拉宾斯、雷尼、斯得拉、佐藤锦/考特	12月底	1月中下旬	4月上旬	1 000
16	临朐月庄5	2连栋大棚	1998	147	9.5	7.2	4	红灯、先锋、美早、拉宾斯、雷尼/考特	12月底	1月下旬	4月中旬	1 500

(续表)

编号	地点	设施类型	建棚时间	结构参数（m） 长度	跨度	脊高	肩高	品种/砧木	扣棚时间	升温时间	上市时间	产量(kg)
17	泰安孙楼村1	单栋大棚	2009	73	26	9	4	红灯、早红宝石、早大果、先锋/考特、吉塞拉	12月底	1月上旬	4月中下旬	300
18	泰安孙楼村2	单栋大棚	2010	100	25	7.8	3.5	红灯、先锋、美早、拉宾斯、萨米豆、雷尼、早大果/考特、吉塞拉	12月底	1月上旬	4月中旬	—
19	泰安朱家庄	5连栋大棚	2010	60	24	6.5	3	美早、红灯、先锋、拉宾斯、萨米脱/考特	12月底	1月上旬	4月中上旬	—

2.2 扣棚、升温时间的确定

由表1可见，大连的扣棚时间一般为10月下旬至11月中旬，升温时间为12月下旬至1月上旬；山东地区最早在11月中旬开始扣棚，大多数扣棚时间为12月中下旬，扣棚后1~2周开始升温，若扣棚时间较晚，则可扣棚后接着升温。扣棚后晚上打开通风口，白天合上通风口，盖上草苫，使温度控制在0~7℃之间，当累积温度达到栽培品种的需冷量后才能升温。

大连地区，扣棚后升温，有前期快速升温法和缓慢升温两种方法，前者，前期高温闷棚，萌芽后马上降温，从升温到花期仅需30d，但风险大；后者从升温到花期需45~50d。两种方法均不需要加温设备。

个别栽培者为使甜樱桃赶在春节之前上市，则在8月份开始扣棚，棚内安装制冷设备，使棚内温度控制在0~7.2℃，当达到甜樱桃的需冷量后接着升温，促进开花坐果。这种促成栽培方法需在采摘后及时补充树体营养，以恢复树体，建议不要连年使用，否则树体衰老很快，缩短结果年限。

由于我国对需冷量的评价模式不一，以至报道的同一品种的需冷量值存在较大差异，对确定升温时间带来了难度。部分樱桃园经常因扣棚时间过早，需冷量不足导致开花不整齐、坐果量低，造成减产甚至绝收，经济损失严重。当前，需冷量的评价模型主要有四种：≤7.2℃、0~7.2℃、0~9.8℃和犹他模型。其中0~7.2℃模型较为常用，该模型能够较好地反映各品种的需冷量。采用0~7.2℃标准，甜樱桃品种的需冷量约600~1 200h。需冷量受多种因素影响，主要由遗传因素决定，同时受栽培地区冬季温度、枝芽特性、树龄、树势等因素影响。栽培者在确定扣棚时间时，既要参考科研结果，又要结合生产经验。

2.3 适宜品种及授粉技术

大连地区主栽品种为美早、佳红、萨米脱、红灯、拉宾斯、雷尼，砧木为大青叶和马哈利。大青叶根系浅，根系的需冷量易满足，日光温室栽培表现好。美早是首选推广品种，主要优点果实个大、硬度大、商品价值高，日光温室栽培，单果重可以达到15~25g。

山东地区主要栽培品种有红灯、先锋、拉宾斯、雷尼、早大果、萨米脱、美早、意大利早红、芝罘红、斯得拉等。临朐、平度大棚栽培表现丰产的主要品种为红灯、先锋、拉宾斯、雷尼。烟台地区的甜樱桃砧木主要为大青叶，临朐、平度等地区主要采用考特作砧木。

主要授粉技术包括蜜蜂授粉、壁蜂授粉、人工授粉、鼓风机辅助授粉。利用蜜蜂或壁蜂授粉时，需在每 $667m^2$ 放蜂 1~2 箱或 300 头左右；

授粉时，空气的湿度和温度十分重要，若温度过低、湿度过大，则花粉不易散出，蜜蜂的活动性较差，授粉受精不良，坐果率低。同时，湿度大，花瓣不易脱落，需要人工辅助脱除花萼，盛花末期利用橡皮锤敲打花枝，震落花瓣，预防病害。

2.4 环境因子调控

影响设施内甜樱桃生长发育的环境因子较多，有光照、水分、湿度、温度和土壤等，但最关键的因子为湿度和温度。

温度是影响设施内甜樱桃生长发育的首要关键因子。在无加温设备的设施内，温度的变化规律为日出前最低，日出后逐渐升高，下午两点左右达到最高，之后逐渐下降，第二天日出前降至最低。设施容积越小，白天升温越快，温度越高，晚上气温下降越快，昼夜温差越大。若升温过快、温度过高，会加快甜樱桃的萌发和开花速度，缩短从升温到开花的时间，但过高的温度会造成花器官败育，导致甜樱桃大量落花、落果；因此，温室扣棚后要逐渐升温，不要提温过快、温度过高。调查表明设施内花期温度，早晨控制在 6~9℃，中午 17~20℃，晚上 5~7℃。

湿度是影响设施内甜樱桃生长发育的另一个关键因子。适宜的空气相对湿度，对甜樱桃的生长发育，特别是坐果至关重要。湿度过大，树体易结露，花粉吸水涨裂失活或花粉粘滞，扩散困难，影响坐果。调查表明，大连、烟台地区一般采用覆盖地膜、通风换气、地面和树体洒水、喷雾或浇水的方式来控制湿度，其中地膜覆盖是设施内普遍采用的控湿方式，它的主要作用是阻止地下水分与地上水气的自由流通，对预防果实裂果效果明显。不同发育期，甜樱桃对湿度的要求不同，从升温到花前湿度控制在 70%~80%；开花期湿度以 45%~50% 为宜，过高过低均不利于授粉和坐果；至果实成熟，保持 50%~60% 为宜；果实变白到着色期特别需要注意降低湿度，否则会引起大量裂果。

2.5 花果管理

花果管理是甜樱桃设施栽培的关键技术措施，直接影响最终经济效益。大连地区花果管理比较精细，产量可达 $2\,000kg/667m^2$；临朐花果管理技术相对成熟，产量一般在 1 000~1 500$kg/667m^2$。本调查表明，甜樱桃促成栽培中花期不一致、坐果率低、硬核期落果严重、果个小、隔年结果等现象较为普遍。

2.5.1 萌芽前喷施荣芽

大连地区，萌芽前喷施荣芽促进花芽萌发且开花期一致，花期缩短至一周，效果明显。

2.5.2 增强树势、控制生长

树势过弱，营养生长过旺是导致落果的主要原因。喷施 PBO 控制新梢生长，喷施叶面肥增加树体营养，提高坐果率效果明显。生产中很多生产者花期不进行叶面追肥和新梢控长措施，引起养分供应不足，导致落果现象发生。花期喷施硼砂（0.3%），盛花后喷施 1~2 次 PBO 200 倍，喷施叶面肥 2~5 次，主要有磷酸二氢钾（0.3%~0.5%）、尿素（0.3%~0.5%）和美果露等。

2.5.3 合理负载

树体负载量过多，易因养分竞争导致芽分化少、质量差、生理落果严重、果实品质下降、丰产、稳产性差。应通过疏花枝、疏蕾、疏果的方式进行负载量调节。一般每个花束状果枝上留6~8朵花。疏果一般在花后14d、生理落果之后进行，每个花束状果枝留3~4个果为宜。生产上可将培育单果重在20g以上，可溶性固形物含量在18%以上的优质大果作为目标，合理负载，显著提高甜樱桃的商品价值。

调查表明，目前生产上大多数栽培者缺乏合理负载的意识，认为花、果越多越好，不及时疏除，导致后期落果严重，或生产的果个太小，含糖量较低，品质低下，因此应积极引导种植户进行疏花疏果。

2.5.4 预防裂果

甜樱桃采收期裂果一般是由水分供应不均衡引起的，因此在果实发育期特别是果实发育后期保证水分供应均衡、防止下雨天设施内积水、保持设施内空气湿度稳定是预防裂果的重要措施。

2.6 揭棚后树体管理

揭棚时间为采后1~2周，可一次性揭除。揭棚后的管理如下：

2.6.1 土肥水管理

由于促成栽培打破了甜樱桃固有的生长规律，揭棚后正是营养积累和花芽分化的关键时期，放松管理将会出现树势衰弱、花芽分化不良、花芽老化等问题，影响翌年的生长结果，必须加强土肥水管理，以尽快补充树体营养、恢复树势。

揭棚后，每隔15d左右叶面喷施绿兴1 000倍液（或活力素等）加尿素0.5%，隔次加磷酸二氢钾0.5%（2~3次），以提高叶片光合能力，增加树体营养积累。8~9月份施基肥一次，基肥种类一般为有机肥，用量为30~50kg/株；扣棚后、花前施化肥或复合肥一次，用量2~3kg/株，以促进花芽分化，多采用放射性环施，沟深20cm左右。

扣棚期间灌水原则为小水勤灌，一般棚内灌水5~6次，分别在扣棚后1~2d、花前、谢花后、硬核前、果实着色期进行灌水，并保证整个果实发育期水分供应均衡。雨季来临时，要搞好棚内排水工作，以防止积涝死树和流胶病的出现。每次灌水和雨后及时中耕松土及除草，深度10cm左右。落叶后清扫棚内杂草和落叶，消灭病原。

2.6.2 修剪管理

甜樱桃成枝力弱且顶端优势极强，应加强夏季修剪，意在增加枝量，防止结果部位外移，促进花芽分化。时间7月上旬以前完成，夏剪方法主要是摘心和拿枝，对主枝延长枝和中央领导枝留30~40cm，其余枝留10~15cm，背上直立枝或强旺枝多次摘心或拿枝。

2.6.3 病虫害管理

主要病虫害包括桑白蚧、红蜘蛛、卷叶虫、流胶病。在萌芽期喷布1次50倍机油乳剂或5波美度石硫合剂，在生长期喷洒10%吡虫啉可湿性粉剂4 000倍液或4.5%士达乳油2 000倍液来防治桑白蚧。开花前使用73%克螨特乳油1 000倍液、1.8%阿维菌素乳油3 000倍液、50%托尔克可湿性粉剂1 500倍液防治红蜘蛛。在卷叶虫的越冬幼虫出蛰期和第1幼虫初孵期喷施20%灭幼脲三号悬浮剂2 000倍液、2.5%功夫菊酯乳油2 000~3 000倍液防治效果良好。不同产区对流胶病的防治措施不同，据泰安朱家庄介绍，流胶病发生后，可在刮胶后涂抹碱面，效果显著，伤口愈合快，不易复发。

3 讨论

通过调查可见，甜樱桃设施结构比较灵活，设施长、宽、高因园片、树高、树形而设计，这也是造成调查结果中烟台、大连和潍坊等地区设施结构的差异的主要原因。

设施内品种及砧木种类选择，扣棚升温时间，温、湿度等环境因子调控，花果管理技术，树体生长控制及土肥水管理技术的正确应用是设施栽培内栽培甜樱桃获得丰产、稳产的前提，以上技术要综合应用，缺一不可。本调查表明，各产区在以上技术的应用上均存在一定的问题，应积极应对，保证稳产和丰产。

设施内栽培甜樱桃上市早、经济效益高，是农民致富的良好途径，具有广阔的发展前景。

参考文献

[1] 刘铭，张英杰，吕英民．荷兰设施园艺的发展现状．温室园艺．2010，(8)：24~33
[2] 王晨，王涛，房经贵，蔡斌华．果树设施栽培研究进展．江苏农业科学．2009，(4)：197~200
[3] 李莉．我国设施果树生产现状分析．山西果树．2010，(6)：41~43
[4] 黄贞光，赵改荣．甜樱桃保护地栽培技术研究．中国生态农业学报，2001，(2)：94~96
[5] 赵德英，刘国成，吕德国等．日光温室甜樱桃限根栽培技术．中国果树．2006，(6)：21~23
[6] 孙瑞红，王涛，秦志华等．山东保护地甜樱桃病虫害发生动态与防治措施．落叶果树．2010，(1)：24~25
[7] 王海波，刘凤之，王孝娣，李敏．中国果树设施栽培的八项关键技术．农业工程技术（温室园艺）．2007，(2)：48~51
[8] 王海波，王孝娣，王宝亮，谢兆，魏长存，聂继云，刘凤之．中国北方设施葡萄产业现状、存在问题及发展对策．农业工程技术（温室园艺）．2011，(1)：21~24
[9] 潘凤荣．甜樱桃温室栽培．农业工程技术（温室园艺）．2001，(11)：11
[10] 潘凤荣．日光温室甜樱桃采收后的管理．河北果树．2002，(1)：49
[11] 孙玉刚，秦志华，安淼．甜樱桃发展30年回顾与展望．烟台果树．2008，(4)：11~14
[12] 孙玉刚，张福兴．甜樱桃栽培百问百答．北京：中国农业出版社，2009